Interpreting Ground-penetrating Radar for Archaeology

Interpreting Ground-penetrating Radar for Archaeology

Lawrence B. Conyers

Routledge
Taylor & Francis Group
LONDON AND NEW YORK

First published 2012 by Left Coast Press, Inc.

Published 2016 by Routledge
2 Park Square, Milton Park, Abingdon, Oxon OX14 4RN
711 Third Avenue, New York, NY 10017, USA

Routledge is an imprint of the Taylor & Francis Group, an informa business

Library of Congress Cataloging-in-Publication Data:
Conyers, Lawrence B.
Interpreting Ground-penetrating Radar for Archaeology / Lawrence B. Conyers.
p. cm.
Includes bibliographical references and index.
ISBN 978-1-61132-216-3 (hbk. : alk. paper)
ISBN 978-1-61132-218-7 (institutional ebook)
ISBN 978-1-61132-653-6 (consumer ebook)
1. Geophysics in archaeology. 2. Ground penetrating radar. I. Title.
CC79.G46C668 2012
930.1028—dc23
2012023956

ISBN 978-1-61132-216-3 hardback

CONTENTS

Illustrations

PREFACE

When I first got involved with GPR and its uses in archaeology in the late 1980s, it was a very lonely world, with only a few collaborators with whom I could compare ideas, share data, and work jointly in developing processing and interpretation methods. I was fortunate to have had Dean Goodman and Jeff Lucius as friends and collaborators, and along with a few others in the United States and western Europe, there was the beginning of a small community of GPR practitioners. Today there are many competing GPR system manufacturers and thousands of users on every continent; the community keeps getting larger and more dynamic with each passing year. It seems at times that the GPR community today is so vast and diverse that it is difficult to keep up with all the new advances. Looking back, I remember producing a "wish list" in 1996 with Dean Goodman, where we listed GPR advances we expected or hoped would occur in the next 10 years (Conyers and Goodman 1997). When he and I reviewed that list just four years later, we were astounded that almost everything on that list had been accomplished, and new advances were progressing at a rate even faster than we could have imagined. Collection and processing of GPR data is now fast, accurate, and very sophisticated, with huge amounts of data collected over areas that are vastly larger than we thought was possible in the 1990s. Many of the standard GPR images, including reflection profiles and amplitude slices, are now commonly produced by all GPR scientists and have become commonplace throughout the industry. However, there still seems to be a good deal of confusion and debate about what all those images reveal about what is in the ground. That is the subject of this book.

All GPR users are now capable of routinely producing potentially usable products from their GPR data, but all of us still struggle with what those images have to tell us about the interesting buried features we are studying. This book is therefore about the interpretation of what is produced from GPR data. There is already a great deal known and written regarding how to collect and process those data (e.g. Conyers and Goodman 1997; Conyers 2004a), and this book does not reiterate those subjects and instead concentrates almost wholly on interpretation. In discussions with some GPR colleagues recently, I commented that there is an interpretive void in GPR knowledge, and many agreed that useful and intuitive hardware and software techniques are no longer what are lacking. All concurred that making sense of the output of these standard techniques, however, still remains problematic.

With that in mind, and knowing that new knowledge and understanding of any scientific method is built on the experience of others, I counted up all the GPR surveys my students and I have conducted in the last 20 years. It quickly became apparent that there might be something I could learn from all that work if I devoted some time to reworking, processing, and thinking about some of those surveys, especially with regard to how those data are interpreted. The number of surveys that my students and I have conducted now reaches more than 600 and seems to be growing grow very fast every year, as collaborators, students, and friends continue to collect new data and test a variety of ideas about the past, the ground, and the archaeological record still buried below the surface. I therefore resolved to categorize the most interesting of the GPR surveys that we conducted in many different environments to produce a book on GPR interpretation. This volume reviews successes, failures, and some major (and many minor) revelations that have come to us over the years, in the hope that others can build upon those results. While no GPR survey is ever the same as another (even the same ground surveyed again during a different season can yield different results), I hope there is enough basic commonality among the examples offered here to be of help to others in their own work.

I often demand that my students take the GPR equipment and go out in the field and make mistakes, learn from them, and then attempt to explain to the rest of us what they have gained from that experience. With that in mind, I have also reviewed all the mistakes I have made in the last 23 years of GPR work around the world, most of which I had not seriously pondered until I started preparing this book. It is the successes and failures together, and an understanding of them, which can push along a general comprehension of how the GPR method can "see below the ground." So, while I have yet to completely understand all of my past mistakes and still have a good deal to learn from what I had always considered to be successes, I did take the better part of a year to reflect on my last two decades of GPR work. My initial plan was to try to derive from these data sets information that was applicable to some basic themes of GPR interpretation, which I hoped would be of interest to the geophysical archaeology community. Having completed this reanalysis, I have personally learned a great deal about GPR that I never appreciated before, and the process of writing down those results has led me to identify many aspects of that past research that had eluded me.

None of the reanalysis of old data sets would have been possible without recent advances in GPR data processing, technology that was not available when I first interpreted the data from many of my earlier surveys. All of the reflection profiles presented here were created with a software program written by Jeff Lucius, in minor collaboration with me, called GPR Viewer. This imaging program allows any format of GPR data to be opened in a window, processed, adjusted for velocity, corrected for gains and time zero, and then further corrected for horizontal and vertical distortions. Interesting features can be "zoomed in" on, colors changed to enhance certain reflection features, and then output as images to be annotated in other programs. GPR Viewer changed my life, and this book would not have been possible without it. The amplitudes from many profiles in a grid were resample using Dean Goodman's versatile and industry-standard software called GPR Slice, and also Jeff Lucius's GPR Process code, with the output of those programs gridded and mapped using Golden Software's Surfer 9 program.

One goal is to make this volume more widely available around the world by publishing it in a standard paper edition (which still seems necessary to give it some legitimacy in the academic world within which I reside), my preferred book format. A second goal is to also make it available widely as a digital book that people anywhere can download from the Internet. Perhaps in that way this somewhat specialized subject can be made accessible to those working in areas of the world that do not have traditional academic libraries containing books of this sort, or the ability to receive deliveries of written material in the typical bound, paper format. This potential nontraditional market of e-books was suggested to me recently by a friend who was traveling in Asia and browsing an out-of-the-way bookstore. He found a very poor-quality "pirated book" of Dean Goodman's and my first book on GPR for archaeology (Conyers and Goodman 1997) for sale on the shelves. I found this remarkable, as it suggests that someone went to some trouble to make and sell a pirated copy of a book I thought might have had a total readership of perhaps a few hundred. If someone went to all that trouble, then surely there might be a need for more easily accessible information on GPR, now that the method has become more commonplace.

Today there are tens, if not hundreds, of software programs that can be used to process and produce images of GPR data. There are also thousands of GPR systems all over the world, and I hear that people in China are buying many systems per month, for largely unknown purposes. Apparently the ability to collect and process GPR data is no longer the problem it was just 15 or so years ago, and it appears that there are very many people with whom I have never had any contact who must be using GPR for various reasons. It also never ceases to amaze me how many e-mails I get every week from other previously unknown researchers from all over the world looking for advice or interpretive help with their data. Dean Goodman also has signed up many hundreds of subscribers to his GPR processing and visualization software, people who renew their licenses every year for "real money," and his subscriber base continues to grow. This suggests to me that the need for developing robust and easy-to-use software has been largely filled. Also, many of Dean's customers are not archaeologists but engineers and geo-technicians of many sorts, so the method is being widely utilized by a wide variety of scientists.

I continue to think that no matter what the reason for using GPR, whether for archaeology, geology, or other specialties, what remains largely lacking is the ability to interpret processed data in a way that can answer important questions. This is especially the case within the broader archaeological communities, specialists who are consumers of GPR results but are only vaguely aware of the method and its interpretive powers and limits in the most general of terms. Others who collect, process, and interpret GPR data more commonly might also find the results presented here to be useful regardless of their geophysical experience and skills.

I have found from my e-mail correspondence that GPR users often already have the software for processing and visualization and are only looking for a second opinion on an interpretation. Or perhaps they have found an interesting feature or have produced a puzzling map about which they have questions regarding their interpretation. While I find these kinds of extracurricular GPR exercises interesting and a good learning experience for all concerned, they also remind me that many practitioners of geophysical archaeology have had little training or experience in data interpretation.

Some of these new recruits to GPR research are also under the erroneous impression that once they have access to the acquisition equipment, they can just go out and collect some data, process it using software that leads them through a number of processing steps, and —presto!—useful images can be magically produced. This reminds me of what we used to do in the old days of film photography, where we took our unexposed film to the one-hour processing center and the pictures were quickly developed for us, ready for our enjoyment, readily interpretable. This is rarely the case in geophysical archaeology; in my experience, I can remember only one or two occasions, with very special data sets, when I was able to interpret results immediately after producing standard GPR images.

Most of the ground that we may be surveying is by nature very complex, and the GPR method, in particular, produces images of the subsurface that are even more complex than those generated by other near-surface geophysical methods. Our typical GPR data sets usually contain an extraordinary array of reflections, all recorded in three dimensions. One small grid of GPR data can contain many hundreds of two-dimensional data sets (containing thousands of individual reflections), which are then commonly processed in a batch to produce other two-dimensional maps (slice-maps) or three-dimensional images such as isosurfaces. These products often contain a bewildering array of "anomalies," which may or may not be indicative of what an interpreter thinks (or hopes) might be there. While these GPR images can often be very interesting and even artistic in a way, they are usually not easy to interpret.

Dean Goodman once suggested that he and I should produce a "coffee-table book" of our beautiful and very colorful images of GPR data, as they are more like art than science. I was both intrigued and appalled at the suggestion, because he is right that some products of GPR processing can look more like modern art than like science. As I thought more about it, it dawned on me that anyone looking at GPR images without an understanding of what goes into producing them cannot possibly see their scientific meaning, but only their interesting shapes and colors—the realm of art. Only a comprehension of the many variables that go into producing radar reflections in the ground and processing them into images will allow accurate or even a marginally useful interpretation of the final products.

Acknowledgments

The list of colleagues, helpers in the field and in the processing of data, students, and funding agencies and organizations for all the projects discussed in this book is lengthy. I have tried to list them all below, in alphabetical order, and apologize if I have left anyone out. Most of the work presented here was done in collaborative ventures, and as with most science, the number of people thinking and working on projects is directly proportional to their success. Many thanks go to John Arnold, Leigh Ann Bedal, Federica Boschi, Tamara Bray, Scott Byram, Maria Caffrey, Richard Carillo, Tom Carr, Bonnie Clark, James Conyers, John Conyers, Katherine Conyers, Doug Craig, Bob Cromwell, Herb Dallas, Mark Demuth, Bill Doelle, Eileen Ernenwein, Jack Fenner, Paul Fish, Suzanne Fish, Kevin Gilmore, Dean Goodman, Michael Grealy, Mike Hargrave, Brian Harrison, Gary Huckleberry, John Hildebrand, Eric Hollinger, Louise Holt, Marin Hopkins, Sally Horn, Dylan Imre, John Isaacson, Al Kinlaw, Michele Koons, James Litel, Sarah Lowry, Jeff Lucius, Corey Malcolm, Daniel Masters, Ian Moffat, Fred Nials, Pat O'Grady, Tiffany Osbury, Rick Pettigrew, Lisa Piscopo, Peter Quantock, Jeff Quilter, Dave Ross, Nan Rothschild, Yossi Salmon, Bruce Schumacher, Julie Shablitsky, Payson Sheets, Jay Silverstein, Susan Stevens, Ed Stoner, Jennie Sturm, Don Sullivan, Mary-Jean Sutton, Paul Swader, Eric Tate, Tiffany Tchakirides, David Hurst Thomas, Jim Thompson, Victor Thompson, Jim Vint, Shelly Wachsman, Chet Walker, Dianne Wall, Lynley Wallis, Candice Wheeler, Jim Weber, Sean Wiggins, Sharon Wilkins, and Doug Wilson. Funding of these projects was made possible by Amache Preservation Society; American Museum of Natural History; Australian National University; California State Parks and Recreation; CERL; City of Portland, Oregon; Colorado Historical Society; Columbia University; Comanche National Grassland; Comstock Cemetery Foundation; Coquille Indian Tribe; Desert Archaeology; Dumbarton Oaks; Flinders University; Four Mile Historic Park; George Bass Institute of Texas A&M University; Haifa University; Harvard University; Heinz Foundation; International Catacomb Society; Joint POW-MIA Command; Leon Levy Foundation; Los Alamos National Laboratory; Mapoon Aboriginal Council; Mel Fisher Museum; Monroe County, Florida; National Geographic Society; Northland Resources; Oak Ridge National Laboratory; Oregon Department of Transportation; Pala Band of Mission Indians; Pambamarca Field School, Ecuador; Reno, Nevada, Transportation Department; RMC Consultants; SERDP, Southwest Heritage Foundation; U.S. Department of Defense; U.S. National Endowment for Humanities; U.S. National Parks Service; U.S. National Science Foundation; University of Arizona; and the University of Denver's Arts, Humanities, and Social Sciences Division.

1
Introduction

Most of the commonly used near-surface geophysical methods, such as earth resistance, magnetics, and electromagnetic induction, produce images generated from instruments that gather averages of measurements related to the various physical and chemical properties of some volume of materials in the ground, with only a moderate amount of depth control. While these data sets can be processed in a way that can estimate depth (or distance), they still represent averages of what are often large volumes of materials. In contrast, GPR generates a data set of reflections from specific materials along the interfaces between units in the ground. Radar travel time is measured precisely, and these measurements can be converted to depth, yielding an often complex three-dimensional data set of reflection amplitudes over a surveyed area. This is the GPR method's utility, but also a potential pitfall, especially if the interpreter does not understand and cannot correctly read the images produced from those raw GPR data. For this reason, my usual interpretation method includes spending a good deal of time at the outset collecting a data set that is calibrated for specific field conditions, and then even more time interpreting the resulting images that are produced in many different ways prior to reaching any conclusions. I spend what others consider an inordinate amount of time analyzing reflection profiles, long before I process them into amplitude maps. I have found that other GPR users often proceed directly to map making when starting their interpretation, often paying little attention to the actual data themselves; in this procedure, the most readily interpretable images might be the amplitude maps, but there could be more information to aid in overall interpretation within the standard reflection profiles. This book will therefore be heavily weighted to the interpretation of two-dimensional profiles, as a way leading ultimately to the interpretation of many of the other types of images, such as amplitude slice-maps and isosurfaces, which are created from multiple profiles in a grid.

Factors Affecting GPR Interpretation

In order to interpret both two- and three-dimensional images of GPR data, users must think geologically, biologically, geophysically, and archaeologically, often at the same time. An understanding of these variables is therefore important (but rarely possible for any individual practitioner):

- Soil changes and types (pedology)
- Soil chemistry (usually the greatest unknown in most GPR studies)
- Stratigraphy of different depositional environments (near-surface geology and geomorphology)
- How energy is propagated, reflected, refracted, and attenuated in the ground (physics and chemistry)
- The types of cultural features that might be present and their geometry, distribution, and origin (archaeology)
- How water is distributed and retained in the ground (hydrology)

- The nature and distribution of other materials in the ground, such as tree roots and animal burrows (biology)
- An understanding of GPR equipment components, how they are powered, and how electronics affects the type of data collected (electrical engineering)

After a GPR data set is collected and ready to be processed, all the above factors must be considered, even if not completely understood. In addition, one must have an understanding of computer processing methods and map making, as well as of image output, before interpretation can begin. That said, few GPR scientists understand and appreciate all these different variables. I grew up with a generation of scientists who knew computers only as mainframes that took our punch cards to process commands, and we obtained our results after waiting in line with others for thick paper printouts. This laborious one-way communication for data processing was obviously not conducive to active interpretation of complicated data sets, and we are fortunate to have long ago moved beyond such methods. I first used a personal computer (an IBM PC with two "floppy" drives, one for software and one for data storage) when I was 30 years old and remember marveling at how fast results could be obtained and then modified on the screen. Because of my very antiquated computer background, I had a steep learning curve in applying computers for GPR processing when I first began in the 1980s. In contrast, my younger colleagues today seem to have been raised with a computer mouse in their hands, and the computer aspects of GPR processing come quite naturally to them. Dean Goodman wrote the first usable GPR processing and image production program on an IBM XT computer similar to the one I first dealt with in late 1988. He told me that he usually had to let the computer run all night to test some portions of his software code, and then in the morning search through lines of error messages to see where the bugs were to be corrected that day. Fortunately, those days are behind us and we can let Dean and other smart programmers deal with our software problems quickly and efficiently.

While a multitude of variables that affect many aspects of GPR collection, processing, and interpretation must be considered, none of us can be expected to be expert in all these subjects. I feel blessed to have had my university training in geology, soils, and geophysics before I started in archaeology, long before I used my first personal computer. My first two college degrees were in geology and geophysics, and only when I was in my thirties did I become exposed to the complexities of the archaeological record within a geological matrix. As a result, when I entered the academic world I often struggled with the details of archaeological interpretation, and my colleagues and students ended up teaching me a great deal about these subjects as I became involved in different projects around the world. I am also hopeless in the subject of electrical engineering and have yet to open up a radar antenna to look at its contents, including the actual antennas and shielding components. My fear is that I would never be able to put it back together again. So, while some GPR experts have in the past tried to engage me in discussions that revolve around capacitors and resisters, I usually try to change the subject quickly so as to not appear as clueless as I really am in this subject, as I know just enough to be dangerous.

The key for any GPR interpretation is therefore to recognize the complexity of the method and at least know something about the variables that might come into play in an area where data have been collected. It is also important to search out other experts as consultants or collaborators when necessary, to help fill in the knowledge voids that will likely exist. I have been very fortunate in my own career to have had two wonderfully talented computer programmers (Jeff Lucius and Dean Goodman) as friends and colleagues. I can let them do what they do best, which is write software code that is useful for the rest of us. I am also constantly requiring the help of engineers at GSSI to keep me out of trouble with my equipment (and fix things when they break), and have relied on hundreds of others for local expertise when gathering data and interpreting results. My warning (and encouragement) to all those who do geophysical archaeology, and particularly GPR, is therefore to know your limits with regard to these variables, constantly ask for help from others who know more than you do, and go out and make mistakes, knowing that this is part of the learning process. On the board in my lab, I keep an archive of "mistakes" that my graduate students make, and we refer to them often; some we know by heart (and we are running out of space on the board!). Then I demand that we all think about other potential mistakes, what might cause other problems, and how to avoid them and learn from them in the future.

In writing this book, I have brought attention to many of my own mistakes, and those of my students, discussing how they were made and what we learned from them. This is not the way most scholarly writing is usually done, as most of us publish only our successes. I can't remember ever reading an article on "mistakes I have made," but with a subject as complicated as GPR I think it is necessary. So, in this book, I am "coming clean" by discussing a wealth of blunders and areas of ignorance, and what was learned from them. Perhaps this will keep others out of the same kind of trouble.

My background in geophysics, prior to GPR, was in seismic exploration. When I saw my first GPR reflection profile in 1989, I remember immediately thinking that I could relate easily to the types of displays that were commonly produced—that is, reflection profiles. I was happy to work with images where all the reflections along a line were visible on one piece of paper, ready for interpretation, and there were depth and distance scales to tell me the location of what I was seeing. This feeling is probably quite rare for others, as most off my students and other newcomers to GPR look at profiles and other images and comment that they "look like a bunch of squiggles." I hope this book will help to make those "squiggles" into something more meaningful and useful.

The Importance of Understanding the Basic GPR Images

My bias with GPR interpretation, which comes from my background, is therefore to begin by interpreting the basic raw data from which images are produced. This means studying and understanding in a general way the components of reflection profiles (the traces composed of reflections within a time-window from one spot on the ground) and then the reflection profiles themselves (composites of traces stacked and displayed next to one another along one transect). Only when these components are understood can other displays such as amplitude maps, isosurfaces, and videos be interpreted. This is often not an easy undertaking, but I think it is the most crucial step in GPR interpretation, and one that is most frequently overlooked by many because of its perceived complexity. In this book, I therefore begin with basic interpretations of individual traces and reflection profiles. Only then do I proceed to the production of amplitude maps produced from many profiles. When this is complete, and there is at least a basic understanding of what conditions are producing reflections in the ground, some of the more sophisticated images produced from GPR data can be meaningfully begun.

I have recently witnessed two interesting (and embarrassing) examples of public displays where very experienced GPR scientists presented incorrect results because of their neglect or inability (or both) to interpret the basic data used to generate those images. The pictures they presented were quite "pretty" and contained much that was fascinating, but the results were flawed because the two-dimensional profiles from which their maps and videos were produced had not been interpreted correctly. Fortunately, very few people in the audience understood their data input either, so there was no real public embarrassment. I didn't point out these errors at the time, because the data from which the final products were produced were presented very quickly and not commented on, and only the results were emphasized. Later I talked to others in the audience and asked them what they thought of the interpretation. Those who understood how GPR data are collected and what produces reflections had the same impression as I did about the erroneous interpretation. For all the others in the audience, the images must have appeared to convey only a very complex data set that was difficult to understand but fun to look at. I discuss some of those interpretation problems in general in the chapters that follow. I am not making this point in order to make fun of others' mistakes but to point out that even very experienced geophysicists can produce erroneous results when the raw data and basic products used to produce more complex images are not interpreted correctly.

Interpretation Themes

Throughout the chapters in this book, I discuss themes that relate to my own experiences in GPR interpretation. These topics are by no means comprehensive, deriving only from the surveys that my students and I have collected and the conditions we were confronted with. Some of the interpretive themes presented here are geophysical and cultural in nature, and others are related more to the geological, chemical, and physical variables that produce complexity in most surveys, such as soil and attenuation variations. There are also modern cultural overprints that can confuse an interpreter, such as recently emplaced objects in the ground, anomalous frequencies of energy that are recorded, or objects that are visible in GPR images but perhaps are not immediately recognizable, such as modern walls, power lines, or even passing cars and trucks. While these anomalous features can sometimes be removed using filtering and other processing techniques, they must first be recognized for what they are. As GPR is a three-dimensional geophysical method, the 3D complexity of the ground can also be both a blessing and a curse. One of my recurring themes, addressed in a number of chapters, is how to think in a three-dimensional way and modify collection and processing procedures to address these three-dimensional variables.

I have devoted a part of this book to a variety of buried cultural features that might be expected in different areas of the world. In Europe and the Middle East, I have been involved in the use of GPR for Classical archaeology, in Roman and earlier sites. Dean Goodman calls GPR at many Roman sites "GPR for Dummies," and in some respects this is true. The Romans built substantial structures that are readily visible with GPR, often composed of beautifully cut, sometimes exotic stones, which can be covered with sediment of a very different composition and texture. The contrast between architecture and the site matrix is therefore dra-

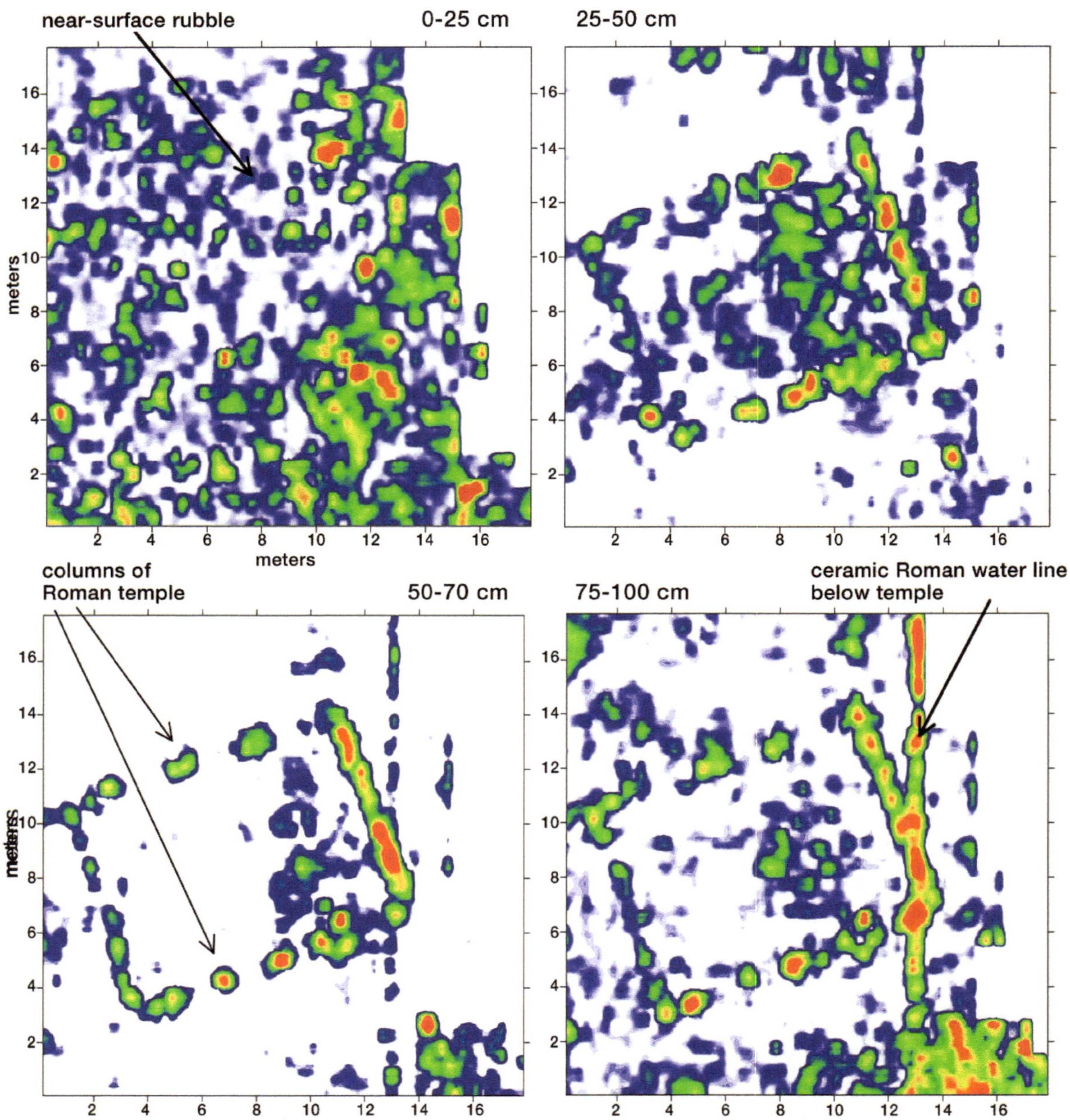

Figure 1-1: Amplitude slices of a Roman temple at Petra, Jordan. The architectural stones and a ceramic water line produce high-amplitude reflections as they are buried by nonreflective windblown sand. Data collected with 400 MHz antennas.

matic, and often archaeological features are so distinct they can be interpreted in "real time" as data are being collected. At Petra, Jordan, I found that the column bases and walls of Nabataean and Roman buildings were so distinct that during collection, as we watched them appear on the system computer screen, we could mark their location with pin flags on the ground surface and produce basic maps of the buried features. I call this (very rare situation) "immediate gratification GPR." Later when all the profiles in a grid were processed into amplitude maps, a temple and other associated features were even more apparent than in the "real time" results (Figure 1-1).

Though less common, the New World has some features that, like Roman architecture, produce strong radar reflections. Some very distinct examples can be seen in historic architecture from the last few centuries composed of concrete or adobe foundations and made into substantially built structures. In southern New Mexico, a Spanish colonial church from the 17th century was constructed with stone foundations and walls composed of adobe bricks. The walls have long eroded away, but the foundations are preserved in the ground (Figure 1-2). At this site, the wall foundation reflections in the profiles were not distinct enough to immediately identify them during collection, but when all profiles in the grid were sliced horizontally, the outline of the church was perfectly shown by the placement of the high-amplitude reflections (Figure 1-3).

Other historic and prehistoric cultural features, such as wooden or earthen structures, are usually not as definitive in GPR images as stone architecture, and interpretation of both profiles and amplitude maps can become more arduous and complex (Figure 1-4). Features of interest may be nonreflective, and layers of eroded architecture are what produce the radar reflections. When this is the case, an understanding of the history of a site, especially after abandonment, is the key to interpretation. Most GPR interpreters are trained to focus on the higher-amplitude reflections, but in some cases it is the low- or non-reflection areas that denote the location of features of interest (Figure 1-4).

Prehistoric sites composed of dwelling floors, hearths, and lithic scatters from all over the world are much more difficult to map with GPR. However, there are some notable successes (and many failures) that I discuss in the following chapters. One example is of two raised platforms on either side of a dwelling floor, which

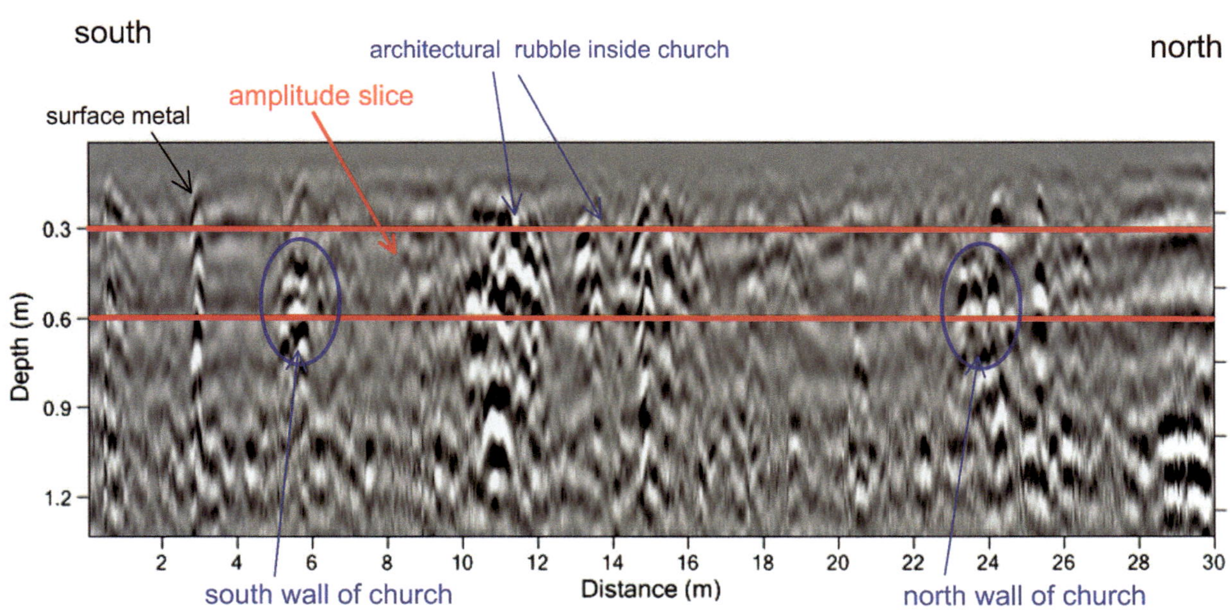

Figure 1-2: GPR reflection profile across a 17th-century colonial church in New Mexico. The stone foundations of the church walls are visible but not readily interpretable in this profile. Only when many profiles were horizontally sliced, as shown with the red boundary lines, and then reflections in those slices processed into amplitude maps (one of which is shown in Figure 1-3) is the outline of the church's foundation visible. Data collected with 400 MHz antennas. *Courtesy of Jennie Sturm and Chet Walker.*

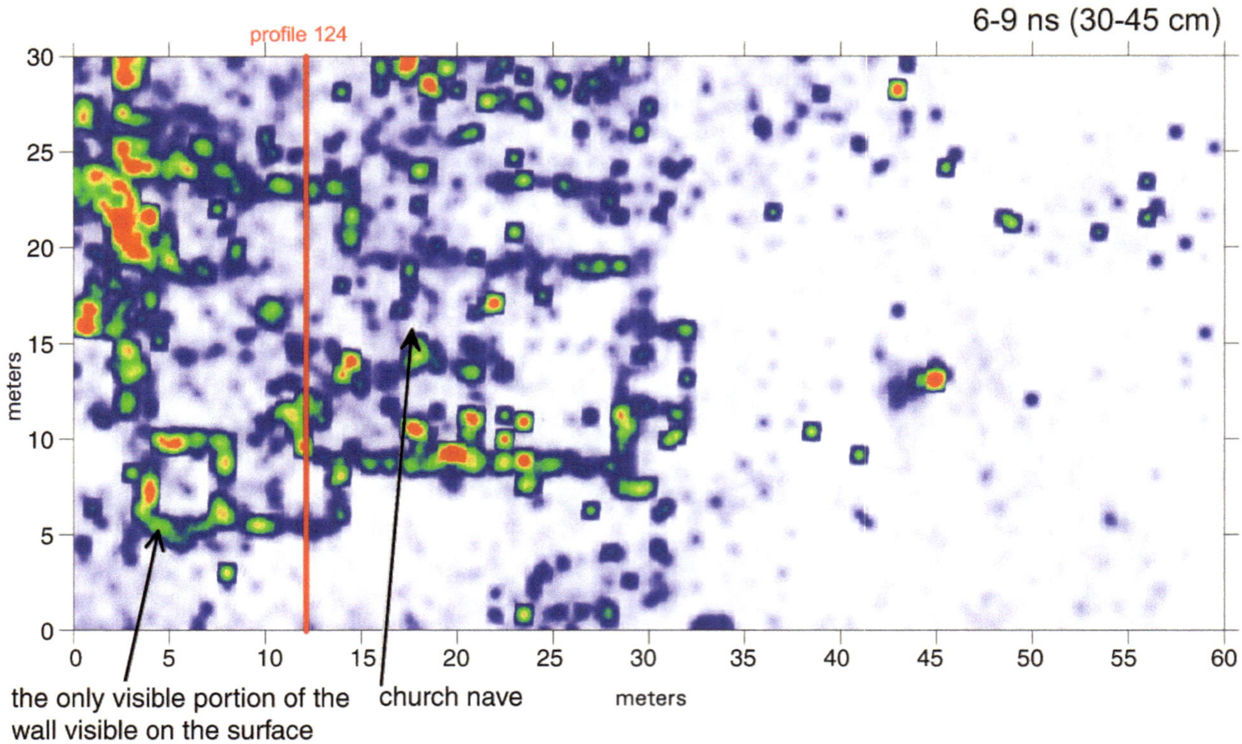

Figure 1-3: Amplitude slice-map of a colonial-period church in New Mexico. The stones in the foundation are highly reflective and show the outline of the church. Data collected with 400 MHz antennas. *Courtesy of Jennie Sturm and Chet Walker.*

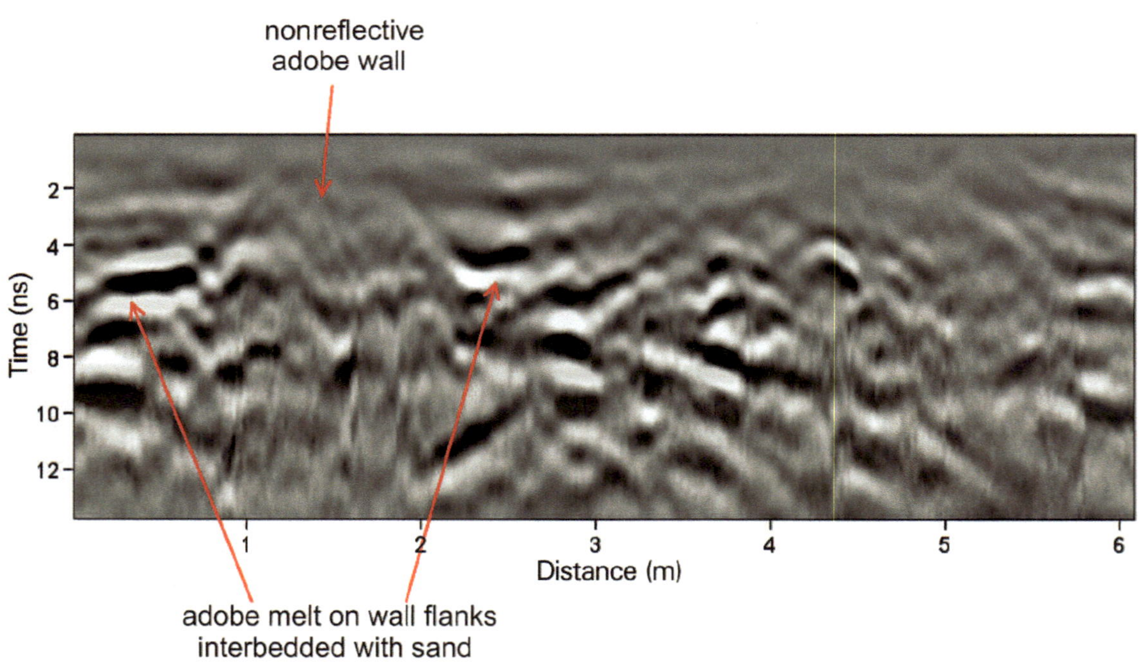

Figure 1-4: GPR reflection profile crossing a thick adobe wall. The wall is nonreflective and not readily visible in this profile. The high-amplitude adobe melt layers, adjacent to the wall, produce the visible layers in this profile. Data collected with 400 MHz antennas in Tucson, Arizona.

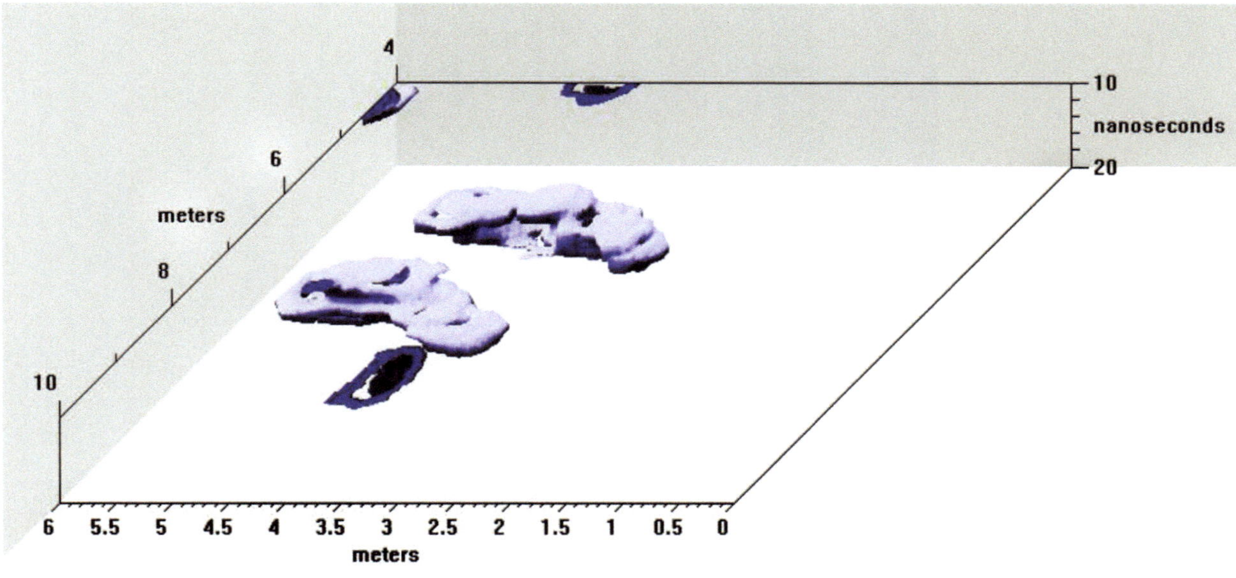

Figure 1-5: Isosurface image of two "benches" of a pit structure buried in coastal sand dunes. The sand that covers these features is nonreflective, and this subtle feature, constructed by hunting and fishing people, is visible in this image because of its distinct difference with the surrounding matrix. Data collected using 400 MHz antennas near Port Orford, Oregon.

was used by hunting and fishing people on the coast of Oregon (Figure 1-5). In GPR images, the features that reflect energy were constructed with clay, and the burying medium was homogeneous sand. The platforms are very distinct in reflection profiles and slices but, in this case, are best visualized in an isosurface image.

I discuss in Chapter 12 some of the other intangible and often interesting and frustrating variables that go along with interpretations, variables that involve perceptions and expectations of others involved in a project. Everyone who has worked collaboratively, especially in projects where other team members do not really understand GPR but are eager to apply the method, have stories to tell about how they have dealt with perceptions, personalities, and communication problems. I devote a small amount of space to this subject, as it relates directly to interpretation and how GPR can be used effectively, or not.

These basic themes are discussed by using the images commonly produced from most GPR data. However, none of the interpretations presented here are possible without a basic understanding of how GPR data are collected and then processed and displayed. The details of the GPR method and the many variables related to equipment, processing, and the production of images are discussed elsewhere (Conyers and Goodman 1997; Conyers 2004a). However, I begin here, in Chapter 2, by outlining some of the basic elements of GPR method and theory and the variables that produce radar reflections.

2

Basic Method and Theory of GPR

Ground-penetrating radar (GPR) data are acquired by reflecting pulses of radar energy produced on a surface antenna, which generates waves of various wavelengths that propagate outward. They spread into the ground in a cone as the waves propagate downward. As these waves move in the ground, they can be reflected from buried objects, features, or bedding surfaces (Figure 2-1). The reflected waves then travel back to the ground surface and are detected and recorded at a receiving antenna that is paired with the transmitting antenna. The two-way travel times of the waves into the ground, to the reflection surface and back to the receiving antenna, are recorded in nanoseconds (ns). As the radar waves propagate through various materials in the ground, their velocity changes depending on the physical and chemical properties of the material through which they are traveling (Conyers 2004a: 45). A reflected wave is generated when waves' propagating velocity changes at contacts between different materials in the ground radar. Some reflected waves will then travel back to the ground surface while the remaining energy continues to propagate deeper and can be reflected again from additional interfaces, until all the energy finally dissipates with depth. Only the reflected energy that travels back to the surface antenna is recorded and visible for interpretation. If buried surfaces that reflect energy are oriented in such a way that the reflected waves move away from the surface antenna, that energy will not be recorded, making those interfaces effectively invisible using the GPR method.

The velocity of radar energy in the ground can be calculated and reflected radar wave travel times converted to distance (or depth in the ground). It is this ability to determine depth that makes GPR capable of producing a three-dimensional data set. There are many ways to calculate velocity (Conyers and Lucius 1996; Conyers 2004a: 99), all of which are estimates of wave propagation speed through packages of sediments and soils. Velocity of propagating waves can vary considerably with depth, usually decreasing as water saturation increases, and also vary laterally because of a variety of other changes in ground composition.

In most GPR data sets, radar antennas are moved along the ground in transects, and two-dimensional profiles of a large number of reflections at various depths are created to produce reflection profiles (Figure 2-2). When data are acquired in a closely spaced series of antenna transects within a grid, reflections from adjoining profiles can be resampled, compared, and then processed into amplitude maps and three-dimensional isosurface images (Figure 2-2). These images produce an accurate three-dimensional picture of buried reflection surfaces (Conyers 2004a: 148), indicating the location of features spatially (in x and y dimensions) and with good depth control (z). An interpretation of the reflections in the ground can then be accomplished using three of the most important GPR images: reflection profiles, amplitude maps, and isosurface renderings.

Production of Reflections, Depth of Penetration, and Resolution

The buried discontinuities where reflections occur are usually created by variations in the electrical properties

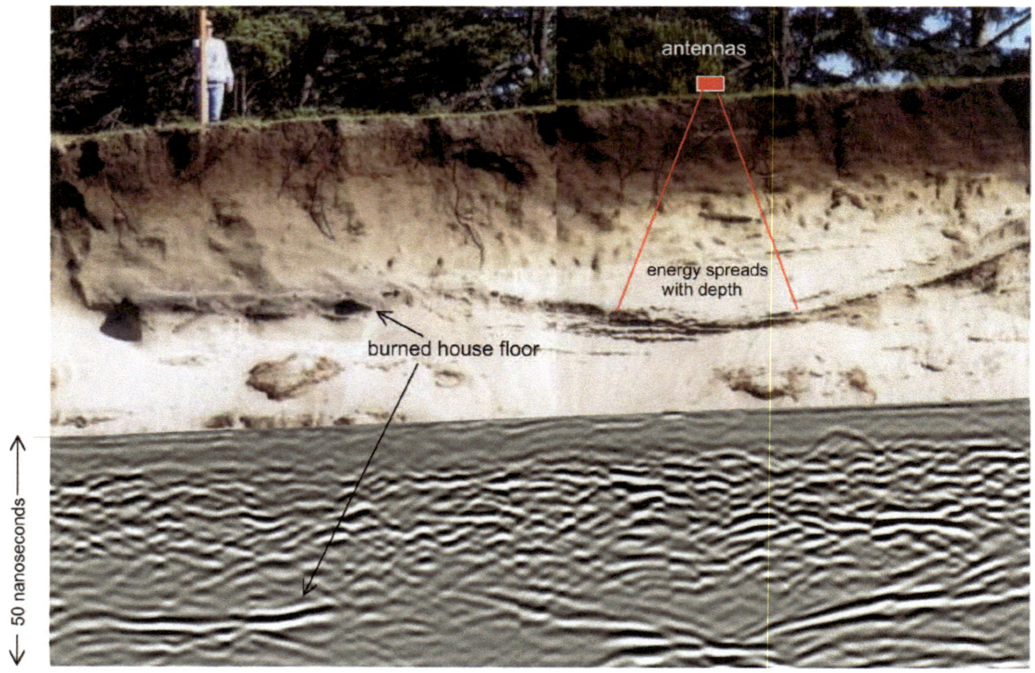

Figure 2-1: Reflection profile of a pit-house floor. This profile was collected over a buried pit-house floor using 400 MHz antennas. A picture of the buried materials is on top, with the GPR profile showing the reflections from these features below. The time-window over which these reflections were collected was 50 ns, corresponding to about 3 m.

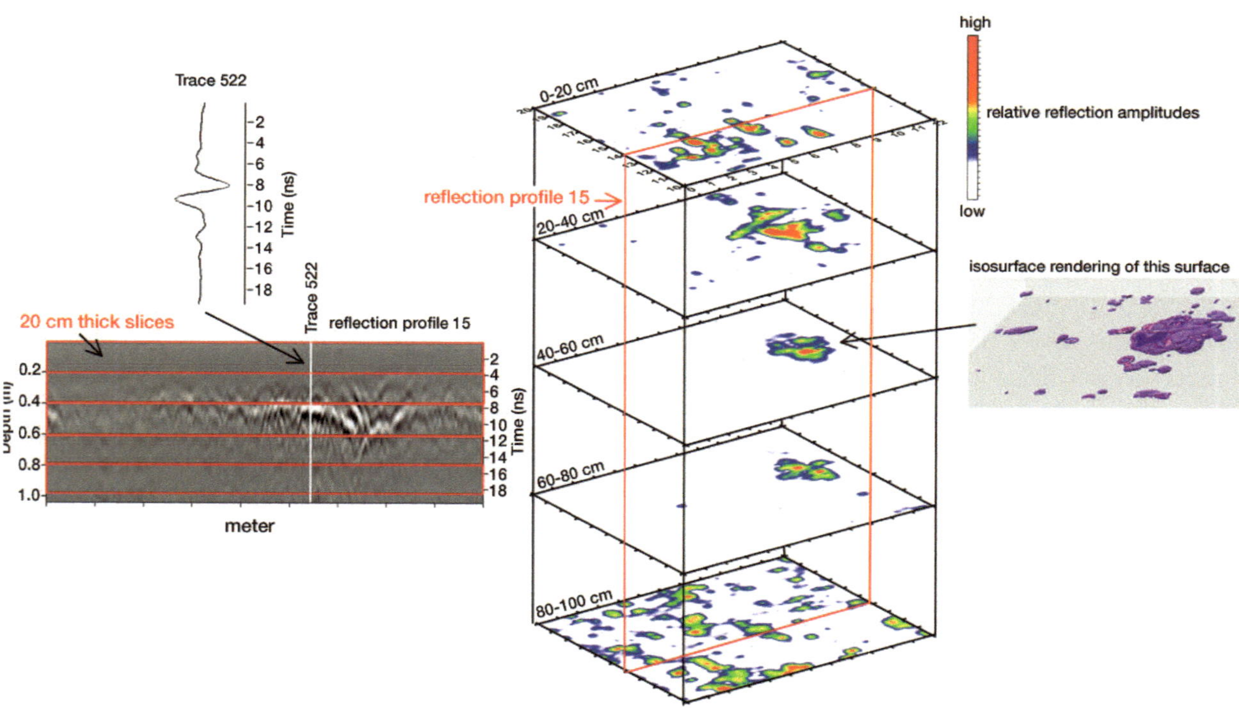

Figure 2-2: Example of the production of GPR images from the basic data to profiles, slice-maps, and isosurfaces. Producing GPR images starts with a basic trace, many of which are stacked to produce two-dimensional profiles. The profiles collected within a grid are then resampled to generate amplitude maps at chosen thicknesses. Many thinly sliced amplitude images or data can further be resampled to generate isosurface images of the features of interest in three dimensions.

of the sediment or soil, changes in the lithology, and differences in bulk density at stratigraphic interfaces. Those measurable (and sometimes visible) differences in materials in the ground create water saturation differences within those buried units, which is what usually produces the velocity changes that generate wave reflections. Reflections can also be created at void spaces in the ground, which may be encountered with burials, tombs and tunnels, or any other variation in buried material that produces an abrupt radar wave velocity change.

The depth to which radar energy can penetrate and the degree of definition that can be expected in the subsurface are partially controlled by the frequency of the radar energy transmitted. Radar energy frequency controls both the wavelength of the propagating waves and the amount of weakening, or attenuation, of those waves in the ground. Standard GPR antennas used in archaeology propagate radar energy that varies in bandwidth from about 10 to 1,200 MHz (Figure 2-3). Antennas usually come in standard frequencies, with each antenna having one center-frequency but actually producing radar energy that ranges a good deal around that mean. Each GPR system manufacturer produces different frequency antennas, none of which are interchangeable with other manufacturers' systems, and each of which has its own interesting and sometimes infuriating pros and cons. I am a user of GSSI systems and have grown to love my 400 MHz antennas. They are the workhorses in my research, as they generate a well-focused transmission beam of radar energy, are well shielded, can resolve features of about 20 cm or so in dimension, and can often transmit energy to 2 m or more in depth. I use my 270 MHz antenna for 5% or less of my research, usually when I need to propagate energy to greater than 2 m or more in the ground. Reflections recorded from these antennas have much less subsurface resolution than those from 400 MHz antennas, so if at all possible I would rather use the 400s. The 270 MHz frequency energy also tends to spread out more from the transmitting antenna, so when these are used I am constantly on the lookout for waves generated from features other than the ones in the ground. Walls, trees, livestock, and people will all produce reflections readily from 270 MHz propagating waves, which can be maddening when trying to differentiate them from the reflections generated within the ground. The 900 MHz antennas generate energy well within the frequency band of many personal communication devices, and I have noticed a good deal of background noise is also collected from cell phone (mobile phone) and various other radio transmissions, along with reflections from within the ground. The 900 MHz antennas are quite good at shallow mapping within a meter or so of the ground surface and have very good feature resolution. But if the 900s are used in even moderately electrically conductive ground, their transmitted energy can be attenuated within 30–40 cm of the ground surface. I have never had the occasion to use the very high-frequency 1.2 GHz or higher antennas, as they are mostly used for extremely shallow concrete

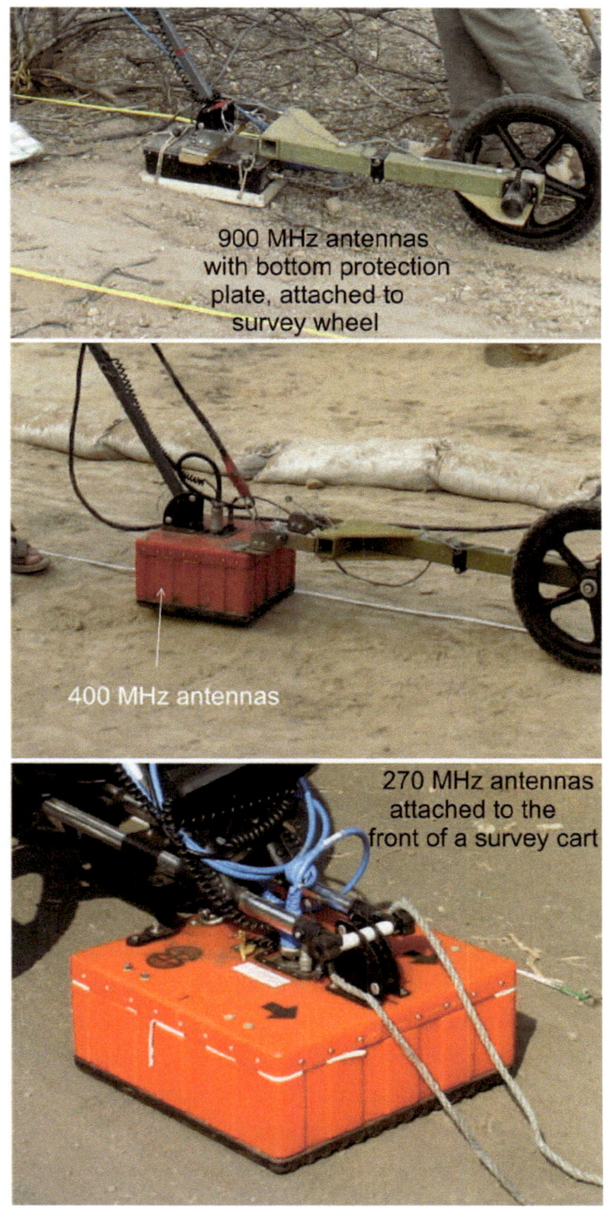

Figure 2-3: Three commonly used GPR antennas. These are the 900, 400, and 270 MHz frequency antennas produced by GSSI.

analyses. The very low-frequency antennas in the 10–100 MHz range are all unshielded, very difficult to move around, and useful only in very electrically resistive ground where depth penetration is important (5–15 m in depth) and the buried features of interest are about the size of a small car. I have never been involved in looking for buried features of that size, especially those buried deeply in ground that is not electically conductive, such as pure quartz sand. The manufacturers Mala Geoscience and Sensors & Software produce similar frequency antennas with very similar resolution and depth penetration as the GSSI units.

DATA COLLECTION AND RECORDING

The most efficient collection method for subsurface GPR mapping is to establish a grid across a survey area prior to acquiring reflection data. Usually, rectangular grids consisting of many reflection profiles are collected within that grid (Figure 2-2) with a transect spacing of 1 m or less. The higher-frequency antennas usually necessitate closer transect spacing than those of lower frequency, as they generate more focused energy that spreads out less with depth of penetration. With the 900 MHz antennas, I usually collect profiles at least 25 cm apart, or less if I have many field helpers and a lot of time. Lower-frequency profiles can be collected with a coarser grid, but I am very distrustful of data collected in transects more than a meter or so apart, as they can leave unsurveyed some buried features located between the profiles that would remain undetected.

Rectilinear grids of profiles allow reflections to be placed within a Cartesian coordinate system, which then can be readily processed using most standard available mapping and gridding software. A few systems are in development that can place randomly collected reflections into space using GPS-collected spatial data. It will not be long before totally random data collection can be processed into a three-dimensional volume, and profiles and amplitude maps can then be produced from a package of reflections chosen by the interpreter. We in GPR are not quite there yet, but the petroleum exploration geophysicists using seismic wave reflections have been interpreting data sets of this sort for many years.

During data collection, the two-way travel time and the amplitude and wavelength of the reflected radar waves derived from the reflected radar waves are amplified, processed, and recorded for immediate viewing and later post-acquisition processing and display. At one specific location along an antenna transect, the waves that are recorded and stacked vertically as they are recorded over some programmed depth is called a *trace* (Figure 2-2). When many traces are recorded sequentially as antennas are pulled along the ground surface in transects, a reflection profile is produced. Distance along each antenna transect is recorded for accurate placement of all reflections in space using a survey wheel, manual distance markers, or GPS receivers connected to the antennas.

Radar energy becomes both dispersed and attenuated as it radiates into the ground. When portions of the original transmitted signal are reflected from buried interfaces back toward the surface, they will suffer additional attenuation by the material through which they pass, before finally being recorded. Therefore, to be detected as reflections, important subsurface interfaces must not only have sufficient contrast at their boundary to abruptly slow or speed up wave propagation, but also must be located at a shallow enough depth where sufficient radar energy is still available to be reflected. As radar energy is propagated to increasing depths, the waves becomes weaker as they spread out over more surface area and are absorbed by the ground, making less energy available for reflection. For every site, the maximum depth of resolution will vary with the ground chemistry and many other geologic and hydrologic conditions, which are discussed in Chapter 5.

INTERPRETATION OF GPR REFLECTIONS

Before making a GPR interpretation based on a visual analysis of the standard processed reflection profiles and amplitude map images, one must have a clear understanding of what causes radar reflections in the ground. This is not necessarily a straightforward process, as there are many variables that affect wave reflection; these variables are a function of equipment employed, ground conditions, collection settings, and background noise, all of which must be accounted for. Over subsequent chapters I will discuss many of these variables, with examples. But most important to interpretation is a basic accounting of what produces reflections of radar waves that are propagating and then reflecting as they move through the ground (Conyers 2004a).

I have heard many GPR interpreters use the word *anomalies* to refer to interesting radar reflections that are visible in profiles and in the mapped amplitude features created from sampling reflections in profiles. Though I occasionally use this word too, I have always found it vague and lacking any descriptive qualities that would help me understand anything at all about the ground and how it has reflected radar energy. In this chapter, I attempt to move beyond the vagueness of *anomalies* and try to describe more effectively the actual radar energy and displays made from the reflections of that energy. The displays of those data might be reflection profiles collected in a grid, slice-maps, or isosurface renderings produced from those profiles, which are then viewed and interpreted. To make a complete and useful interpretation of most data sets, all potential images must be analyzed in a number of ways to understand how those reflections were generated, based on suppositions of ground conditions, which are often usually only partially understood (Conyers 2011b). Using an analysis of reflection profiles in two dimensions yields very different images from those in horizontal amplitude maps, which tend to create "composite" reflection features of many complex reflections in a three-dimensional space. In this section, I use both reflection profiles and amplitude slice-maps to demonstrate what each can produce when interpreted alone, and what can be determined when they are used in conjunction for interpretation.

I refuse to allow my students to identify—or even worse, interpret—any of the reflections or the processed features produced from GPR images as *anomalies*, which only means "something different from the norm." In today's world of GPR, when there are plenty of ways to produce many different displays of reflection data, we should not have to depend on this meaningless interpretive crutch, as it only suggests that something exists in the ground that is different. What might the "anomaly" be different from? Different from no reflections at all? Users of this word are usually just making a cursory attempt at GPR interpretation or perhaps are confused about, or have not attempted to determine, what actually might be producing the recorded reflections in the ground. We can surely do better than this. *Anomaly* is a word to use in GPR interpretation only as a last resort, when meaningful description and interpretation fail.

My students are required to describe radar reflections in many different ways, whether analyzing them in two-dimensional reflection profiles or in other processed images produced by three-dimensional analysis of many profiles. Their descriptions must first begin by describing reflection features of interest in geophysical terms, when viewed in profile. For instance, a common feature in a reflection profile might be described simply as a "high-amplitude point-source reflection hyperbola with an apex at 15 ns, or about 90 cm in depth" (Figure 2-4) or a "moderate-amplitude undulating planar reflection between 32 and 35 m in profile 6, at a depth of 1.1–1.2 m (20–23 ns)" (Figure 2-5). Such basic geophysical descriptions are the first step toward defining reflections in a way that other people with some geophysical understanding can begin to visualize and think about with regard to reflection origins (Conyers 2011b). These types of descriptions are much more than just describing some reflections as "anomalies."

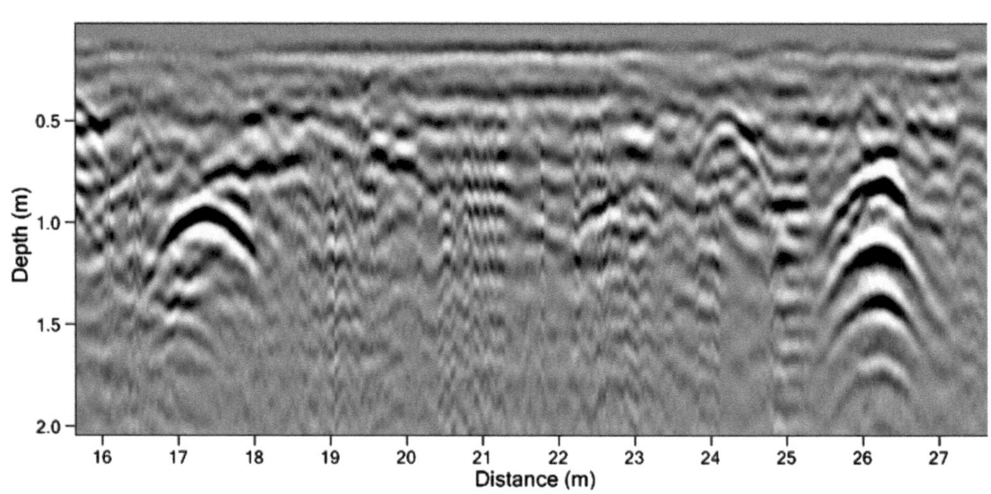

Figure 2-4: Point-source hyperbolic reflections. These reflection features are created from the tops of buried Roman columns. The feature on the right has stacked column pieces that produce multiple hyperbolic-shaped reflections from the interfaces between each stone. Collected with 400 MHz antennas at Ashkelon, Israel.

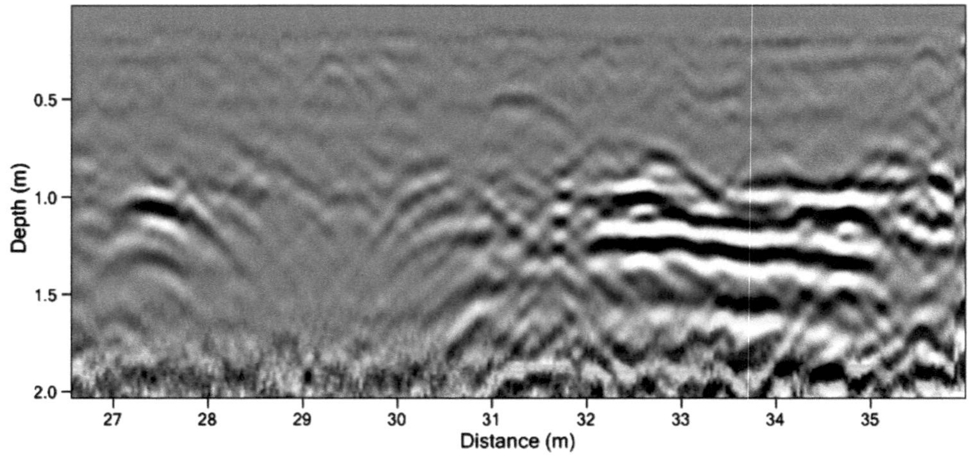

Figure 2-5: Undulating planar reflections. These were generated from a buried floor in a basilica. Collected with 400 MHz antennas at Ashkelon, Israel.

DESCRIBING REFLECTIONS GEOPHYSICALLY

Once we have first described our reflection features geophysically, we can then usefully resample many adjacent reflection profiles within a grid and process them into amplitude maps of certain slices in the ground (defined in either radar travel times or depths); this in turn allows other types of reflection features to be described that yield even more of an understanding about their origin. For instance, horizontal amplitude maps of point-source reflections can produce "aerially distinct high-amplitude reflections from individual objects" in some areas of a grid, while the planar reflections might be described as a "4 × 6 m oblong high-amplitude reflection feature from 18 to 23 ns (1–1.2 m in depth generated from a horizontal layer)" (Figure 2-6). Using both the profiles and plan-view maps, the reflection features are then visible in multiple planes and are describable even more precisely in geophysical terms. Of course, many of these features might also be "anomalous," but at this point they are much more than that, as we have learned a lot about their reflections in terms of shape, reflection strength, orientation, and distribution.

The next step is to take both spatial and geophysical descriptions of reflections and examine them in archaeological or geological terms. This is only possible with an understanding of the conditions of burial and preservation and how potential cultural features were constructed, used, modified, and ultimately abandoned over some period of time. Often these aspects of a site are not well known at the time of interpretation, particularly if GPR data are being collected well before excavations take place. Sometimes the purpose of a GPR study is to identify features for excavation, so archaeological and geophysical interpretations of this sort must sometimes be tentative at best. But often by asking people with experience in an area, doing other research in advance, and even looking around for nearby exposures of soils and sediments, we can find important clues about what is below the surface, and an integration of all this information can produce very accurate interpretations of GPR images. The final product of this analysis and interpretation can be integrated from many different sources and could well lead to an understanding about the people who constructed and used these features. Those interpretations might be able to illuminate aspects of cultural change or events of historical importance, or lead to testing hypotheses of anthropological significance (Conyers 2010; Conyers and Leckebusch 2010).

To be able to make these types of initial geophysical descriptions, which can later lead to geological and archaeological interpretations, it is important first to consider what causes radar wave reflections. I refer readers to previous books on the basics of GPR (Conyers and Goodman 1997; Conyers 2004a) and the plethora of other articles on what causes GPR reflections in the ground. What is covered here are only some basic conclusions related to GPR data collection, processing, and image production so that we can move along quickly to the actual interpretation of GPR images.

Typical GPR systems collect and record only these data, from which interpreters must work:

- The two-way travel time of each reflection recorded. (Actually each trace is defined by a certain number of samples recorded within a programmed time-

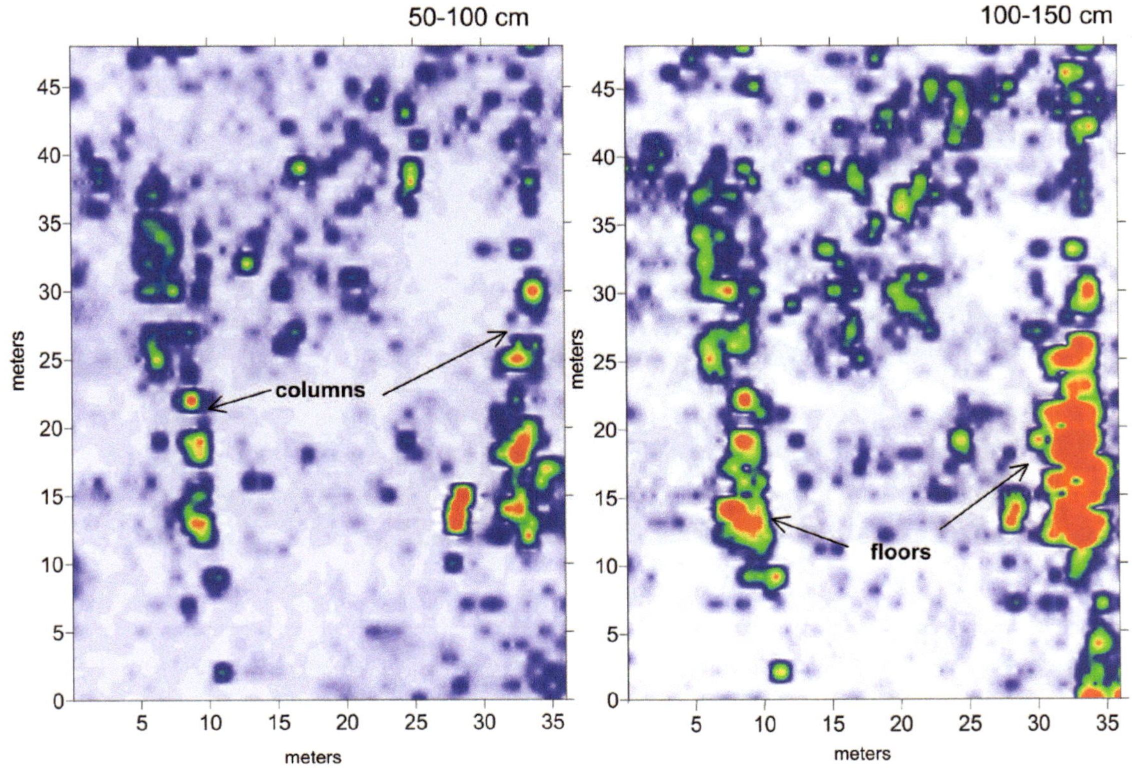

Figure 2-6: Amplitude maps of planar and point-source reflections. The column tops, which produce point-source hyperbolas in reflection profiles, can be visualized as aerially restricted reflection features in amplitude maps. Floors that appear as undulating planar reflections in profile yield more extensive reflection features when resampled and plotted in map view. These maps are of a basilica at Ashkelon, Israel, collected with 400 MHz antennas.

window and then those samples are assigned a two-way travel time in most software programs.)

- Amplitudes of all recorded waves within a dynamic range (8, 16, or more bits).
- Information about multiple frequencies (using broad-band antennas) of recorded waves where information about various frequency band waves can potentially be arrived at with some specialized data processing.

From these basic measurements, images are produced, including reflection profiles, amplitude slice-maps, and other three-dimensional images. Variables that affect recorded data—for example, equipment variations, setup parameters, background noise, and modern cultural overprints—are not taken into account here but are discussed in later chapters when presenting examples of various interpretations.

All reflections generated from radar waves propagating in the ground are created at interfaces where differing materials are in contact along a boundary and are different enough that the velocity of moving waves intersecting the interface changes abruptly (Conyers 2004a). An example of a material change that affects velocity in this way might be where a clay floor sits on an underlying sandy soil, and where these materials are then buried by some other different material. That is, where the base of the clay floor contacts with the underlying soil and where the top of the floor intersects with the covering sediment are both interfaces that could generate wave reflections. The radar waves propagated from the ground-surface antenna would be moving at a fairly rapid rate in the overlying material, slow abruptly as they passed into the clay floor, and increase again as they passed out of the clay floor into the underlying soil. Each abrupt velocity change would theoretically create a reflected wave (Conyers 2004a: 65) that then travels back to the surface to be recorded as a sine-shaped wave with both positive and negative deflections (Figure 2-7). In contrast, a gradational change in materials over some distance would not

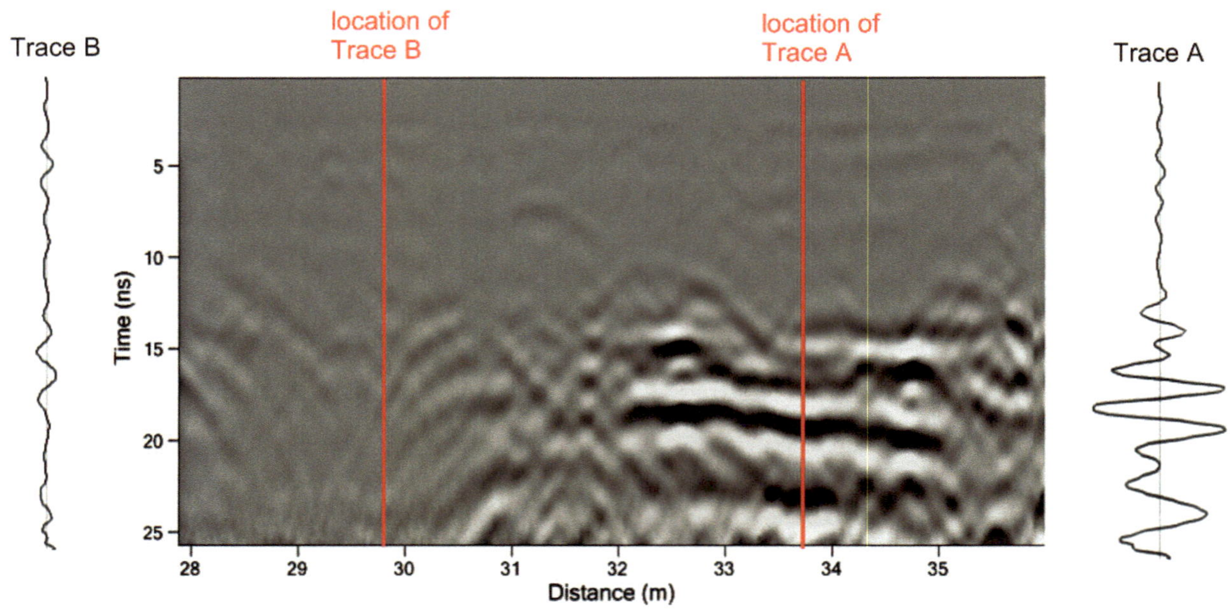

Figure 2-7: Production of reflections from buried interfaces. Traces show how reflections are created from the compacted floor on the right (Trace A), and in homogeneous sand fill on the left (Trace B), which contains no material changes that could generate velocity changes. Data collected with 400 MHz antennas.

produce a reflection, as there would be no abrupt change in radar velocity and thus no reflected wave would be generated. This kind of gradational change might be found when the sediment in one layer changes from silt to sand over a distance of a meter or so (Figure 2-7). In general, the greater the change in velocity across a boundary, the greater the amplitude of the wave that is reflected back to the surface and recorded.

Reflection *profiles* are nothing more than a sequential stacking of many hundreds of reflection traces, each consisting of many reflected waves from different depths in the ground. Each trace is recorded at a discrete surface position along an antenna transect, and the display of all these is used to produce a two-dimensional vertical slice in the ground. All sine waves in all traces are given certain colors or shades of gray to denote relative amplitudes, and the two-way travel times of those waves and the distance along a transect make up the *x* and *y* axes of the profile.

DESCRIBING REFLECTIONS BASED ON THEIR GENESIS

When viewing the radar reflections produced by changes in materials at various interfaces in the ground, most interpreters of GPR reflection profiles usually describe them in terms of constituents of the materials on either side of the interface—as I did in describing the materials in that simple three-layer system of a clay floor bounded above and below by soil and some sandy covering layer (Figure 2-7). However, long experience in comparing GPR reflections to known materials in the ground, combined with many hours of laboratory analysis of those materials, has led me to deduce that it isn't really the sediment type differences that are producing the reflections of radar waves at all. Instead, it is how the material differences in the ground on either side of an interface either retain or distribute water (Conyers 2004b). I have concluded that in many (but not all) contexts where GPR is commonly used, it is actually water content changes in those buried units that are producing the radar velocity changes that generate the reflected radar waves. Therefore, any interpretation of sediment (and soil) and other constituent changes, as well as archaeological component variations within packages of ground, must take into account how those materials retain and distribute water.

I began to consider my interpretations of GPR images differently after collecting a number of grids of data where I could immediately compare reflections in profiles with what I could see directly in outcrops or from excavation trenches. Often I saw many more reflections

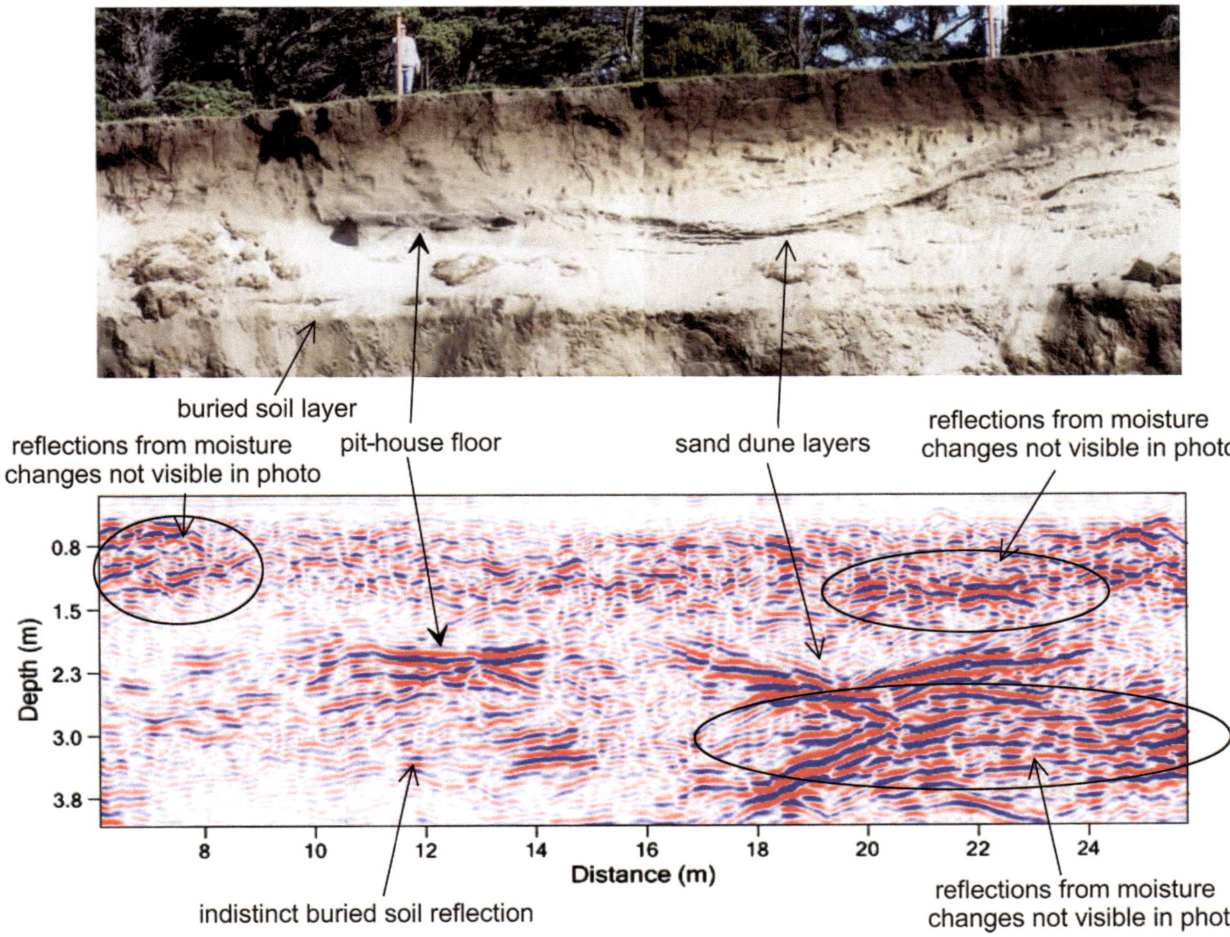

Figure 2-8: Comparison of a GPR reflection profile with exposed materials. A direct comparison of a GPR reflection profile on the bottom with a picture of the exposed materials shows which interfaces are producing some of the reflections. The GPR profile was collected on the edge of the cliff. The pit-house floor is readily recognizable, as are some sand dune layers. But other layers visible in the picture, such as the buried soil horizon at the base of the sequence, are visible only as weak reflections in some areas of the GPR profile. Many other stratigraphic horizons not visible in the picture generate reflections because they retain more moisture. GPR data collected with 400 MHz antennas near Port Orford, Oregon.

in profiles than I could see in the actual exposure of sediment, or I saw what I thought were important material differences in the outcrop that did not produce radar reflections visible in reflection profiles. This difficulty in interpretation is demonstrated in Figure 2-8, where distinct archaeological and geological layers visible along a cliff could be directly compared with the GPR reflection profile collected along the edge of that exposure. In this case, a house floor encased in sand dunes was the target of the GPR study. The exposure along the cliff showed that this windblown sand package lay directly over a distinct, dark brown, buried soil horizon. A well-formed surface soil had formed near the present ground surface. The house floor and the most distinct dune strata within the sand are visible as high-amplitude reflections in the GPR reflection profile (Figure 2-8). But a closer look at the GPR reflection profile shows that there must be many other sediment variations in the dune deposit that are producing radar reflections but are not visible to the human eye in the outcrop. Also, the lowest dark soil horizon, which is clearly visible in the photograph, did not produce a continuous distinct reflection in the GPR profile as might be expected. A cursory sediment analysis made in the field showed that the lowest soil unit is not appreciably different in composition from the overlying sand dunes, except in containing some minor additional clay and organic matter that gives it a distinctly different color. Apparently there is not a significant enough change in material at this interface to produce a high-amplitude reflection.

In this case, the GPR image (Figure 2-8) displays many high-amplitude reflections that must correspond to sediment changes not visible in the exposure. While no detailed sediment analysis of each unit in this exposure was made, some of the laminations in the sand dunes were noted to have had slightly more clay than others, which would likely have retained more water. These variations were not immediately visible to the human eye but are distinctly visible in the GPR reflection profile. If the GPR profile alone had been used for interpretation, some important layers, such as the basal soil unit, likely would have remained undetected, along with a number of thin sand dune layers not readily visible in the exposure. The differences between what was visible in the exposed sediments and what can be seen in the GPR reflection profile are solely related to the varying amount of water retained by the materials on either side of bed contacts.

The house floor is visible in the GPR profile as a high-amplitude reflection, as it was composed of imported clay, was compacted, and had been partially burned by fires (Conyers 2011b). It retained much more water than the sand above and below it and therefore generated a distinct reflection. The lower, dark brown sandy soil is only partially visible, because its composition is not appreciably different from the overlying sand, and while there is a distinct color difference between the two units at their interface, one retained no more water than the other, and therefore no distinct reflection was produced.

This comparison of radar reflections to what is visible in an exposure demonstrates that GPR profiles can sometimes produce a good representation of what is in the ground while not matching up with what you can see directly in the field. This mismatch is related to compositional differences that affect the amount of water retained in various layers in the ground. Thus, in order to make a detailed and accurate interpretation of ground conditions, all the factors that lead to the production of reflections must be considered. In the example in Figure 2-8, it was fortunate that the goal of the survey was to map the house floors in an ancient village built within these dunes, which were visible with GPR because they were composed of such different material than the sediment matrix. However, if the goal had been to understand sediment changes within the windblown sand layers or the underlying dark sandy soil, much more work would have been needed to explain the reflection complexity in those units. In most archaeological contexts, it is rare to find details about the distinct composition of each unit in the ground, and rarer still to be able to directly compare a reflection profile to a well-exposed section of material of this sort.

WATER RETENTION AND DISTRIBUTION AS THE KEY TO GPR REFLECTION ORIGINS

Admittedly, a few examples of the sort shown in Figure 2-8 are not enough to make a strong case that water's differential distribution and retention in the ground is the most significant variable in the production of radar reflections. I therefore offer support for that hypothesis with analysis and observations made in the laboratory and also some basic calculations of velocity in different materials. Some disclaimer is also necessary, as the water discussion presented here is probably only applicable to common sediments, soils, and natural building materials such as clay floors, which might be found in usual archaeological contexts. There are a number of other buried materials that do not necessarily rely on water to produce reflections, such as metal objects, void spaces, very hard rocks or stones with little porosity, and also modern objects such as plastic and asphalt. Many minerals and other exotic materials in the ground, not normally found in most near-surface archaeological contexts, also have very different properties that can potentially produce differential properties that affect the transmission and reflection of radar energy (Olhoeft 1981).

Exceptions aside, my hypothesis holds that the greatest changes in radar velocity are a function of the ability of a material to hold water. The radar wave velocity in freshwater is about as different from radar wave velocity in dry sand as is physically possible. Wave velocity in a void is about 30 cm/ns (the speed of light), and slower (about 17 cm/ns) in dry sand. Radar will rarely travel slower than it does in freshwater (about 3 cm/ns). For the sake of this argument with regard to radar velocities in usual ground conditions, we can consider very dry sand and freshwater as two ends of a radar wave velocity continuum. All usual materials in the ground of archaeological or geological importance will therefore have velocities between those of dry sand and freshwater, with only void spaces sitting outside these two end members. Therefore, all velocity changes are nothing more than the relative proportions of water that are retained in a buried material.

Some basic calculations can demonstrate this concept. *Relative dielectric permittivity* (RDP), while actually a very complex calculation that includes many variables and constants, is often used as a proxy measurement for velocity in GPR studies (Conyers 2004a: 48). The higher the RDP assigned to a material through which radar passes, the lower the velocity. Freshwater has an RDP of 80 (very slow), whereas dry quartz sand has an RDP of about 3 (fast). The velocity of radar in a void space is defined as an RDP of 1. Therefore, the greater the difference in RDP (as a proxy for velocity) between two materials at an interface, the higher the amplitude of the radar wave generated. There is an equation that attempts to quantify the amplitude of reflected waves, termed *coefficient of reflectivity*, which some have used as a way to determine various amplitudes at unit contacts (Conyers 2004a: 49). I have never had the occasion to use this equation in any useful way, and instead I look at general differences in RDP across a boundary to determine if a generated reflection will have high or low amplitude. A difference in RDP between 7 and 8 across a boundary would likely generate only a very low-amplitude radar wave reflection, if any at all, whereas a difference in RDP between 7 and 40 at an interface would generate a high-amplitude wave. Very high-amplitude reflections are also generated at the interfaces of a void space (with velocity of the speed of light) and any surrounding lower-velocity materials in tunnels and caves. This is just what can be seen in reflection profiles (Figure 2-9) over these features. Metal objects are perfect for radar wave reflections and generate very high-amplitude reflections, while plastic pipes (depending on whether they contain water, are empty, or perhaps are being used as conduits for wires) usually create lower-amplitude reflections (Figure 2-10).

To illustrate the concept of reflected wave amplitude with regard to water, a simple model was constructed (Figure 2-11). In this model, a sand layer overlies a clay unit, with a sharp interface between them. Using RDP values for these materials, a simple set of calculations allows for a basic understanding of how water differences relate to reflection generation and amplitude. The first model determines the amplitude of a radar wave generated at the interface when both sediment types are totally dry (Figure 2-11:A). The generated wave amplitude is a function of velocity change, but the model uses RDP (as a proxy for velocity) for both units, so a few calculations are necessary.

- If the sand is totally dry and has a porosity of 30% (about as high a porosity value as possible for a typical uncompacted sand), then 30% of the sand has an RDP of 1 (when all pore spaces are filled with air), and 70% of the unit (the sand itself) has an RDP of 3. The calculation to determine RDP of this dry sand unit as a whole is therefore $(0.3 \times 1) + (0.7 \times 3) = 2.4$.

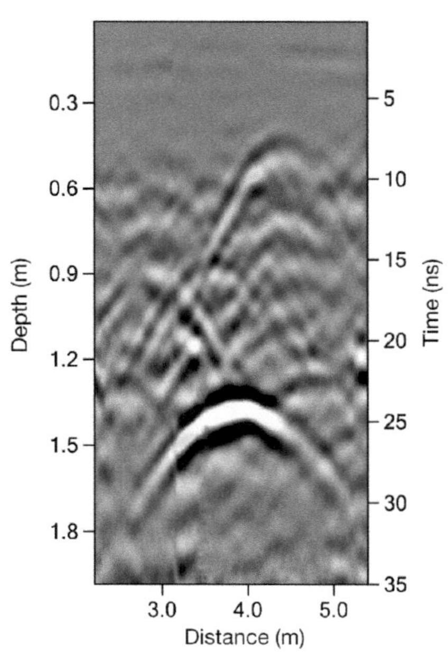

Figure 2-9: Reflection profile across a tunnel lined with concrete. The void space produces high-amplitude reflections. GPR data collected with 400 MHz antennas.

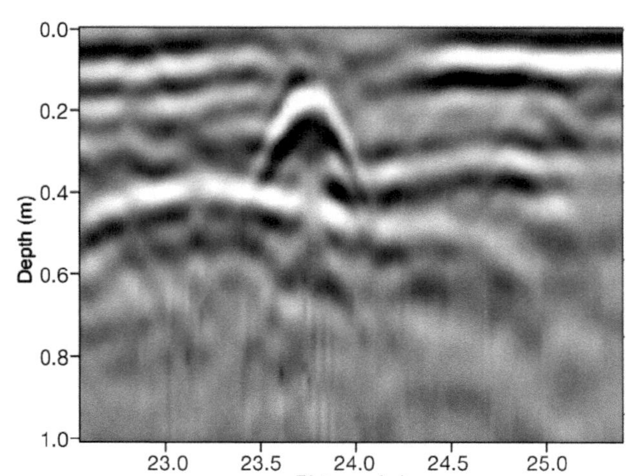

Figure 2-10: Reflection profile of a plastic fiber-optic conduit. This feature produces a high-amplitude hyperbolic reflection. Collected with 900 MHz antennas, University of Denver campus, Colorado.

- A totally dry clay layer has a porosity of 50% (typical for uncompacted clay). The RDP of the clay particles in that layer are modeled as having an RDP of 4 and the void space an RDP of 1. The total clay layer RDP is therefore (0.5 × 4) + (0.5 × 1) = 2.5.

The model shows that there is essentially no difference in the RDP (and therefore no velocity change) between the two units when they are totally dry, and almost no reflection amplitude is generated at the interface (Figure 2-11:A).

The model now assumes that water has been added to the system. If the clay were totally water saturated, the following calculation determines its overall RDP:

- 50% of the clay's pore spaces are filled with freshwater (RDP = 80), and the other 50% of the unit is clay, with an RDP of 4: (0.5 × 80) + (0.5 × 4) = 42.

Using these RDP values, the model generates a reflected wave at the interface between the saturated clay on the bottom and the totally dry sand above (Figure 2-11:B). The difference in RDP is 39.6, which corresponds to a very high-velocity difference at the interface, and a high-amplitude reflected wave is generated at the interface. In nature, this would be a very unusual situation, as the overlying sand would likely also retain some water.

Finally, both units are modeled as totally water saturated. The saturated sand RDP can be calculated this way:

- Sand porosity is 30%, filled with water (RDP = 80), and the remainder of the sand has an RDP of 3: (0.3 × 80) + (0.7 × 3) = 26.1.

The model shows a moderate amplitude reflection produced at the sand-clay interface when both are saturated with water, as the RDP difference between the two is 15.9 (Figure 2-11:C).

The purpose of these calculations and models is to demonstrate that the differences in RDP (and therefore velocity contrast and the resulting reflected wave amplitudes generated) are only a function of the amount of water retained in the pore spaces of each unit. The units when totally dry are so similar with respect to the velocity that radar moves between them with no reflection generated. What is actually generating the wave (and varying its amplitude) is the amount of water that is held by each in its pore spaces.

These modeled conditions of these units when totally saturated can be seen in GPR reflection profiles collected in freshwater lakes (Figure 2-12). In a lake in Colorado, a high-amplitude reflection was generated at the water-sediment interface (where the difference in RDP was between a value of 80 for water and perhaps 30 or so for the saturated silt sediment on the lake bottom). All other reflections generated from layers in the totally saturated sediment layers below the lake bottom were due only to the varying amounts of water retained

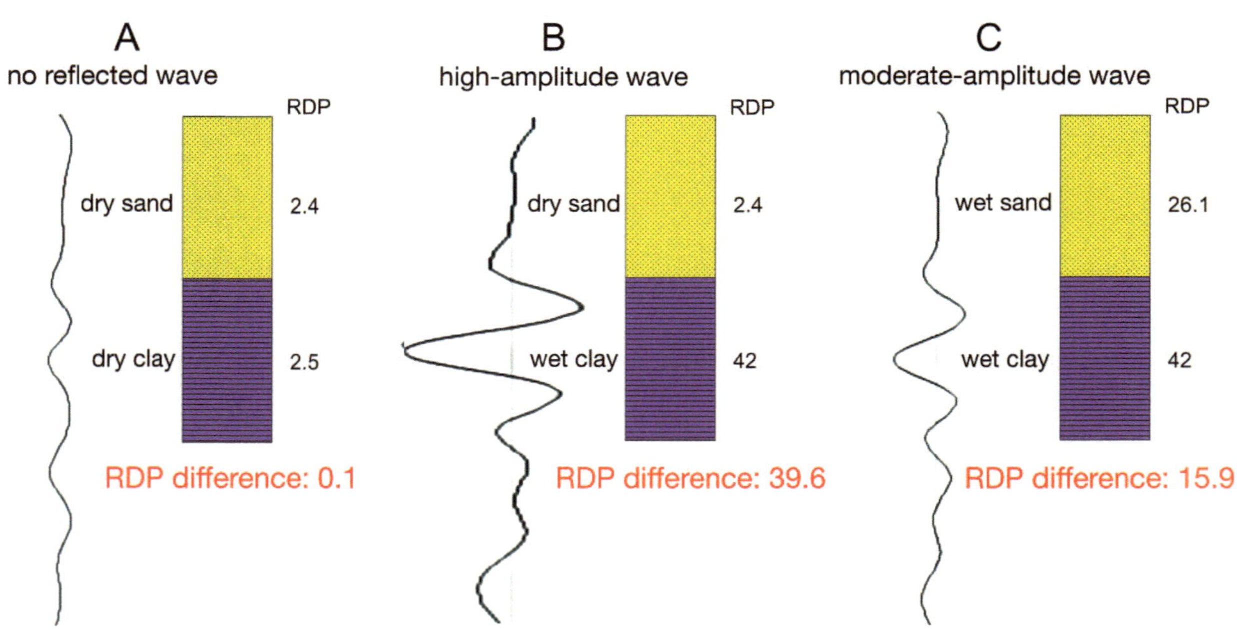

Figure 2-11: Models of reflected wave amplitudes in three different moisture conditions. A is dry sand over dry clay, B is dry sand over wet clay, and C is wet sand over wet clay. The difference in wave amplitude is a function only of the amount of water in each unit.

in the pore spaces of those different sediment layers, as all were totally water saturated. The highest-amplitude reflection was generated at the interface of the upper silt layer and the lowest clay unit.

In an attempt to quantify the RDP of many different units at archaeological sites where we had collected GPR data, and compare the reflection amplitudes generated from those layers of interest, my students and I collected samples from many test sites and analyzed them in the laboratory. We also obtained values for water saturation of the buried units, both in the field during collection and later during laboratory analysis. This process produced a large amount of data from different soil layers, sands, clays, silts, and interbedded cultural features of archaeological interest (Conyers 2004b). We found that every sample of every sediment and archaeological unit we analyzed had, when totally dry, an RDP between 3 and 4. It was only when we added known quantities of water to each sample in the laboratory and then re-collected our measurements that RDP varied between samples. These tests showed that changes in RDP, as measured both in the field and in the laboratory, varied only as a function of the amount of water that was retained in each type of material. This is another example of how changes in RDP are only a function of retained water in materials, and not of the types of material themselves.

From 2001 to 2004, a time when I was obsessed with trying to determine the cause of the varying radar reflections in different areas, my students and I collected a huge amount of GPR data at two archaeological test sites (Conyers 2004b) where it was easy to quantify these types of measurement. At both sites, a number of archaeological-like features had been constructed in excavated pits and then buried, after documenting these materials and their location and depths. One site in Illinois was built within a clay matrix, and the other, in Washington State, within sand (the CATS and Hammer sites, respectively). After collecting many tens of grids of GPR data with various antennas and GPR systems over a period of three years, we found that most buried features were visible (or invisible) only as a function of the amount of water retained by those features. Some archaeological features generated high-amplitude

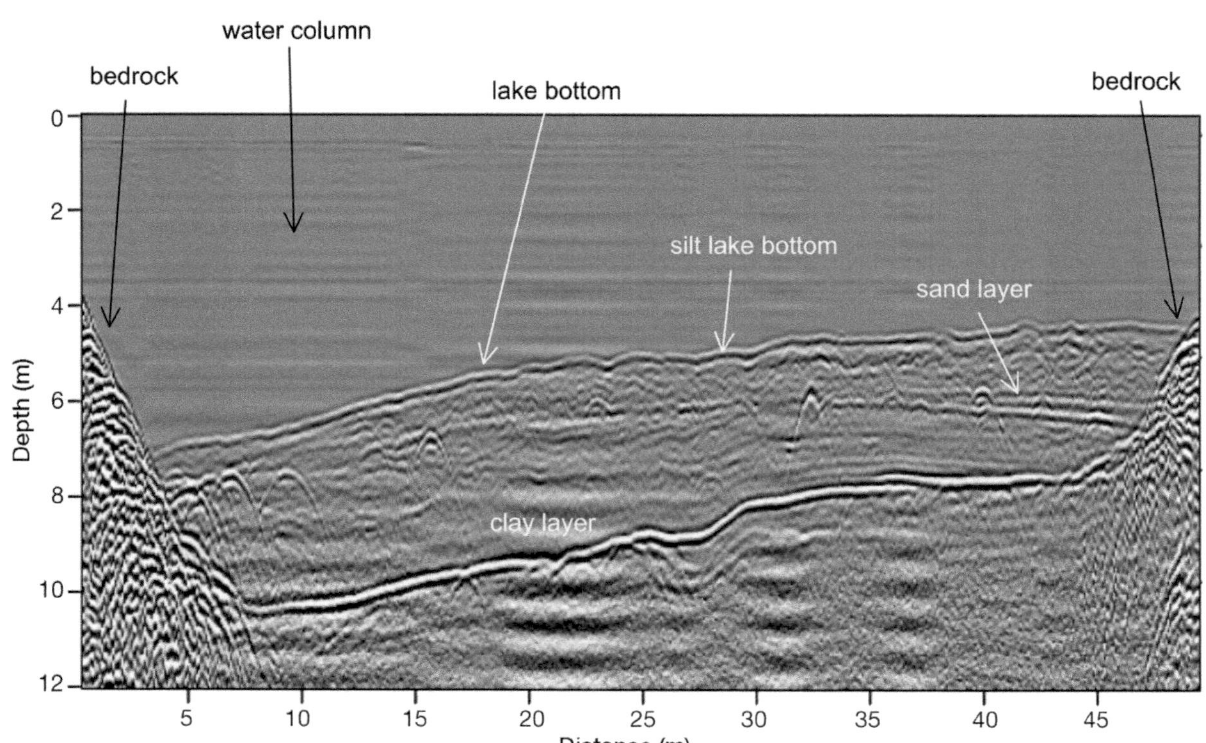

Figure 2-12: Reflection profile collected from a boat on a lake. The sediment on the lake bottom produces a high-amplitude reflection, as does a deeper clay unit where it is overlain by sandy silt. A sand layer within the sandy silt sediment package is visible as a weaker reflection. Collected with 400 MHz antennas at South Mesa Lake, Colorado.

reflections when the ground was dry and others when it was wet. Frozen ground conditions appear to have produced only some small differences in the coupling of radar energy with the ground and therefore in the amount of energy that could travel to the buried features. It appears that frozen ground did not play a direct role in the amplitude of reflections from the buried features of interest, but did affect the amount of energy that moved into the ground to be reflected.

Figure 2-13: Burned clay floor during construction at the CATS site. This test facility is located in Champaign, Illinois. *Courtesy of Mike Hargrave and CERL.*

In this study, a burned clay floor at CATS (Figure 2-13) generated only a very weak radar wave reflection when the ground was dry (it had not rained for about two months prior to GPR data collection and the ground was desiccated). Our measurements showed that the constructed clay floors had retained almost the same amount of water as the clay overburden at the time of data collection, and almost no reflections were generated at the interface. After a very strong summer rainstorm, we collected the GPR data again over the floors, using the same collection parameters as when the ground was dry. We found that downward percolating rainwater had pooled water along the burned floor's surface, producing a very high-amplitude reflection (Figure 2-14). Other nearby floors, which had been constructed of the same material as the site matrix, were not compacted when constructed. They did not provide an interface to retain water and did not therefore reflect radar energy, whether

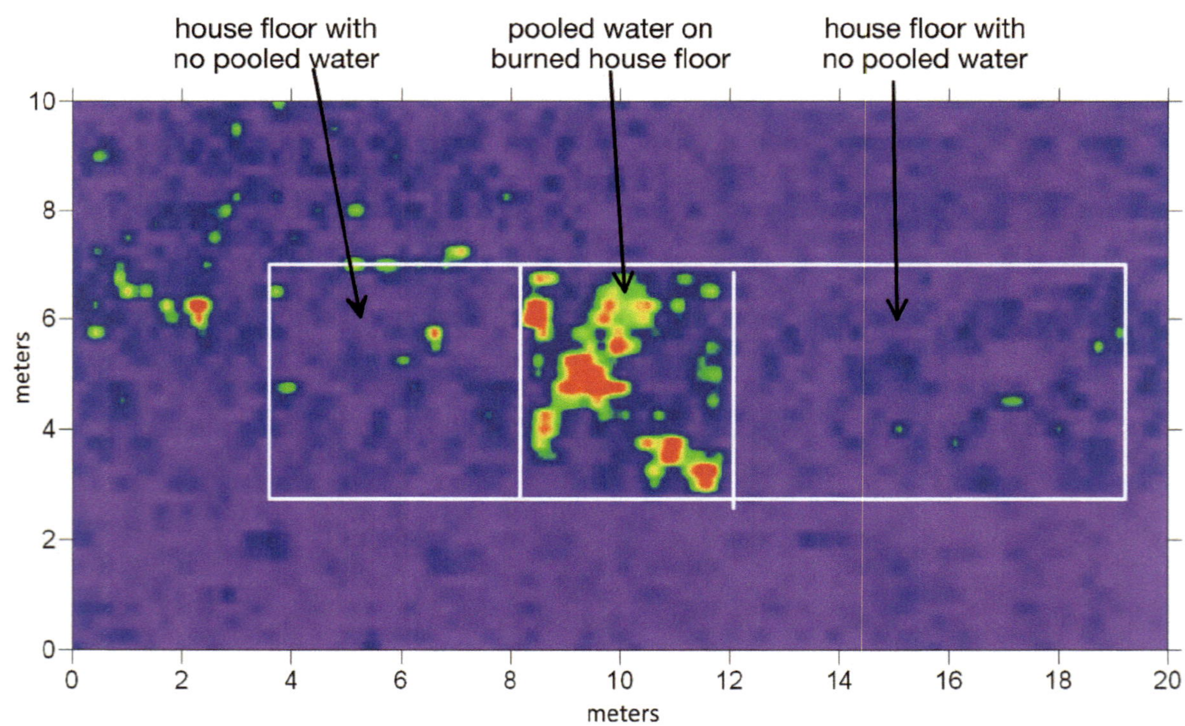

Figure 2-14: GPR amplitude map across constructed house floors at the CATS site. The profiles from which this image was produced were collected after a very heavy rainstorm. The only floor visible is the burned feature, because it collected percolating water on its surface. The other floors are not visible in reflection amplitude maps, as they contain about the same amount of water as the surrounding material. Data collected with 400 MHz antennas, Champaign, Illinois.

the ground was wet or dry. Only the burned floor pooled water on its surface after water was added to the system, and therefore only it produced a high-amplitude reflection. In this case, amplitudes of reflected waves from these floors were totally a function of each unit's ability to retain water.

At the Hammer site, where the ground was very sandy, features such as hearths and various objects such as stones and metal readily generated reflections. Some wooden logs (Figure 2-15), however, produced no radar reflections when the ground was dry, as those materials had an RDP that was not much different from the surrounding sand (Figure 2-16). When we applied a large amount of water to this site (using a fire hose, as it rains only rarely at this location), and then re-collected the GPR data, the wooden log became one of the most prominent buried features visible in both reflection profiles and amplitude maps. The log had retained water in its pore spaces like a sponge, and the remaining water we had added to the ground percolated into the surrounding sand and moved downward. In this case, the wooden log was visible because of its high water content and its contact with the quickly draining sand, which created a large difference in velocity.

Many GPR users from all over the world have commented on the differences they see in reflection amplitudes depending on whether the ground is wet or dry.

Figure 2-15: Construction of the Hammer site, with wooden logs in a sandy matrix. This site is located in sandy ground in central Washington. *Photo courtesy of Darby Stapp.*

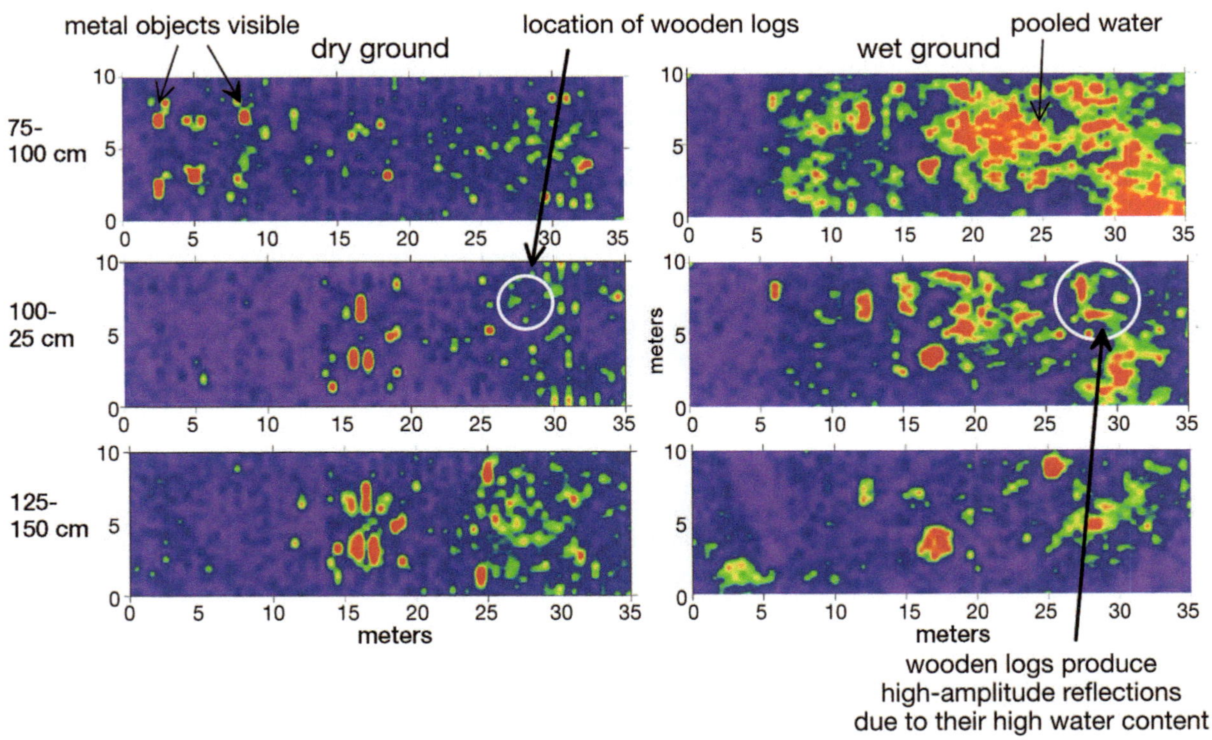

Figure 2-16: Amplitude maps at the Hammer test site. These show the difference between reflections in three slices when the ground was dry and after it was saturated with water. Only metal and stone objects produced reflections in dry conditions. When the ground was saturated, the wooden logs in the 100–125 cm slice produced high-amplitude reflections because they retained water. Data collected with 400 MHz antennas in Richland, Washington.

For instance, I re-collected a data set over a kiva at Bluff, Utah, the day after a large rainstorm. When those reflection data were processed, all that was visible were random high-amplitude reflections, which were not at all related to the circular archaeological feature we knew to be present. The reflection profiles and amplitude maps were producing only images of the reflections of differentially retained water in the sediment fill, obscuring the buried archaeological features of interest. Dean Goodman told me about a survey he conducted in Japan where a Kofun-period tomb was invisible when the ground was dry and only produced high-amplitude reflections after a typhoon dropped a large amount of water. In this case, the water retained on the archaeological feature was what was visible, not necessarily the archaeological feature itself. In both cases, it was the addition and distribution of water within the materials being studied that was the largest variable in the production of radar reflections.

It was with a good deal of apprehension that I presented these conclusions about water, and how almost everything we are seeing in the ground with GPR might be related to water distribution, at the Tenth International Conference on GPR in Delft, The Netherlands, in 2004 (Conyers 2004b). Would the world's top GPR scientists in the audience berate me for being ignorant of what they had long known to be the case, or approach me later to agree that my conclusions were an interesting new idea about GPR and perhaps even agree that this idea fit with their own observations? Neither of these reactions occurred, and my conclusions were met with silence from the audience. Since that time, after having published these data often (Conyers 2004b, 2006c, 2009) and having stated these conclusions in many public talks, the GPR community has continued to hold its collective tongue. No one has yet contradicted my hypothesis about water distribution, so perhaps others had been thinking of this idea already, or everyone else already knew of this concept (and just never published it), and perhaps I am a latecomer to this idea. Whatever the case, I am not going to worry overmuch about what others think and instead will continue to present the results of those studies that my students and I have conducted, and readers of this book can test the idea themselves with their own data.

The case I am making here is that there are often very distinct GPR reflections, which can be readily related to buried features or other materials of interest. The higher the generated amplitude of reflections, the greater the radar wave velocity contrasts along the interface. While we all try to conclude our studies with an interpretation that relates radar reflections visible in profiles and other images directly to the actual features themselves, what is actually producing the reflections is likely more complicated. Our conclusions may state that the GPR images show a buried house floor, while in reality many other variables that relate to water and its distribution with respect to that floor are probably more likely the cause. In the remainder of this book, I will continue to directly interpret GPR images by pointing to features in terms of their archaeological or geological origin, knowing that it is actually water variations within them that are producing the reflections.

BASIC DATA PROCESSING STEPS FOR INTERPRETATION

Most of the images presented in this book were generated from reflection profiles that were processed in just a few basic ways and therefore would be considered by many a form of "raw" data. Basic post-acquisition processing procedures, which are usually employed to view GPR data and begin interpretation, are quite simple and do not usually remove or modify reflections that were generated in the ground. These basic procedures are time-zero adjustments, profile axis calibrations for depth (using a calculated velocity) and horizontal distance, regaining to highlight or enhance certain amplitudes, and use of basic background removal and frequency filters. Many interpreters do not venture beyond these simple basic steps, and for most interpretation applications these procedures are sufficient. There might be good reasons to proceed to more complicated processing and filtering methods, but it is always important to have a well-thought-out reason for making those more complex modifications to a data set. For instance, I feel good about seeing point-source hyperbolas in profiles, as they can quickly point to the location of individual objects, and I also use hyperbola shapes to quickly determine velocity (Conyers 2004a: 117). The hyperbolas can later be removed using migration processing, but this would then make it more difficult to see individual objects in profiles.

I am not in favor of arbitrarily processing data in the hope of randomly producing results that might reveal more interesting "anomalies." Some people recommend a standard procedure of processing steps, which may or may not have worked in the past, for all data sets. In my view, following a complex processing procedure of many steps is like taking a random blindfolded walk in the dark and then attempting to figure out where you are once you can see again. It might lead you to an interesting location, but you would be confused about your surroundings, and it would be difficult to know how to proceed from there. I have read a good deal of literature on the types of GPR processing steps that some software packages make possible and have concluded that there are a few that can be very useful in basic interpretation procedures, but many I just don't understand. Some of those advanced processing procedures have been transferred to GPR from seismic data procedures, and others were written by software coders who have incorporated a plethora of "wish list" processing steps, some of which have very specific applications for disciplines that most of us never dabble in. I am therefore hesitant to experiment with complex waveform processing procedures, as I don't know exactly what I am doing or how to interpret the final product. Life as a GPR interpreter is complicated enough for me when I think I know what I am trying to do, and becomes almost impossible when I have no idea what I might have done to alter my data.

I always start with individual reflection profiles in any interpretation procedure (Figure 2-17). After those raw reflection data have been adjusted on the horizontal and vertical scales to a size that is useful for viewing reflections (often with some vertical exaggeration), I can usually see most of the features I care about (Figure 2-18). When slicing the reflection data from many profiles in a grid to make amplitude maps, even these initial steps

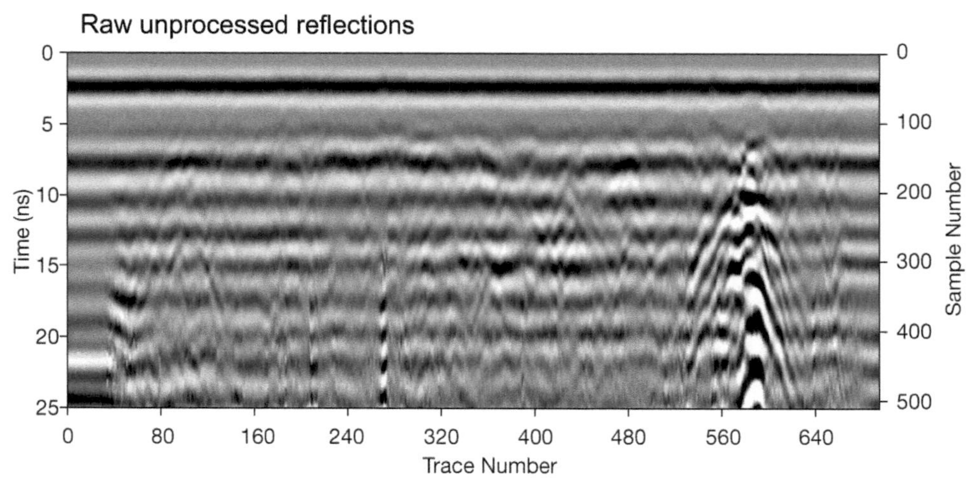

Figure 2-17: Reflection profile showing all raw reflections. Data collected at Petra, Jordan, with 400 MHz antennas.

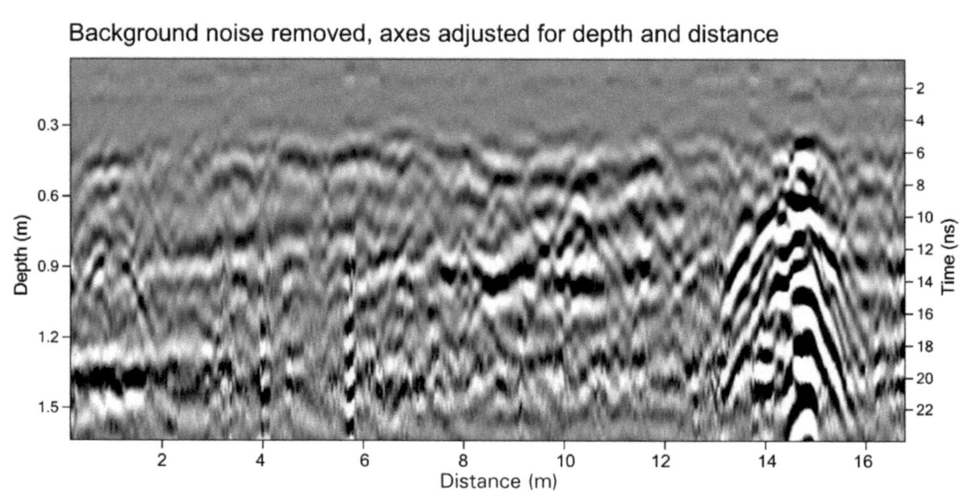

Figure 2-18: Reflection profile seen in Figure 2-17, after background noise was removed and axes adjusted for distance and depth (using an average velocity). Data collected at Petra, Jordan, with 400 MHz antennas.

are often not necessary. The resampling procedures used to produce these images are set up to sample the relative amplitudes of reflections in all profiles using certain input parameters, and the resulting maps produce images of the spatial placement of variations in those amplitudes in a horizontal map view. If the slicing thickness is appropriate for the features to be imaged, colors of the displayed amplitudes can be adjusted for varying amplitudes, which is analogous to adjusting the gains of reflections in profile view. The resulting stacked horizontal amplitude maps that yield a three-dimensional view of all reflections in a grid are then ready for at least a first round of interpretation. In this process, as long as the data have been collected correctly, little processing of the individual profiles is necessary, as the final product is only displaying variations in reflection over a large area.

Once I have produced my first set of amplitude maps, I usually return to profile viewing software to interpret individual profiles in order to make sure I know what features visible in profiles are producing the amplitude features I see in map form. If I have sliced too thick or thin, or interpolated the amplitudes too much, or not enough, I return to the amplitude map production software and try different resampling and gridding procedures. In these iterative steps, it is always important to think about what in the individual profiles has been producing the amplitudes in the maps. At this point in an interpretation task, the basic processing and image production steps can be modified so that buried features of interest are enhanced or delineated in a way to make them easier to identify and understand. Those basic processing steps, using both two- and three-dimensional images in concert, are sufficient for many interpretation tasks.

MORE ADVANCED DATA PROCESSING STEPS

There are a few other data processing steps that are becoming increasingly common in GPR interpretation, which can be readily employed without worrying too much about modifying the reflection data in incomprehensible ways. One is *trace stacking*, which can help to overcome problems with antenna coupling on uneven ground. These produce more "average" reflection profiles that can tend to remove coupling changes in reflection amplitudes. The risk with this step is that the smaller reflection hyperbolas will be filtered out during the production of these images, and only average images of the more extensive planar reflections remain. I made the mistake of stacking traces during collection when confronted with the very bumpy ground in newly plowed fields in Ethiopia (discussed in Chapter 5). Only later in the day did I realize I was also removing all reflection hyperbolas, which precluded any velocity analysis from them. I blame that mistake on jet-lag, as it was my first day in the field after a very long trip. Another important step is to zoom in on features of interest in profile, which can help identify what is producing specific reflections in the ground (Figure 2-19).

A data processing step that is commonly used prior to the production of amplitude maps is *migration*, which can enhance the amplitudes at the apex of most reflection hyperbolas while removing their axes. To perform this step, velocity estimates of the ground must be determined so that hyperbolas can be effectively removed, as their geometry is a function of velocity. This is always difficult, because in most ground conditions, velocity changes with depth, usually slowing. A velocity that was determined from a shallow depth, when applied to deeper reflections will often not completely migrate the hyperbola axes (Figure 2-20). Migration processing yields a set of reflection profiles in which reflections are very discretely placed in space (in both depth and distance) and individual objects in the ground are not as blurred or distorted by hyperbolas. This is a particularly beneficial processing step, as profiles that are unmigrated

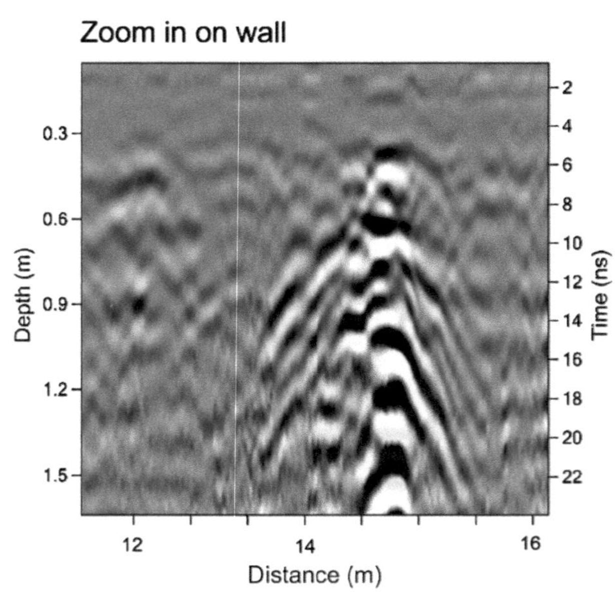

Figure 2-19: Reflection profile of a Roman wall. This profile is zoomed in on the Roman wall displayed in Figure 2-18. Data collected at Petra, Jordan, with 400 MHz antennas.

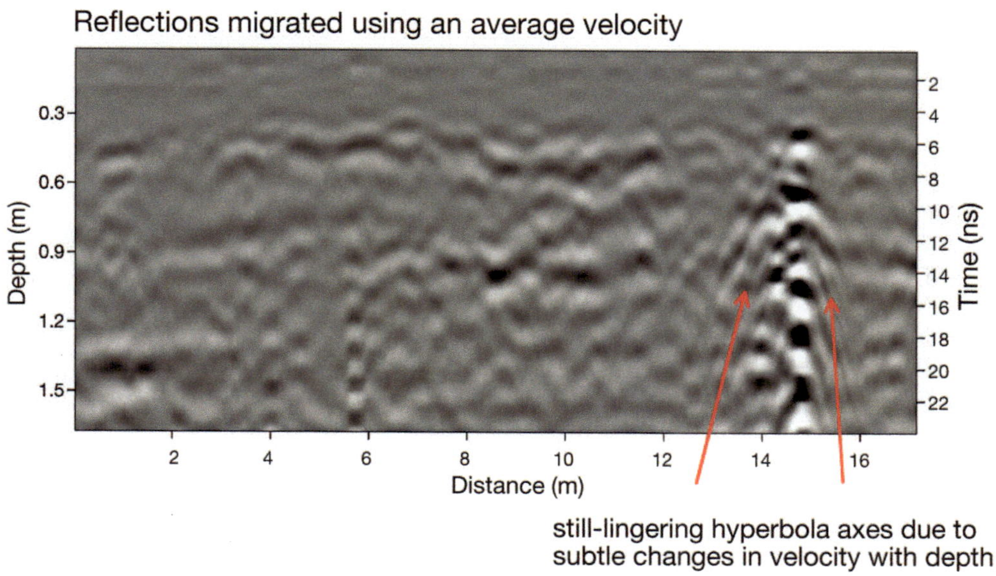

Figure 2-20: Reflection profile seen in Figure 2-18, after reflections were migrated using an average RDP of 4.6. Data collected at Petra, Jordan, with 400 MHz antennas.

and contain many hyperbolas will produce somewhat blurred reflection features in amplitude maps, as the software is sampling the amplitudes of the axes of the hyperbolas, which are recorded deeper (as measured in time) and spread out from their point-source origins. This processing procedure can also make the migrated profiles more difficult to interpret, as all hyperbolas are removed, but it is a very useful tool to use prior to producing amplitude slice-maps. The resulting amplitude maps produced from migrated reflections can illustrate, with no distortion or blurring, individual stones and walls. This is especially useful in data sets where there is an abundance of hyperbolic reflections from buried features of interest such as graves, walls, or stones in general.

Another useful processing step, which can be applied to data prior to producing amplitude maps, is *frequency filtering* (Grealy 2006). Most GPR antennas are "wide-band" transmitters and receivers, which are always producing and recording a range of wave frequencies around a mean (Conyers 2004a: 39). Therefore, when an antenna is defined as having a frequency of 400 MHz, this really means that it is generating energy on either side of an estimated "center frequency," and there are actually many waves produced and collected of many frequencies. The rule of thumb is that in broadband antennas, the frequencies vary between one-half and two times the center frequency, which means the 400 MHz antennas are generating waves between about 200 and 800 MHz (Conyers 2004a: 40). Most of the energy produced is varying only a small amount from the mean, but a few of the other frequencies are still moving into the ground and being reflected and recorded back at the surface. Each antenna also has its own subtle variations, so it is always difficult to know exactly what frequencies of energy are really being produced and recorded. Also, at any one location, all the waves of varying frequencies that are recorded are a function of a wide range of chemical and physical variables within the materials through which the energy passes. As a result, what is displayed in a reflection profile, or what is resampled from profiles to produce amplitude maps, is averages of all the waves recorded in many thousands of individual reflection traces. To frequency-filter data, all traces in a profile must be resampled using processing software that can remove some frequencies while retaining others to produce a whole new set of reflection profiles (Figure 2-21). Amplitude maps can then be generated from those processed profiles, displaying only the reflections that were generated from certain frequencies of energy. The higher-frequency energy is shorter in wavelength and therefore will usually reflect from smaller buried features or objects in the ground with better spatial definition, but the energy will have had a shallower depth penetration. Lower, longer-wavelength frequency energy will theoretically travel deeper in the ground, but it is only reflected from the larger objects or thicker layers

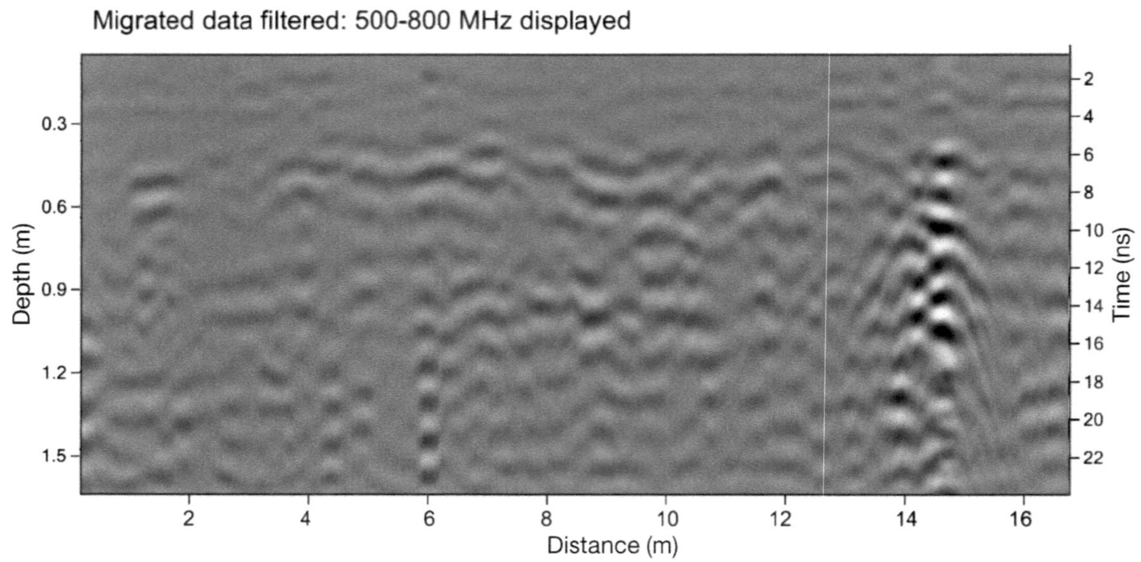

Figure 2-21: Reflection profile seen in Figure 2-17, showing reflections both migrated and filtered so that only 500–800 MHz data are displayed. Data collected at Petra, Jordan, with 400 MHz antennas.

(Conyers 2004a: 65). This processing step can be quite beneficial if one finds, during interpretation, that smaller (or larger) buried features need to be defined and only one frequency of antenna was used to collect the data originally.

Depending on the questions asked during interpretation about buried features of interest, some of these processed-data images can be produced to answer those questions. They should not be employed, however, until the raw data are first interpreted in the more standard ways outlined above. Then, a second or third version of the same data can be produced and compared with the first interpretations in a way that can help define certain buried features of interest. Interpretations of this sort take some time, as all these steps must be performed in a knowledgeable and deliberate way, and the outcomes can often be quite surprising and informative.

As an example of how some of these processing steps can be used, I reprocessed reflection data to produce a series of amplitude maps of a buried Roman temple in Petra (Conyers et al. 2002). The unprocessed 400 MHz profiles, when horizontally sliced in 4 ns thicknesses, produced a good image of the back wall and the bases of what appear to be pillars of this structure. This "raw data" amplitude map is quite good and displays all the buried features that might be necessary for a general-purpose interpretation (Figure 2-22). The back wall of the temple is visible as a mostly continuous reflection feature, which appears just slightly discontinuous in places.

When all the profiles in this grid were migrated, the point-source reflection hyperbolas axes were removed and migrated back to their apex. The resulting profiles were then resampled, and an amplitude map was produced using the same slicing parameters (Figure 2-22; migrated data). In this image, the back wall becomes more clearly defined, showing the individual stones that make it up. I then took those same raw-data reflection profiles and both migrated the reflections and then filtered the resulting reflections so that only the 500–800 MHz energy remained. All those profiles were then resampled and sliced again, using the same thickness and interpolation parameters as before. This process produced an amplitude map that even further defined the stones of the buried feature's back wall (Figure 2-22: migrated reflections, filtered 500–800 MHz).

In that final, highly processed image, what had been originally interpreted as the continuous "back wall" of the temple can be seen to be composed of small individual stones, which do not make up a continuous wall at all, but only its foundation. The final maps also show that little remains in the ground of the structure other than a few individual stones outlining what had at one time been a substantial temple. This feature was excavated, and this is exactly what was found, much to our initial surprise (Figure 2-23). Our initial interpretation was based on unprocessed reflection data, which tended to smear all the buried features, and we thought this "temple" would be much more substantial than it was.

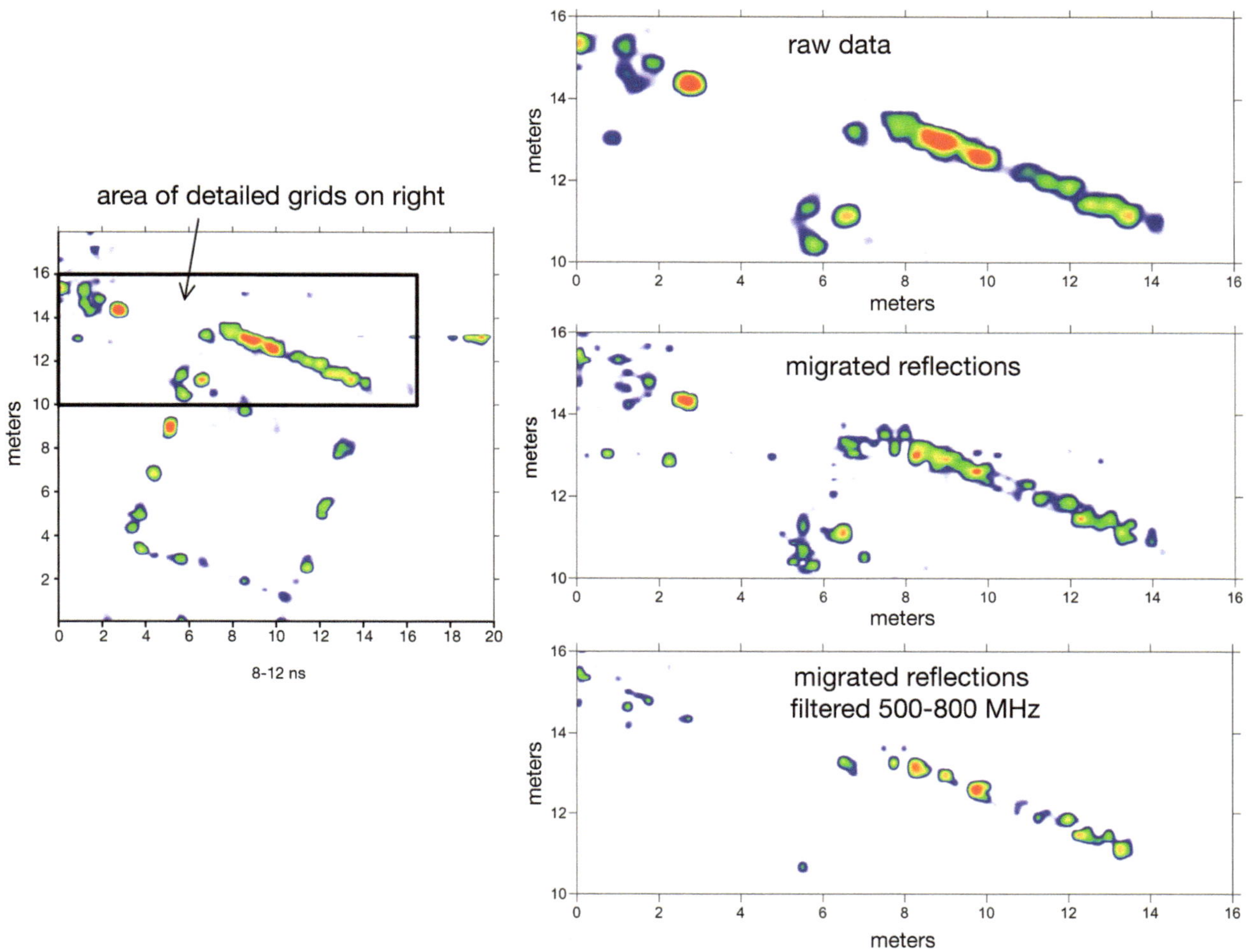

Figure 2-22: Amplitude map showing various images using incremental processing of the reflection data. On the left is the amplitude map of the 8–12 ns slice of the whole Roman temple feature at Petra, Jordan, using 400 MHz data. The back wall of the structure is then shown when the raw data are sampled and displayed (top right image). When these same reflections are migrated (middle right), the wall becomes more distinct, and when those migrated reflections are further filtered to contain only the 500–800 MHz frequencies (lower image on right), individual rocks making up the wall become visible.

We had visions of carved marble columns and other fancy architectural elements of a temple dedicated to the goddess Venus, or something like that. What we really found was the foundation stones of the temple, composed of building materials that were not considered important and had been left in place when this structure was dismantled. It is likely that the well-cut stones were recycled into other buildings elsewhere sometime during the last 1,500 years.

Figure 2-23: Excavations of the back wall of the Roman temple shown in Figure 2-22. These are the individual rocks that made up the foundation of the building.

Only a considered series of processing steps, which are understood and documented during interpretation, should be used for these types of data processing and associated interpretation. In these processes, some data are removed and others enhanced, with certain aspects of the radar waves incorporated into completely new profiles. The example illustrated here included five basic processing steps and three more advanced steps, to produce the final outcome. That outcome answered important questions about the buried feature, and therefore all these steps were justified and could be well documented. Just blithely processing data using all these steps, without knowing what each procedure is accomplishing, could have led to results that were at best uncertain. In our example, that was not the result, which the excavations demonstrated.

3

A Personal History of GPR Interpretation

Early Processing and Interpretation Methods

In the early days of GPR (prior to about 1992), the standard data set with which to make interpretations was comprised only of reflection profiles, most of which were images of unprocessed and unfiltered reflections. These were often printed out on paper, and we marked them with colored pencils to annotate reflections we thought were important. Locations of those reflections and interpretations of them were then manually transferred to a base map and finally interpreted by attempting to understand the shape and possible origin of reflections and their depth below the surface. For instance, in what is now considered a very antiquated interpretation method, reflections were described in such terms as "hyperbolic" or perhaps "high-amplitude western dipping reflection" or other similarly vague and marginally useful language. Interpreters usually placed the location of these visible reflections on a base map and tried to understand what might be in the ground based only on that information. It was often less than informative but, at the time, was considered the best that could be accomplished with what we had.

This is exactly what I did with my first GPR data set from the buried Mayan village site of Cerén in El Salvador (Conyers 1995). This impressive site consists of well-preserved Classic-period farmhouses constructed of earth and built on a distinct ash layer (Sheets 1992; Conyers and Spetzler 2002). The village was buried during the eruption of a small nearby volcano in a matter of a few hours or days, and today the cultural remains are located between 2 and 7 m below the surface. Some of these Mayan houses have been excavated and provide excellent models from which to interpret those that remain buried and are visible only using GPR (Figure 3-1). My task in the late 1980s and early 1990s was to find those buried houses using GPR and map them in order to understand more about the social organization and other aspects of the lives of the people in that ancient village by studying the buried cultural landscape.

The GPR data set at the Cerén site was collected in 1979 using an analog GSSI SIR-7 system with a monostatic 80 MHz antenna, prior to the development of digital GPR systems. Reflection profiles were printed out on thermal paper during collection, and these printouts remained the only output, so there was no possibility of any type of sophisticated data processing (Figure 3-2). It was laborious and tedious work to interpret these rudimentary images, and when finished, we had only a vague idea of what might be in the ground. I could see reflection "anomalies" and was able to basically describe them, but was still quite confused about what might have produced them in the ground. Some reflections were planar, aerially extensive reflections that appeared to change their orientation in the ground. It appeared they might be ash layers draped over house floors or living surfaces, which were barely visible within the plethora of other complicated reflection features created from covering ash layers and more ancient buried living surfaces. Other reflections appeared to be of high amplitude and were interpreted as house floor reflections, which were visible in many profiles (Conyers 1995). A third type of reflection feature was the aerially discrete hyperbolic reflection, which I hypothesized was generated from house platform corners or perhaps upright clay columns, built on an aerially extensive high-amplitude buried living surface. Many other complex reflections were also visible that defied categorization and were assumed to have been generated from the layered volcanic tephra overburden. Those basic descriptions, which were a rudimentary interpretation at best, were all that was possible using those very raw reflection images.

With help from Jeff Lucius of the U.S. Geological Survey, I was later able to convert those analog data to a digital format and then conduct very basic data processing steps, including background removal and horizontal distance normalization on the resulting profiles. This was possible because the analog data of reflected radar waves, consisting of discrete traces that made up the profiles, had been recorded as minor voltage variations on magnetic tapes, which had been stored in a closet and forgotten for more than 10 years. I returned to Central America in 1991 and 1992 to collect new data with a digital GPR system (GSSI SIR-2 system) and used those data to calculate velocity (Conyers and Lucius 1996), convert radar wave travel time to depth, and more precisely map buried features with new 300 and 500 MHz data.

Figure 3-1: Cross section of a buried Mayan house at Cerén, El Salvador. The clay platform and columns of the farmhouse were built on an ancient living surface of white ash, which can be seen on this picture directly above the lowest dark clay. GPR produced distinct reflections from the clay platforms, columns, and the underlying living surface. *Image courtesy of Payson D. Sheets.*

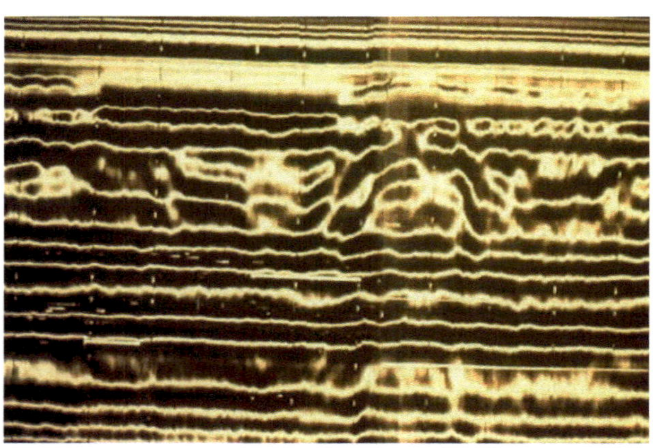

Figure 3-2: GPR reflection profile from 1979 vintage analog GSSI SIR-7 system. This profile was printed on paper from the monostatic 80 MHz antenna at Cerén, El Salvador.

The only output from all those data sets remained reflection profiles, but because they were digital, at least we could do some rudimentary processing to remove background noise, adjust scales, and assign reflections to a location in space. The interpretation of the processed profiles took many months and a lot of trial and error; and the origin of some of the reflection "anomalies" in some profiles remained questionable. Fortunately, I also had excavation information that allowed me to directly compare features visible in the GPR reflection profiles with what was known from the ground. It was only after thinking more broadly about the conditions in the ground (both the archaeology and the surrounding geological matrix) that I was able to produce a GPR interpretation that made sense and could help to clarify aspects of the Mayan site that were not otherwise apparent (Figure 3-3). House floors of various sizes and orientations, agricultural fields, a central plaza, and a complex buried landscape on which all were constructed could be mapped, and the interpretation indicated a more extensive and complicated village than was understood from the limited excavations. The resulting interpretive maps were still very "cartoon-like" and based only on a visual interpretation of many reflections visible in many hundreds of reflection profiles.

About the time that I was struggling to determine the origin of reflections in reflection profiles, Dean Goodman taught me to use his newly written GPR Simulation program (Goodman 1994), which he had developed for his own research in Japan in the early 1990s. The purpose of this innovative software was to produce two-dimensional models of known archaeological features using information about burial conditions and the composition of various building materials. The results were computer-generated synthetic reflection profiles that could be directly compared with reflection profiles

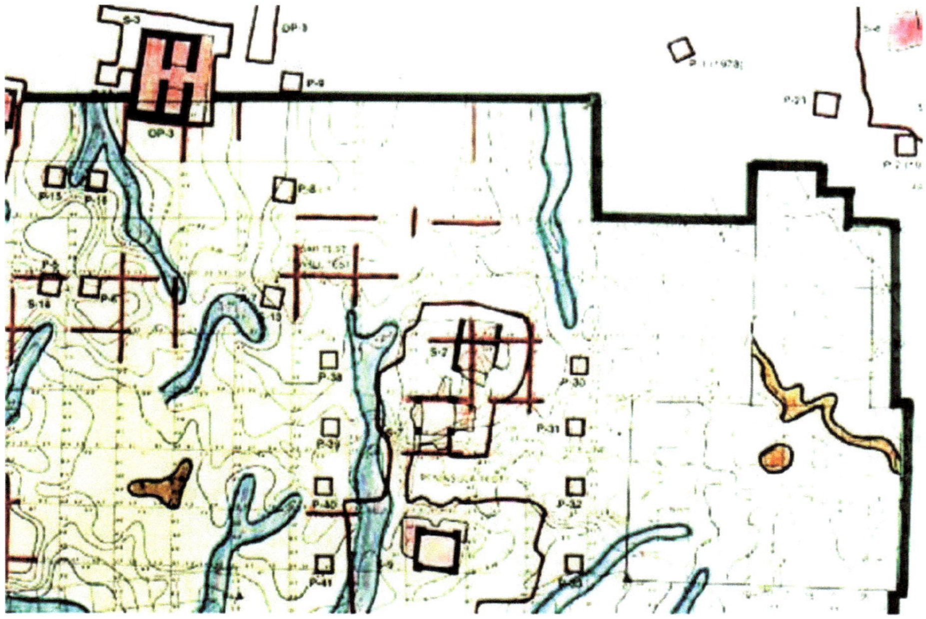

Figure 3-3: Hand-contoured map of GPR reflection "anomalies" at the Cerén site, El Salvador. This map was made by visually interpreting many reflection profiles in a number of grids.

generated from data collected in the field and thus serve as an aid to interpretation. I was particularly interested in determining what Mayan farmhouses might look like in GPR reflection profiles, and what the variations in architecture could tell me about the ancient village, long ago buried in ash. We modeled house floors with standing columns and found that under certain feature orientations and burial conditions, reflection hyperbolas were created from certain architectural components. The models also indicated that other types of houses that were less substantial, or had been flattened by volcanic ash flows, would appear almost invisible in GPR reflection profiles (Figure 3-4). These models, developed in accordance with the categorization of house types known from excavations, which *might* be visible in GPR profiles, could then be directly compared with the actual reflection profiles as a way to increase interpretive integrity.

There also turned out to be a good deal of complexity in the ground at the Cerén site due to variations in water saturation, which, because of its role in attenuating radar energy, tended to obscure some buried features in some locations. Radar wave velocity across the site varied with depth and also laterally (due to variations in retained water within the volcanic ash overburden), so that some features appeared deeper or shallower than expected when only an average velocity was used to convert time to depth (Conyers 1995).

The goal of the project was to use the GPR data as a way to interpret the buried archaeological features and ultimately arrive at some understanding of human behavior and ancient life in a Mayan farming village. In this first GPR study, reflection anomalies were "transformed" into buildings, and the placement and orientation of this architecture on an ancient landscape were used to model the layout of the village and some aspects of Mayan behavior in a rural agricultural setting. It was not an easy process and necessitated a good deal of manual placement of reflection features on base maps (Figure 3-4), but the final product became a three-dimensional map of the ancient Mayan village, which comprised more than just a collection of mostly unknowable "anomalies." All this interpretive work was done by hand, and the resulting maps were hand contoured. When finished, I remember thinking that there had to be a better way to interpret GPR data.

About this same time, Dean Goodman was writing software for a number of other GPR processing functions, the most important being the now common method of spatially imaging amplitudes in slice-maps using three-dimensional grids composed of many two-dimensional vertical profiles (Conyers and Goodman 1997). With this method, interpreters could dispense with the laborious plotting of reflections on maps from visual interpretations of paper copies of each and every

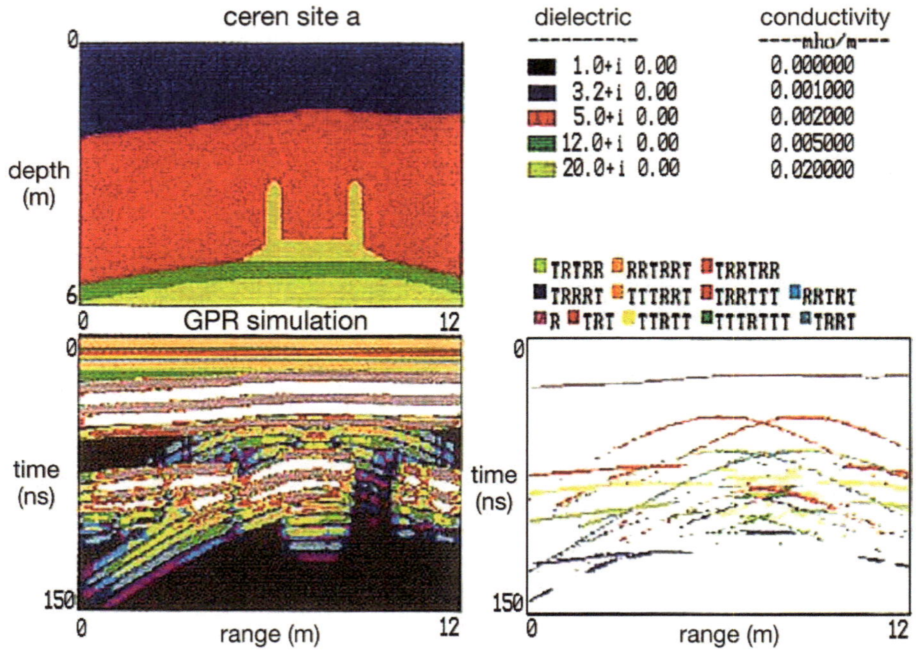

Figure 3-4: GPR simulation model of a Mayan structure at the Cerén site, El Salvador. This model shows what reflections would likely be generated in a reflection profile using known physical and chemical parameters of the architectural components and surrounding matrix.

reflection profile. By 1992, Dean had written the first GPR amplitude analysis software that could take multiple reflections and produce horizontal slice-maps (Goodman et al. 1995). I remember being very excited about the possibilities of this newly developed GPR processing method, but my enthusiasm was not universally shared. One of my early advisors, Gary Olhoeft, then at the U.S. Geological Survey in Denver, Colorado, commented that Dean's concept of producing horizontal amplitude slices was "not advisable." Nonetheless, as it turned out, amplitude analysis of this sort has transformed GPR processing and interpretation from the "clunky" and labor-intensive method that it was. This basic image-producing step is the most widely used interpretation process among all GPR users and has allowed GPR to become the important geophysical method that it is today. The ability to take batches of complicated individual reflection profiles, which would previously take days to process and interpret individually, and create useful images from them in only a few minutes, has transformed the GPR method. But Gary's comment about this method's advisability was also correct with regard to the complex data set I was then working with, as its geological complexity ended up producing amplitude slice-maps that were totally misleading and led to some erroneous conclusions on my part. At the time I was so enamored with the ability of Dean's program to process huge amounts of reflections in just a few minutes that I ignored Gary's warnings and began producing many amplitude maps from my Mayan village site anyway. Amazing "anomalies" showed up on all of them, and all were processed in only a few minutes at most. I was totally enthralled with the power of this computer processing that had transformed so much complex GPR data into readily interpretable images.

The horizontal slices I produced using the amplitude mapping program and the "anomalies" in my maps turned out to be misleading at best. I didn't recognize this at the time (about 1993). Amplitude slice-maps are produced by resampling all reflections in all profiles within a grid into discrete slices, usually defined by upper and lower values of radar travel-time increments. For instance, the program can be set to measure all the amplitudes (whether positive or negative deflections in the recorded waves in all traces) in 5 ns thick packages across a grid. Those amplitudes values are given a location in space (x and y coordinates in a surface grid), and then the data are gridded and mapped to produce horizontal amplitude maps at the programmed depths, as defined by the sampled times (Conyers and Goodman 1997). As I had calculated velocities across my study area at Cerén, I could convert two-way travel time (TWTT) to depth, and the strength of reflections in the ground at the depths I cared about could then be visualized. Perhaps I was too eager to save interpretation time and too enthusiastic about this new processing method, but in hindsight, my first maps using this method proved to be almost meaningless and mostly misleading. It took about a year of this type of processing and interpretation before Gary Olhoeft's prescient comment (that it was "not advisable") began to sink in.

What I had done in my zeal to save time was to partially sample reflections in the ground from a variety of ash layers, which encased and bounded the archaeological features I was really interested in. Those layers, which were often mounded over some of the buried buildings but also were deposited on a topographically variable living surface, sometimes happened to fall within one or another of the arbitrary slices I generated. The amplitude maps I produced turned out to be very complex, as they were horizontal (parallel to the flat ground surface today) whereas the layers in the ground that were producing the reflections in profiles dipped and arched across my GPR grids (Figure 3-5). Reflection anomalies were therefore produced everywhere that a highly reflective volcanic layer, or the underlying ancient living surface, crossed into or out of my arbitrary horizontal slices. Looking at it reflectively, it was a terribly innocent and simple mistake to make, coming out of my eagerness to avoid the tedium of manually processing such a huge data set. Not until after I had used some of these flawed interpretations (and other valid ones too, which were the ones I did originally using manual placement of reflections) to obtain a Ph.D. did I recognize that some of my most striking amplitude maps were practically useless. In fact, those maps were only marginally serviceable geological maps that offered nothing about the buried archaeology within this complex layered system. The slice-maps mostly showed linear high-amplitude features parallel to the strike of the dipping volcanic beds, with horizontal slices displaying one high-amplitude layer in progressively deeper maps, which is just what the software was programmed to do (Figure 3-6). In 1995, I finally figured all this out.

I had made the common mistake of horizontally slicing across dipping beds and producing maps that were showing the varying depths of a complexly layered system—and not the archaeological features contained within this geological package. Even as recently as last year, I saw some results of similar amplitude maps presented at a conference, which were described by the authors as "unknown anomalies." A quick look at the reflection profiles from which the slices had been made, together with a basic interpretive knowledge about the geological matrix of the site, could have avoided this erroneous interpretation.

Perhaps I can be somewhat philosophical about my own initial interpretation errors because, in the long run, the development of the amplitude slicing method was

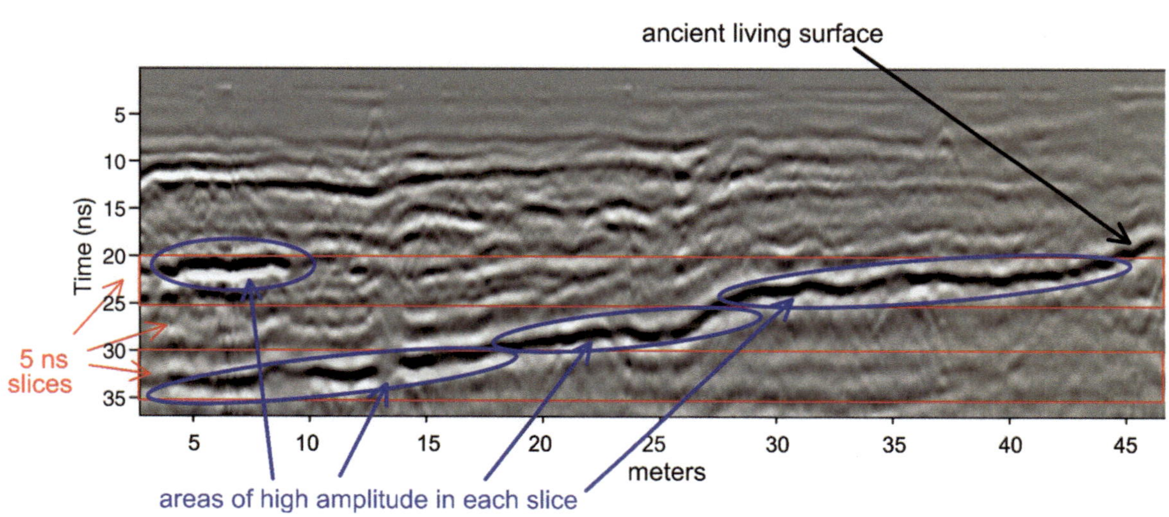

Figure 3-5: Reflection profile sliced and then sampled in 5 ns thick slices. The continuous high-amplitude reflection is the ancient living surface at the Cerén site, El Salvador. As the layer that produced the high-amplitude reflection is not horizontal, it is sampled in three different depth slices. Data collected with a GSSI SIR-2 system and 300 MHz antennas.

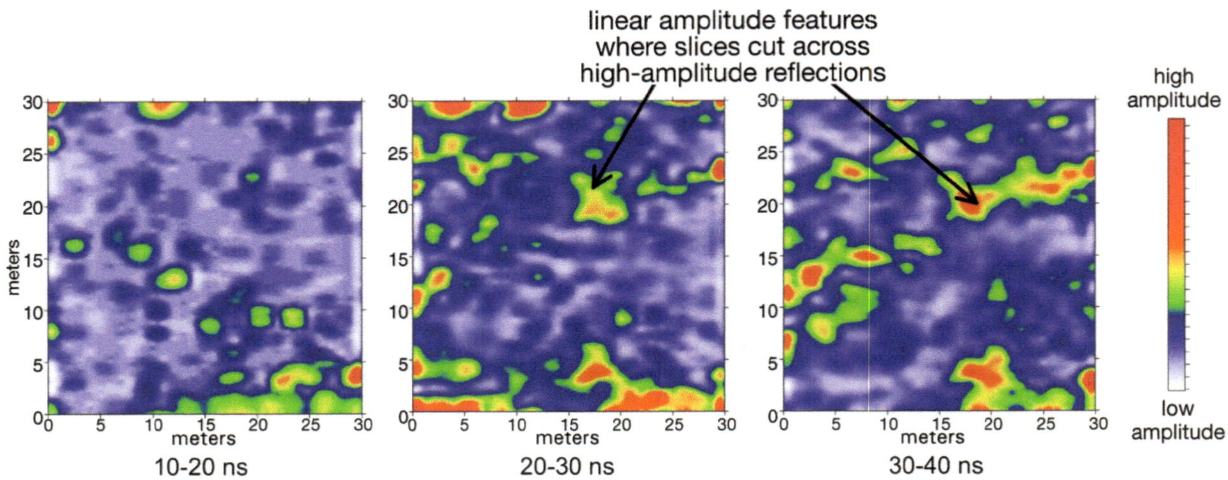

Figure 3-6: Amplitude maps produced by sampling many reflection profiles horizontally, when layers dip. When many reflection profiles (such as in Figure 3-5) are sampled and the amplitudes are plotted in map form, they produce linear features at varying depths, as the high-amplitude reflection is located within progressively deeper slices. This map displays the location of only one horizon in progressively deeper slices, which is not at all useful information.

more important than those initial erroneous results of mine from the early 1990s. But there are two lessons to be drawn from this story. The first is, don't do what I did; make sure you know what you are sampling and how amplitude maps are being produced. The second is, by all means use Dean Goodman's invention, in its many forms today, but do so judiciously. This will produce maps of GPR reflections in just minutes, a process that used to take weeks of manual plotting and hand contouring in order to produce any usable results.

AMPLITUDE MAPS BECOME STANDARD TOOLS FOR INTERPRETATION

For the most part, GPR practitioners throughout the 1990s continued to be slow in incorporating amplitude maps into their usual processing steps, perhaps because many thought, as Gary Olhoeft did, that it was "not advisable." Many continued to manually interpret reflection profiles, often little processed and poorly understood. There was a slow transformation in the GPR community to the amplitude slice-mapping method as an important processing step for most data sets; and by about 2000, it was the first and sometimes only step in GPR processing and interpretation, used by many. As would be expected, GPR users who were asking geological questions relied more on the manual interpretation of two-dimensional reflection profiles because of a profile's ability to illustrate complex stratigraphic interfaces of interest (Baker and Jol 2007; Bristow and Jol 2003). Conversely, archaeologists working in less geologically complicated areas, where the targets of GPR mapping were distinct archaeological features that could be readily discriminated from geological features, quickly moved to the amplitude slice-mapping method as their basic interpretive step. My initial forays into GPR processing incorporated both interpretation techniques, but often I didn't really know why I was using one or the other, and therefore I used both, as each appeared to be telling me something different about the ground.

By 1996, Dean Goodman had written his amplitude analysis software for Windows (the first versions were in DOS) and was starting to make it commercially available. The idea for his GPR software was inspired by the tool that seismic geophysicists had been using for years for petroleum exploration (Dean's Ph.D. research was in deep marine seismic geophysics). His initial targets of GPR mapping for archaeology were Japanese burial tombs and other distinct archaeological features, such as kilns, in a geological matrix that was often homogeneous volcanic ash or well-mixed soils. As he experimented with the amplitude mapping method, he produced, and then published, some striking maps of tombs, moats, and associated burial features with great clarity (Goodman and Nishimura 1993) and demonstrated that this technique was superior to manual reflection profile interpretation in those burial conditions. Dean's data from Japan were mostly collected with low-frequency antennas (200–300 MHz or so) for maximum depth penetration, and as a result, he only collected reflections from the most pronounced of the buried features. This collection method and the features he was interested in were perfect for amplitude mapping, as there was a minimum of associated reflection "clutter" from the surrounding geological matrix that might be confused with the buried cultural materials.

In 1996, I began collecting GPR data in the American Southwest and had some initial success at the amplitude slicing method at some Pueblo sites (Conyers and Cameron 1998). My sites were not as conducive to amplitude mapping as Dean's Japanese burial tombs, as they were often stratigraphically complex. Also, I was interested in shallower deposits, and I assumed I needed higher-resolution antennas than Dean had been using, such as the 500 MHz. In southeastern Utah, there are large circular archaeological features called *kivas* that were built into the ground and today are largely covered by sediment and soil (Conyers and Cameron 1998). These were constructed beginning about A.D. 200, and some similar structures are still in use today by the descendants of the Pueblo people. One of these ancient kivas was hypothesized to exist in a shallow circular depression near Bluff, Utah, and I collected a GPR grid across it to test the amplitude mapping method. The amplitude slice-maps showed a circular area in the middle of the shallow depression, which contained nonreflective material, with high-amplitude material surrounding it (Figure 3-7). All aspects of the GPR amplitude slice-maps indicated this feature was indeed a buried kiva. The GPR amplitude maps were more complex than I had hoped for, and I learned the reason for this when this feature was excavated. The kiva had been constructed by digging down into bedrock and then lining the walls of the large circular room with the excavated stone. After abandonment, the walls of the structure partially collapsed into its interior, and these collapse features generated reflections within the kiva in a random pattern. Both the stratified bedrock outside the kiva and the random collapse within it produced high-amplitude reflections that were displayed in the slices. Also, because I was using the higher-resolution 500 MHz antennas, all these materials produced distinct reflections. When I analyzed the reflection profiles, the nature of the kiva structure was easily identified, as its wall could be easily differentiated from the adjacent bedrock (Figure 3-8). But those subtle differences between reflections produced from the walls of the kiva and those from the surrounding bedrock and wall collapse were sampled by the slicing software in the same way. They all produced high-amplitude reflections, and there was no differentiation between the two in the amplitude maps. They could only be differentiated by interpreting the individual reflection profiles. This experience demonstrated to me that both the amplitude mapping and reflection profile interpretation methods were necessary for any complete analysis of complicated sites of this sort.

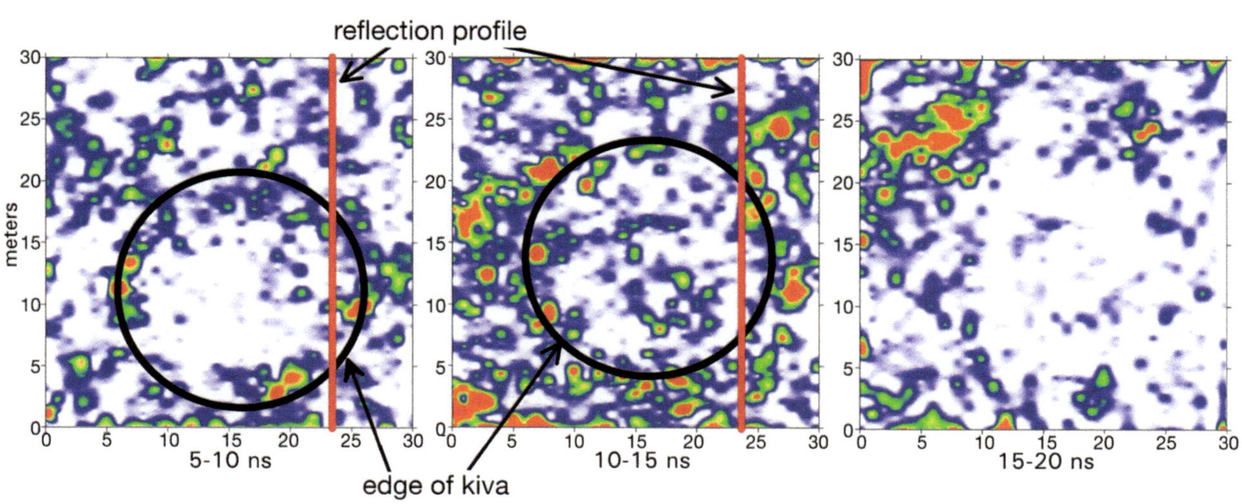

Figure 3-7: Amplitude maps of a circular kiva. These horizontal amplitude maps show the buried circular kiva at Bluff, Utah. The interior of the kiva is generally visible as a circular area of low-amplitude reflections where the structure is filled with fine-grained sediment. The walls of the kiva are difficult to differentiate from the surrounding bedrock, as both produce high-amplitude reflections. The reflection profile shown in Figure 3-8 is shown in red on the first two slice-maps. Reflection data collected with 500 MHz antennas.

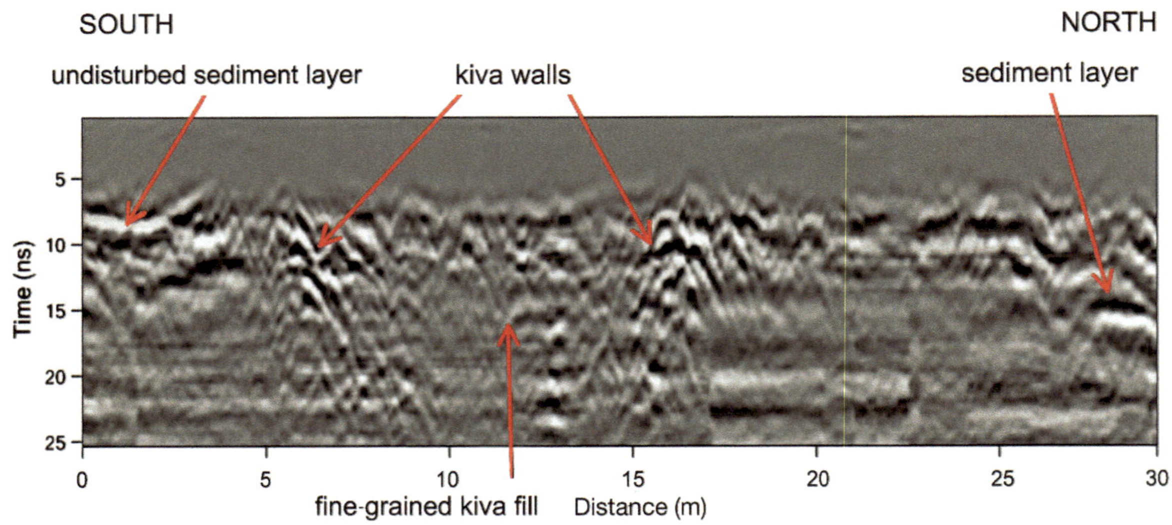

Figure 3-8: Reflection profile across a kiva. This kiva at Bluff, Utah, shows the walls of the kiva as distinct hyperbolic reflections. These cultural features generate the same amplitude reflections as the surrounding bedrock units, and therefore both show up as undifferentiated high-amplitude features in the amplitude maps in Figure 3-7. Data collected with 500 MHz antennas.

AMPLITUDE MAPPING IN CONJUNCTION WITH PROFILE INTERPRETATION

Soon after this kiva study in Utah, I collected a data set nearby that was very easy to interpret using either slice-mapping or individual interpretation analysis. A local landowner had told me of a location in the floodplain of the San Juan River where subtle depressions in an open field were filled with pottery and stone tool fragments. We collected GPR data across all these depressions, and no buried architecture was visible in any of the profiles; likewise, the amplitude maps showed only random reflections from subtle sediment beds in the fluvial material (Conyers 2006a). These same low-amplitude reflections from the homogeneous river silt were also visible in the reflection profiles. When we were almost ready to give up on his project, in the far corner of the grid (under a very large and wicked thorn tree), there appeared a very distinct floor within the sediment package (Figure 3-9). Here was the pit-house floor we had been looking for. Amplitude mapping showed an aerially distinct floor (which turned out to have been partially eroded) of a pit structure (Figure 3-10). We cored it and found it to be composed of a compacted clay floor, with abundant pottery fragments preserved on its surface. In this case, the amplitude mapping was much more precise at showing the aerial extent of this floor. However, the subtle features on it, and the geological nature of the erosion event that incised through it, were visible only in the individual reflection profiles (Figure 3-9). This study again demonstrated to me that a combination of both amplitude mapping and reflection profile interpretation is necessary for most studies.

When Dean Goodman wrote the first amplitude analysis program for GPR, he recognized that stratigraphic complexity can sometimes be overcome in amplitude analysis to produce useful maps. One option his software provided to deal with the complexity of layered systems was the ability to create "horizon slices" (Conyers and Goodman 1997): using a computer mouse, a practitioner can follow one reflection of interest in a profile by manually tracing in each profile (or allowing the software to follow) one particular wave peak or trough (a horizon) throughout a profile. This produces the amplitude variations within one layer, no matter its depth in the ground, which can then be sampled and mapped. This process has now become standard in most GPR processing programs.

A great deal of variation in processing and interpretation methods continued during the early 2000s, with some GPR users sticking to two-dimensional analysis of profiles and others beginning to process data almost wholly in amplitude slices, with the final product often being many maps of many depth-slices in the ground. The reflection profile interpreters tended to be cautious about slice-maps, perhaps because they realized that their data sets were geologically complex and slice-maps often produced a plethora of anomalies that were difficult to interpret while frequently showing no archaeological features at all. Amplitude slice-mapping interpreters often came to the GPR method without a geological background or were focused solely on the search for large buried archaeological features. If the work site was in a geologically simple area and the site matrix was not of great interest, or if

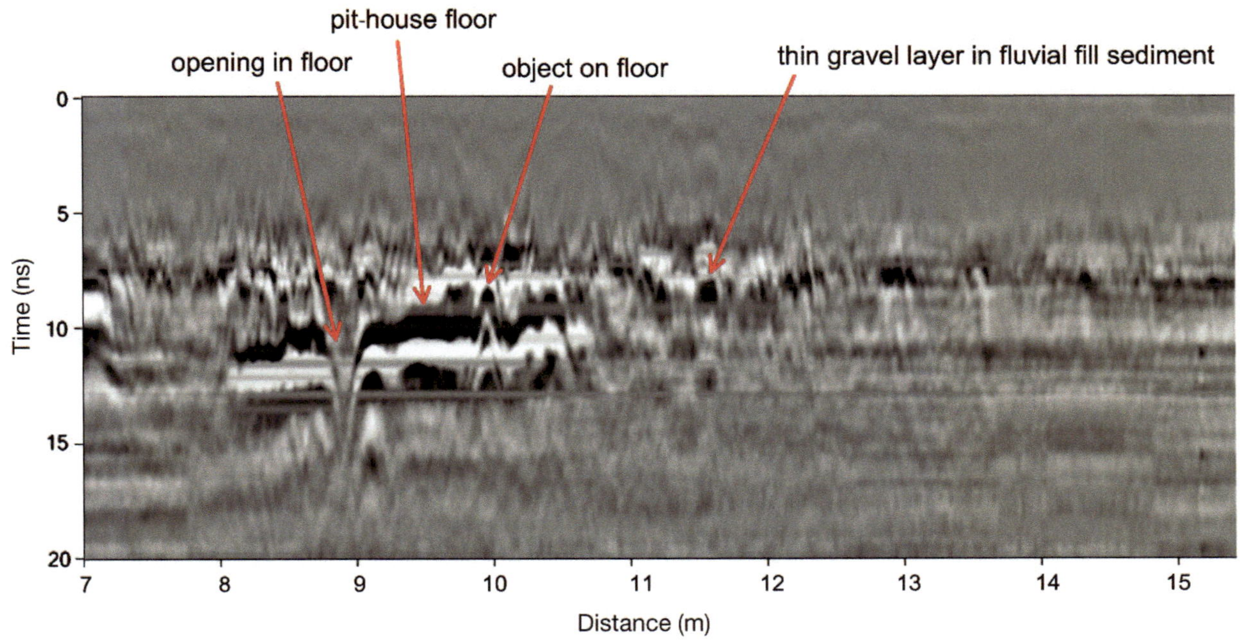

Figure 3-9: Reflection profile of a pit-house floor. This 500 MHz reflection profile shows the high-amplitude floor of a pit house in southeastern Utah. An object on the floor produced a distinctive point-source hyperbola, and an opening in the floor, perhaps an opening to a subfloor storage cistern or burial, can be seen on this profile. Horizontal banding is background radio noise.

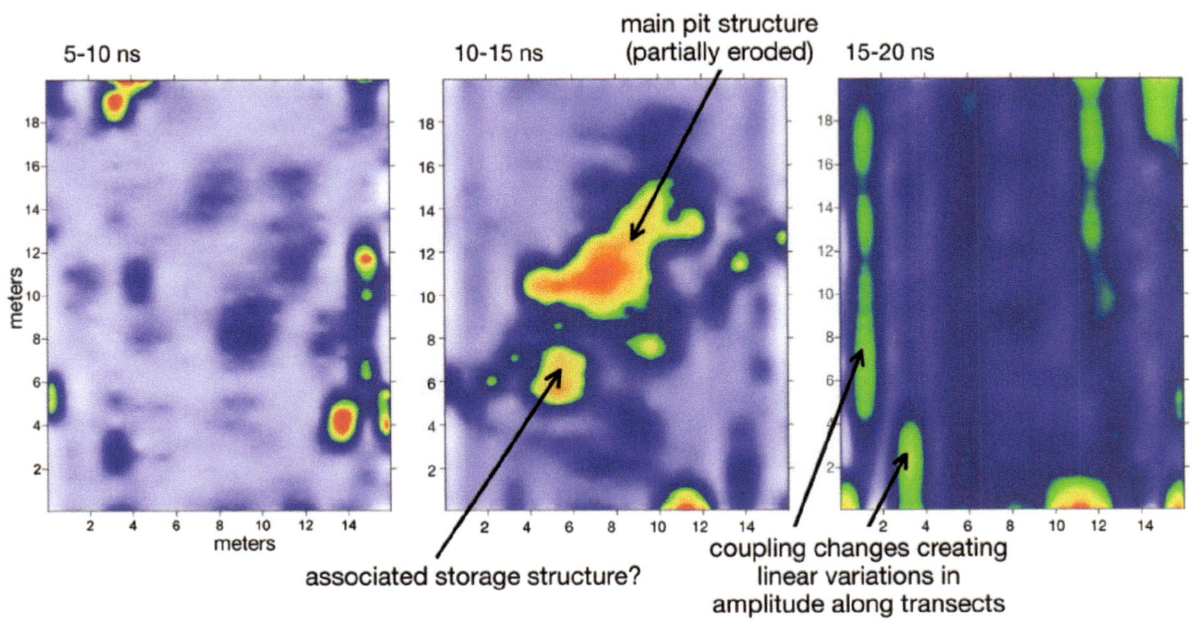

Figure 3-10: Amplitude maps of a pit-house floor. This is the feature shown in profile in Figure 3-9. The linear streaks in the lower slice were caused by coupling differences between antenna transects.

it produced only low-amplitude reflections while the archaeological features were high amplitude, slice-mapping was the logical processing method and interpretation was relatively simple. In such cases, there was no need to interpret each and every profile, as the buried cultural features of interest "jumped out" of the amplitude slice-maps (Figure 3-10). In my early years in GPR, I tended to work in areas that contained both simple and complex conditions, and I remember being conflicted as to what was the best image production method for data processing and interpretation. In the end, I used both, but at the time this was mostly because I was not sure exactly what I was dealing with in the ground (both geologically and archaeologically) until after the data had been collected and processed and interpretation had begun. These days, I always incorporate both slice-mapping and individual profile interpretation in all my projects, and I insist that my students use both methods simultaneously. In this book, I make the case that the two basic methods should always be used in conjunction in order to understand what often are complex burial conditions at most sites.

MANY INNOVATIVE IMAGE TYPES INVENTED FOR INTERPRETATION

In 2001, Dean Goodman and I attended a conference in Japan on GPR for archaeology, hosted by Yasushi Nishimura, where we were amazed to see what the University of Austria GPR scientists had developed. Wolfgang Neubauer and his team demonstrated their method of taking large grids of data, often collected with very fine transect spacing of just a few tens of centimeters, and then slicing them with their own proprietary software and producing videos of large tracts of ground (Neubauer et al. 2002). They had produced *isosurfaces*, a three-dimensional visualization method (which Dean also had developed in his own software package) from which they produced useful and very high-quality images, including videos and shaded relief surfaces of materials in the ground. This started a competition (perhaps only in my mind) to catch up with what the Austrians were doing with their amplitude images, as we all became aware that GPR data had the ability to go far beyond our standard displays of the day. All these methods of visualization have now become commonplace and are incorporated into all GPR programs. Dean Goodman has even developed a way to "fly through" three-dimensional GPR images of the ground, as if playing a video game. These displays give me motion sickness, but I suspect they are very useful for others.

Today, with the hundreds of GPR users, equipment developers, and software programmers, we are witnessing some dramatic changes in our ability to use this method for a wide variety of applications (Conyers and Leckebusch 2010; Viberg et al. 2011). There are now multichannel systems capable of collecting massive amounts of data over huge tracts of land very quickly (Gaffney et al. 2012). An ability to quickly place reflection data in space with GPS has arrived and now is incorporated into most software packages. Every time I talk to Dean Goodman about these new innovative GPR methods, he almost always tells me that some new invention or other "will revolutionize the way GPR is done," and I inwardly groan, as this appears to mean that I will have yet another huge learning curve to overcome. But when I think about it, any GPR data collected, no matter how, still needs to be interpreted using the basic data components that are still only visible in individual reflection traces and reflection profiles. So, the basic need for interpretation remains. I am content to let those who are much more skilled than I in these technical arenas work on these new hardware innovations and the software that must be written to process the huge data sets that result. In the near term, I can perhaps add to the discussion of GPR in the interpretation sphere, which can often lag behind hardware and software advances.

4

Geological Complexities

Inevitably during the analysis of most GPR data sets, interpreters searching for archaeological features are confronted with images that contain many reflections of geological origin, which must be identified and understood. When, as is usually the case, it is not immediately clear what interfaces might have produced the reflections in the ground, the first task is to identify which reflections are geological in origin and which are archaeological. The recognition of geological layers also can help with an understanding of ancient landscapes, by placing the identified cultural features within an environmental context. In addition, geological analysis of GPR data can be a great aid in illuminating post-occupational events by helping to determine the burial context of a site and other events that might have altered a site through geological processes.

When viewing GPR images on the screen during collection, or soon after, I often fall prey to "interpretation anxiety"—that is, I am anxious to satisfy onlookers and colleagues who ask for immediate interpretations. Usually I am wrong in my initial assessments, and long experience has taught me to be deliberate in my interpretations, as it hard to be sure what is geological and what is archaeological. Too often, I find out later, after integrating my GPR results with an analysis of the geological features, that the reflections I had identified as being produced from cultural features were actually soil units, stratigraphic horizons, or bedrock reflections. The same can be true even after data have been processed into both reflection profiles and amplitude maps, so I have become much more wary about making uninformed or hurried interpretations. Moreover, regardless of how much work has gone into an interpretation, if I fail to make it clear to others which reflection features are anthropogenic in origin and which are geologic, it can lead to confusion, disappointment, and possible disillusionment with GPR, especially if excavations show that what were interpreted as archaeological features turn out to have been only geological.

Of course, it can be both fun and instantly gratifying to blithely produce amplitude maps from many reflection profiles in a grid, which will always generate "anomalies," some of which might appear to be cultural features. Those anomalies may or may not be of any importance to the goals of the study, and if the resulting maps are given to excavation planners without detailed explanations, confusion can arise. A thorough analysis of the origin of reflections in GPR amplitude maps, with a specific identification of which are geological and which archaeological, is required if an interpretation of the "anomalies" is to be serviceable in archaeology. Only once have I ever produced an amplitude map where the buried archaeological features immediately "jumped out" at me, and that was in Petra, Jordan (Figure 2-1), where it was difficult to miss the buried Roman temple. Buried features of this sort, which show up distinctively in amplitude maps, are quite rare.

I made the mistake long ago of producing amplitude maps of great complexity and then handing them over to archaeologists who, having no knowledge of what produces reflections in the ground, proceeded to use my "anomaly maps" as if they were images of archaeological features. After excavating some of the anomalies, my collaborators became perplexed when only geological features were uncovered. In the past when I made this mistake, I might get a report back from the excavation team regarding their results based on my maps, with comments about the validity of my GPR interpretations such as, "We dug where your maps showed something and there was nothing there." In this case, "something" was nothing more than an anomaly, and "nothing" meant that their excavations did not uncover whatever it was they had hoped to find. Another painful comment of this sort might be, "The features that you claimed were of archaeological interest proved to be nothing." As this type of comment is quite bruising to the ego, I determined never again to produce images for others without a detailed analysis of the origin of all the reflections in whatever images I produced.

I have often thought about whether one could dig in the ground and "find nothing" and have determined that, short of finding a void space, it is not possible. Of course, what actually occasions such emphatic frustration is when excavation plans are based on an incomplete interpretation of GPR maps and profiles, one that has no explicit and well-explained geological interpretation. When those excavations uncover the soil changes or other geological features that produced the reflections, as far as those excavators are concerned it is as good as finding nothing. Archaeological excavators are on the lookout for archaeological features; they are often not trained in sediment or soil stratigraphy and are not likely to be searching for how geological variations might retain different amounts of water to produce the radar reflections—so in their mind they have found "nothing." Mike Powers, a GPR colleague, once told me that "GPR doesn't lie," a statement I think about often. Our GPR images are telling us *something* about the ground, and it is our obligation to figure out what it might be. Therefore, any interpretation of images must encompass what is being illustrated in terms of geological as well as archaeological features.

Geological complexity can seldom be avoided, except in rare cases where the matrix of a site is some homogeneous material that produces no reflections whatever. In all other cases, sediments and their structures, bedrock units, and soil horizon reflections will come into play. In this chapter, I present a few of the more definitive examples of interesting geological features I have encountered when mapping archaeological sites. Often these features were not immediately visible in the raw data during collection; only after reviewing profiles and maps, and perhaps digging a few holes or drilling to obtain samples, was an accurate interpretation achieved. Here I present examples in fluvial and alluvial settings, lakes, glaciers, beaches, sand dunes, and volcanic terrains, from coastal settings to high mountains. While the selection is far from comprehensive with regard to geological environments that one might confront, it includes some relevant examples of how geological interpretations can help to resolve archaeological questions and place cultural features within an environmental context. There are abundant references available in the literature for many other common terrestrial environments, which can be used as further examples of these and other common environments that can be studied using GPR (Bristow and Jol 2003; Baker and Jol 2007).

SAND DUNES

I begin with an example from coastal sand dunes, which I used earlier in discussing how GPR reflections can be misleading when an interpreter fails to take into account what causes the reflections (Figure 2-8). Here I have produced the longer reflection profile along the top of that sea cliff to illustrate the complexity of the geological strata that produced radar reflections at this site along the Oregon coast. When viewing the 55 m long reflection profile, the abundance of reflections can be overwhelming and difficult to interpret. There are many sand dune beds visible as dipping reflections, as well as portions of the buried soil horizon that forms the lower boundary of this sediment package. A recently buried roadbed is visible in one area, and the target of this survey, the clay floor, can be easily mistaken for other high-amplitude planar dune bedding reflections. Zooming in on that floor helped in its identification (Figure 4-1), but I might have missed it had I not been able to directly compare the floor reflection visible in the reflection profile with what I was able to see with my own eyes in the sea cliff exposure. Once I knew that this planar, aerially restricted reflection was generated from a floor, the hyperbolic reflection generated from a hearth constructed on that floor (not visible in the exposure, as it is still buried some distance into the preserved sediment) also became noticeable. If many profiles of this sort had been collected within a grid and then sliced horizontally to produce amplitude maps (which I did not do), the result would have been a set of very complicated maps, as all the geological reflections as well as a few generated from

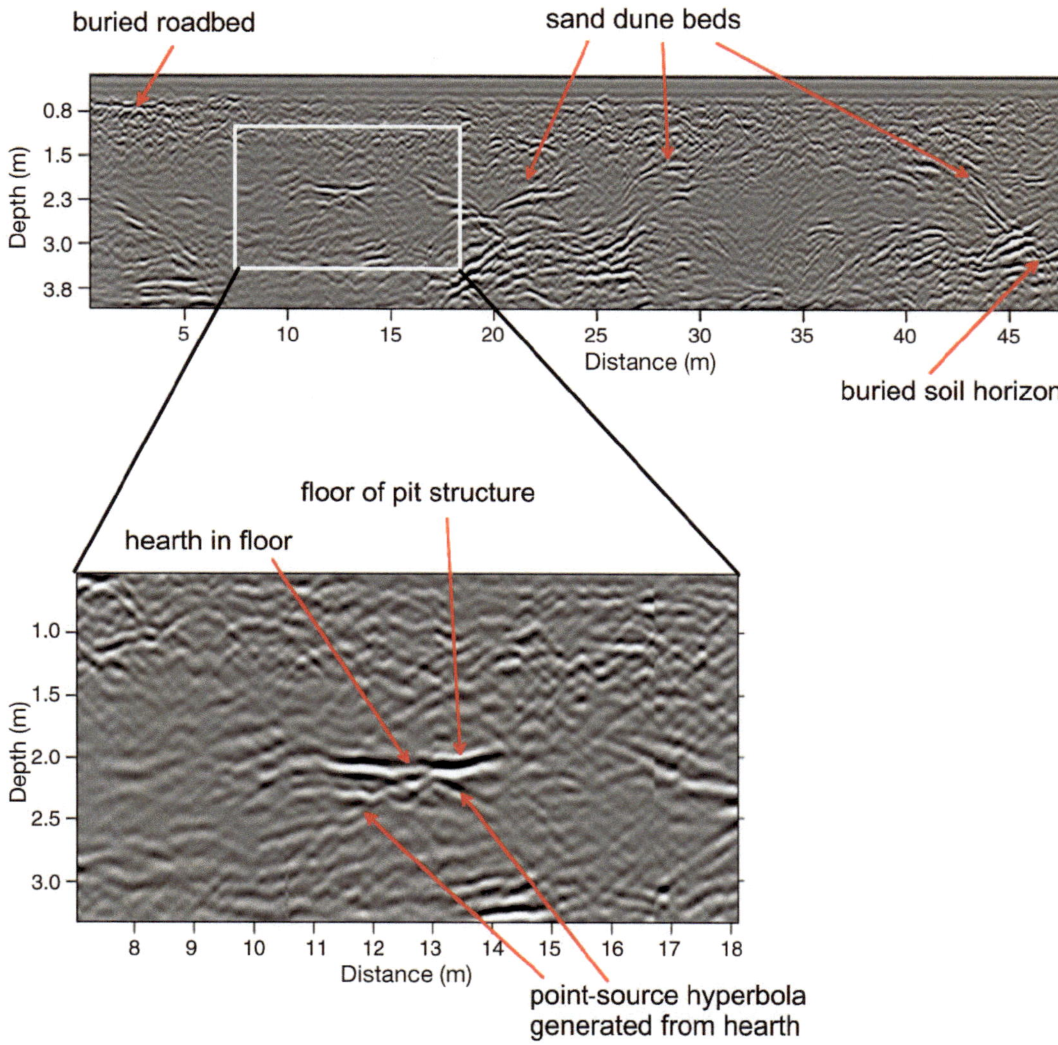

Figure 4-1: Reflection profile along the top of a sea cliff. This is the same feature discussed in Figure 2-8, showing the complexity of geological and archaeological reflections. Data collected with 400 MHz antennas near Port Orford, Oregon.

the one house floor would have produced many reflection "anomalies." Only an interpretation of the individual reflection profiles can hope to differentiate which reflections are of geological origin and which are archaeological. Perhaps somewhere in those complicated amplitude maps there might have been a floor reflection "anomaly," but it would likely have been obscured or not easily differentiated from the many geological reflections.

Sand dunes in general are easy to see in GPR profiles, as the cross-bedding units will usually reflect radar energy and produce sloping surfaces of various orientations. Finding other features within those complex packages of reflections takes some skill at differentiating the orientation and distribution of both the geological and archaeological reflections. Some confusion is inevitable, as these environments are reflection rich.

RIVERS AND RIVER TERRACES

Along the California coast near Lompoc is one of the Spanish missions founded in 1787 called La Purísima. It was largely destroyed by an earthquake in 1812 and abandoned soon after. Today it is a California state park, and in preparation for the construction of a new visitor's center, large grids of GPR reflection profiles were collected in open areas around the mission building, where nothing was known about any possible buried cultural remains. In one grid, a very prominent stream channel was discovered, far away from any active fluvial drainage today. When all the profiles in the grid were processed into amplitude maps, the orientation of the stream was immediately apparent (Figure 4-2), especially when viewing the uppermost slice, from 0 to 5 ns (within about 25 cm of the ground surface). However, when viewing this channel in reflection profiles (Figure 4-3), the stratigraphic variations that produced such a distinct linear channel in the uppermost slice map were not visible. Apparently there are variations in the amplitude of reflections in that uppermost slice, which were recorded in the digital data but are not visible to the human eye in standard profile images. This is common in many data sets where the near-field zone of surface antennas tends to obscure actual amplitude changes in standard profile images, but the digital data contain information that can be sampled and made into usable maps (Ernenwein 2006). In this case, the river channel visible here had enough variation in reflection amplitude across this grid to produce a strikingly linear channel in the shallowest amplitude slice. The material that filled in the uppermost part of the channel, which is likely sand, was probably deposited during the last filling episode before the channel migrated elsewhere.

The progressively deeper amplitude slices show very different images of the channel, which appears to have been slowly filled with sediment over time (Figure 4-2). They delineate amplitude features of the banks of a channel that narrows with depth, and the amplitude maps display only the highly reflective banks, which are of varying distances apart with depth. Two reflection profiles within this grid illustrate the nature of this channel and why these complex amplitude slice-maps were produced. While the base of the channel was below the penetration depth of radar energy, the reflection pro-

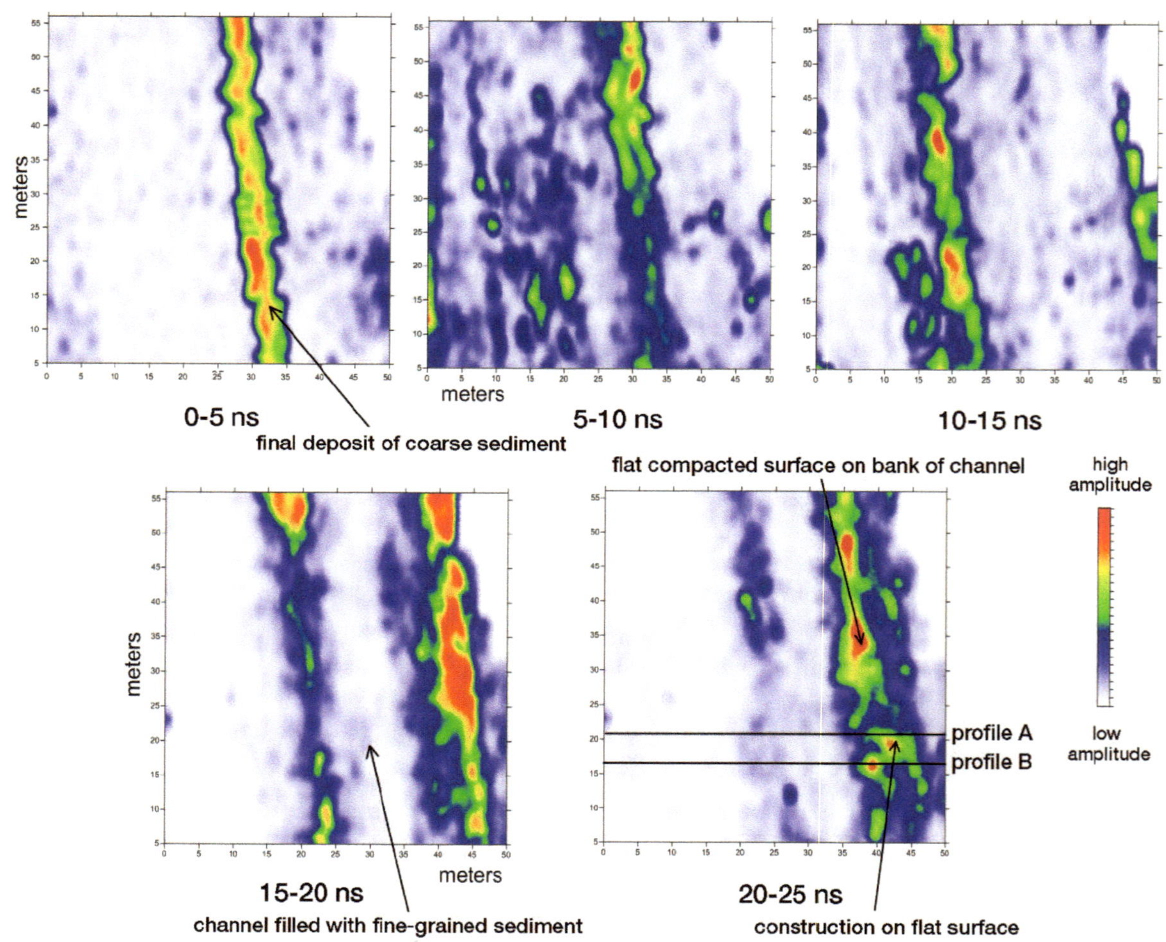

Figure 4-2: Amplitude slice-maps of a river channel. This is a small channel at the La Purísima site, California. Each slice, which sampled different units deposited during a long period when this channel was being filled with sediment, shows a very different picture in the amplitude maps.

files illustrate that at one time the channel's banks were composed of some material that today retains water (likely clay) and differs in radar wave velocity from the material that later filled that channel.

A core taken in the middle of the channel showed it to be filled with coarse sand. This material is largely nonreflective and homogeneous, as can be readily seen in the reflection profiles (Figure 4-3). The channel's banks prior to filling with sand were likely receiving clay or some other fine-grained sediment that today holds water and produces this very high-amplitude reflection in the profiles.

Along the eastern bank of the channel, a number of high-amplitude reflection hyperbolas were recorded directly on top of a very reflective surface. While the nature of these point-source reflections is not known, they are likely cultural in origin, as they appear only in one small area along the eastern bank of this stream channel and sit directly on that surface. This high-amplitude planar reflection in some locations is perfectly flat (Figure 4-3: profile B) and appears to have been generated from a compacted or clay-floored work area, or perhaps some other type of water-retaining living surface.

When this area was inhabited, perhaps during the time the mission was occupied, this small stream was active. A flat work area and associated human-built features were constructed on its eastern bank. The nature of those features is not known, but this modified area on the stream bank is distinctly different from other areas along the stream channel where the channel is symmetrical. The unmodified areas of the stream bank show a surface sloping into the stream, which was filled with sand, presumably after the work area was abandoned (or perhaps coincident with the filling episode during a flood event). The association of this small cultural area with the stream is significant, as it suggests some activity took place in this location that was associated with water. While the amplitude maps generally

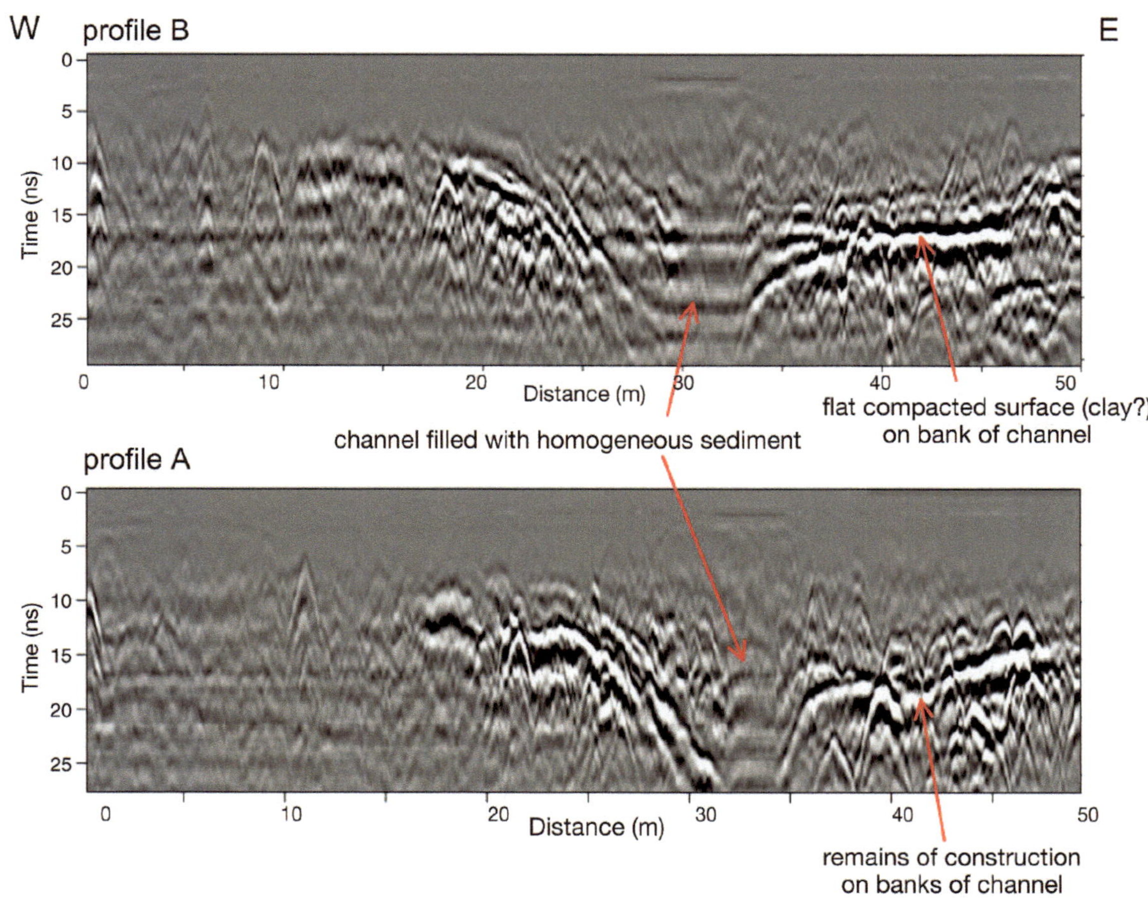

Figure 4-3: Reflection profiles collected across a river channel, with associated archaeological features. Data collected with 400 MHz antennas at La Purísima, California. The location of these profiles on the amplitude maps is shown in Figure 4-2. The remains of a work area and possible construction materials built on that work surface are visible on the eastern side of the channel.

show the orientation of the stream channel and the work area on its eastern bank, they are each quite complex, as the horizontal slicing levels cross the complex stratigraphy of the channel. An integration of the archaeological features with the geological context of this small site shows how powerful GPR images, with limited excavations, can be when analyzing cultural features within their environmental context.

Also in southern California, along the eastern bank of the San Gabriel River, are the remains of the small ranch of Pío de Jesus Pico IV, better known as Pio Pico, the last governor of what was then Mexican Alta California. When California was annexed by the United States after the Mexican-American War, Pio Pico remained in the area, opening a hotel in Los Angeles and beginning other business ventures. He died in 1894. The remains of his house, constructed about 1853 when the area was an active cattle ranch, is today a California state park. A number of GPR surveys were conducted surrounding the house to locate buried cultural features associated with his family's occupation.

The Pio Pico house was constructed on a river terrace just above the San Gabriel River, and the terrace scarp is visible today just to the west. A grid of GPR data was collected directly below the scarp of the terrace on which the house stands. The grid began beneath that terrace riser (the scarp between the terrace on which the house sits and the lower terrace where the GPR survey was conducted) and continued out into what was the floodplain during the 19th-century occupation. It was apparent in GPR profiles that there were concentrations of small point-source reflection hyperbolas along the terrace scarp, which were hypothesized to be a possible trash midden. Each small reflection was likely produced from individual objects in the thick package of historic trash (Figure 4-4). Excavations were undertaken based on this interpretation, and the trash deposit was found to contain many whiskey bottles but also other interesting objects such as porcelain doll parts and many household objects from the period when the Pio Pico ranch was occupied. Most of the reflection profiles that crossed this feature only showed random high-amplitude reflections with no discernible base to the trash deposit. In a few areas, however, the base of what was an ancient channel of the San Gabriel River was visible in reflection profiles, with trash directly in it. The river channel (Figure 4-4) was likely used as a dump for trash, which was a common practice at that time in rural locations such as this. The channel's base, while visible only in some profiles, can be seen as a discontinuous planar reflection because of the abundance of small point-source hyperbola reflections that were produced from the objects that sit directly on it.

Some unusual, but very distinct archaeological features were found within complex fluvial sediments in Arizona. In the Tucson area, the ancient Hohokam people constructed elaborate irrigation canals, some dating as far back as 4000 B.P. (Conyers 2010). These canals are often interbedded with river-deposited sand and gravel, and can be difficult to differentiate from the braided river channels. In one study area, along the south bank of Tanque Verde River, a buried living surface with an organic Ab soil horizon (buried topsoil layer) was identified in an outcrop along the riverbank. This soil contains pottery suggesting it was formed between about A.D 1200 and 1450, when a village just a few hundred meters away was occupied. One hypothesis is that this soil unit identifies the ancient agricultural fields that were farmed along the bank of the river during that occupation. Today this hypothesized agricultural field is covered by a golf course. In an attempt to map features on this buried soil horizon, GPR profiles were collected just adjacent to the river on the golf course. Reflection profiles identified a distinct reflection at the same elevation as the buried soil horizon, visible nearby along the riverbank (Figure 4-5). The soil horizon visible in the reflection profile produced a high-amplitude reflection at its contact with the underlying fluvial units. Overlying this soil is a mostly nonreflective layer, consistent with a homogeneous gravel-sand flood deposit that covered the soil sometime after about 1450.

Incised into the underlying fluvial sediments is a distinct channel about 1–1.5 m in width and 40–50 cm deep, which is about the dimension of known Hohokam canals that have been excavated elsewhere in the area (Figure 4-5). The top of this canal is at the same elevation as the buried soil horizon, which is consistent with a water distribution canal in an ancient agricultural field. The base of the canal generated a very high-amplitude reflection, as it was likely lined with clay to prevent water seepage, and also accumulated fine-grained sediment during its use-life. That very reflective cultural horizon is distinctly different from channel units in the braided river sands and gravels that make up the matrix of the site, which contain little clay.

An interesting high-amplitude reflection feature was recorded below the irrigation canal. Rather than illustrating actual bedding features at that depth (Figure 4-5), it was produced by complex radar travel paths in the

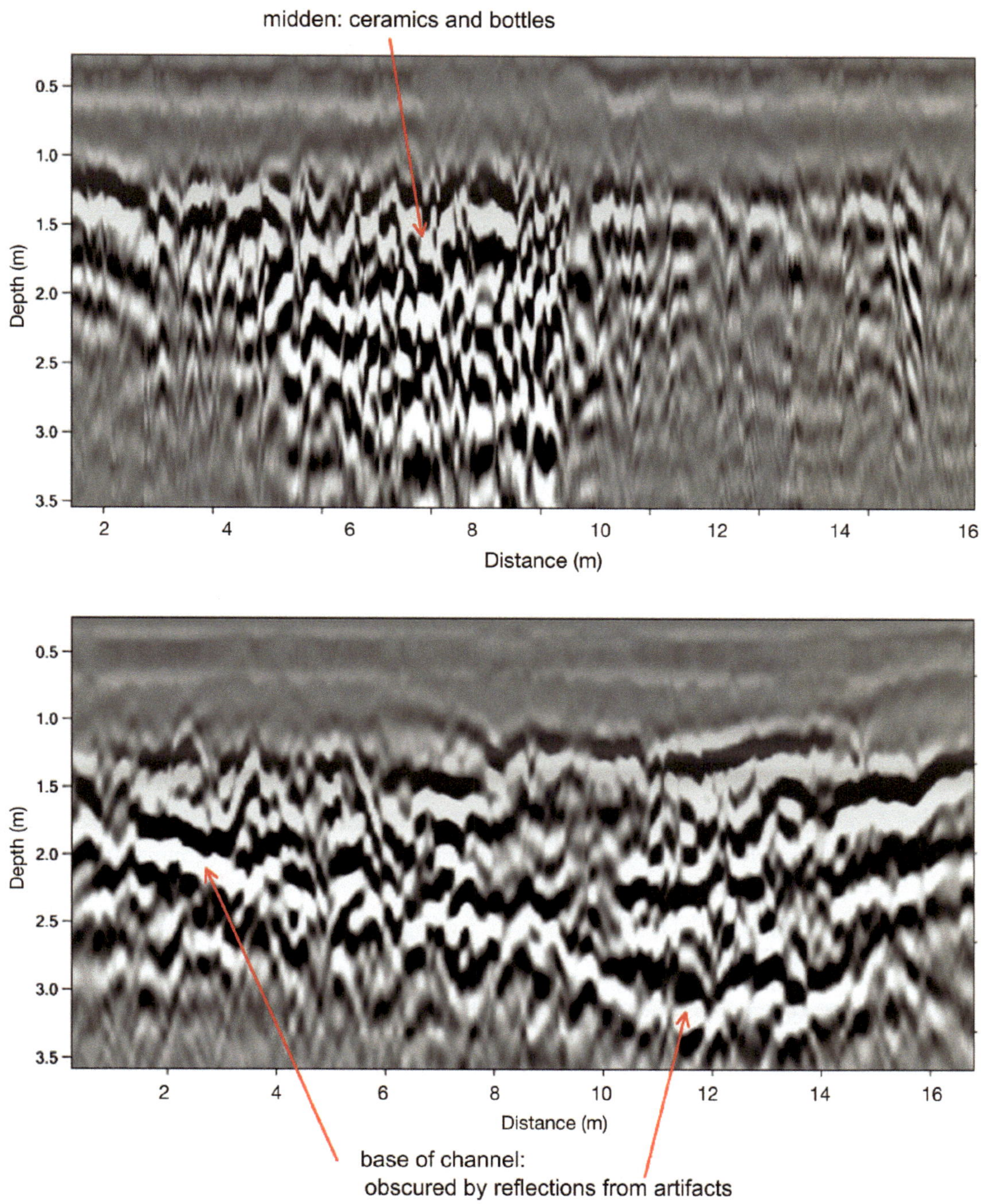

Figure 4-4: Reflection profiles of a historic trash midden in California. The larger artifacts in this midden are visible in profile with data collected using the 400 MHz antennas. The upper profile shows abundant high-amplitude reflection hyperbolas, which were generated by glass and ceramic artifacts thrown in a river channel. The lower profile shows a discontinuous planar reflection generated from the base of that channel, which is somewhat distorted by the hyperbolic reflections produced from the artifacts deposited in it.

ground. In general, "bow-tie"-shaped reflection features of this sort are produced in narrow features like canals with steep banks, small river channels, or other features incised into the ground. Radar energy spreads as it moves into the ground from the transmitting antenna (Figure 4-6), and energy is reflected off the opposite bank of the

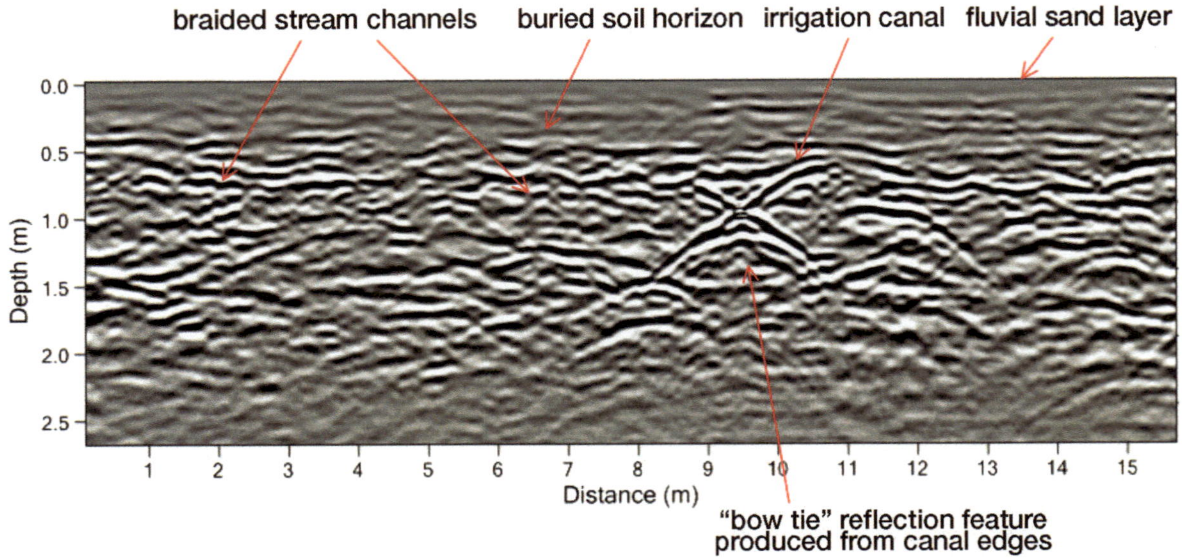

Figure 4-5: Reflection profile showing a buried soil horizon with an irrigation canal incised into underlying braided stream sediments. The canal edges appear as high-amplitude reflections along its edge, with a complex "bow-tie" reflection below the canal. Low-amplitude sand layers cover the canal and the ancient living surface. Data collected with 400 MHz antennas along Tanque Verde River, Tucson, Arizona.

channel and back to the surface antenna. That reflected wave is recorded as if it came from directly below the antenna, and the elapsed travel time recorded for it plots the reflection at a depth below the actual channel itself. Some energy also travels directly to and from the antenna vertically, producing a "real" reflection from the canal edge directly below. As the antennas are moved closer to the channel, the elapsed time for the radar waves moving at an angle in front of the antennas becomes shorter, and the resulting reflection is plotted closer and closer (as measured in time) to the reflections from the channel that traveled a more direct path. The same phenomenon occurs as the antennas travel away from the channel, producing a vertical bow-tie reflection where only the upper portion of the tie is producing a correct image of the channel.

The river and river terrace examples shown here illustrate the complexity of these environments in GPR interpretation. River channels do not always look like channels in reflection profiles, due to the complexity of radar wave travel paths in the ground. When many profiles are sliced into amplitude maps, they tend to look even less like channels (Figure 4-2), as it is often the features within the channels that are reflecting energy, not always the channel itself. Braided stream systems can look much different than meandering river systems, and I have not

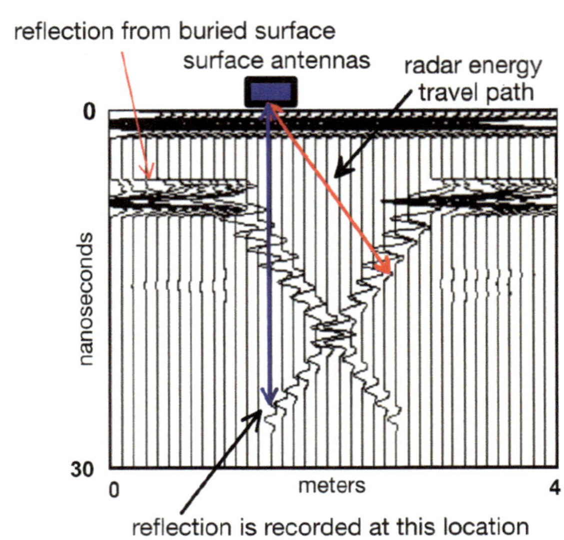

Figure 4-6: Model of a canal showing resulting radar reflections. This model shows what would occur as energy is transmitted in front of a moving surface antenna, reflected off the opposite side of the canal but recorded as if that reflection occurred directly below. As the antennas move across this feature, reflections from the opposite side of the canal are recorded, producing a bow-tie-like reflection feature where only the upper reflections of the "tie" are a true representation of the canal edges.

even begun to address those complexities here. Only a very deliberate analysis of the reflections as they appear in profiles will allow for a correct interpretation of the environment as a whole. That interpretation must begin with a basic knowledge of the types of depositional systems that might be present in a study area, and an understanding of their geometry and sedimentary constituents.

BEDROCK

An understanding of the depth of bedrock at a site, and how its exposure or depth influenced the human habitation or other activities, is always important. People tend to build structures on bedrock and sometimes make foundations within it, but rarely dig far into bedrock if it can be avoided, usually choosing less consolidated ground. Bedrock reflections tend to be continuous and of high amplitude in GPR reflection profiles, but reflections from a bedrock surface or layers within the bedrock can often be very complex in many ways, as is demonstrated below from a few examples in different environments.

In Key West, Florida, the local bedrock is a Pleistocene-age well-cemented calcareous layer called the Miami Oolite, which was deposited about 120,000 years ago during the Sangamon Interglacial period, a time when sea levels were higher than they are today. Following that period during the last glacial period, which lasted from about 110,000 to about 12,000 years ago or so, sea level was much lower than today, and this carbonate layer was exposed, weathered, consolidated, and cemented into what the locals call "hardpan," the local term for bedrock. Hardpan is very difficult to dig through and tends to be a physical barrier, below which there was little past human activity. Over much of the island, this bedrock layer is overlain by windblown sand and, adjacent to the coast, beach deposits.

An identification of the Miami Oolite reflection was important in a GPR study where the target of the surveys was graves of African slaves and pirates, buried during the 19th century in what was called the South Beach area of Key West. The city and county had proposed a redevelopment of the area, now a popular beach for tourists, and wanted to make sure that the proposed construction activities would avoid human remains, which were inferred to exist from historical documents. Today the ground surface is almost flat and has been modified with imported beach sand, sidewalks, planted areas, and other modern features, so there are no surface indications of graves or topographic features that might have existed in the past (Figure 4-7).

Sedimentary structures within the Miami Oolite bedrock unit generally produce a number of complex reflections in GPR reflection profiles, but its upper interface with the overlying carbonate sand is usually quite distinct (Figure 4-8). The amplitude of the bedrock-sediment interface reflection varies depending on the porosity of the bedrock and its retained water, and the type of sediment overlying it. To complicate matters, the bedrock has been greatly modified in the past by wave action and floods during strong hurricanes that often pass over the island. These dramatic events are known to produce a good deal of erosion, and the surface mantle of sediments can be removed and redeposited offshore, or sometimes moved into newly formed gullies. After such storms, the city of Key West always fills in the newly created erosion channels, leveling the ground surface for roads and parks, and, as a result, little is known about the landscape in the 19th century except what can be gleaned from some historic photos and low-resolution maps.

A number of GPR surveys were conducted in an area that historical documents indicated might have been the old South Beach burial area. It became immediately apparent by an analysis of the GPR profiles that in much of the survey area, bedrock was within only 50 cm or so of the ground surface. In addition, the GPR profiles displayed complex images of a number of erosion channels that had cut through bedrock, illustrating a very incised and rugged landscape prior to filling and

Figure 4-7: Collecting GPR data along the beach at Key West, Florida. Here the 400 MHz antennas were used to discover what is below this imported beach sand, which covers a complex buried topography incised into bedrock.

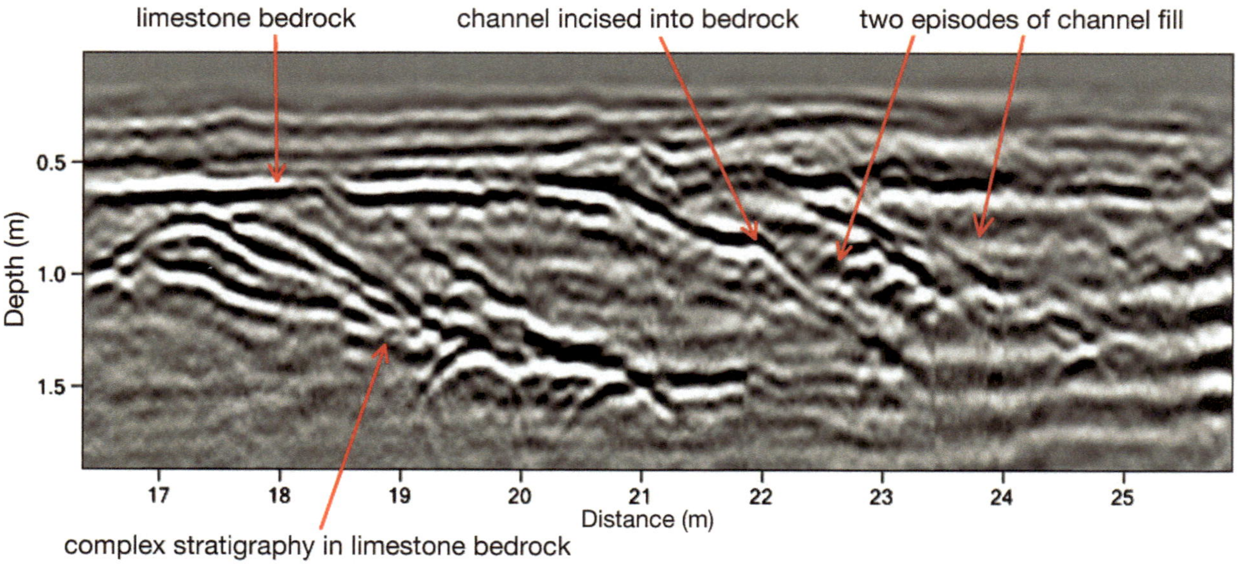

Figure 4-8: Reflection profile from Key West, Florida. This profile illustrates a very high-amplitude reflection generated from bedrock, and other high-amplitude reflections from natural strata within the bedrock. It also shows a channel that was incised into bedrock and then was filled with carbonate sand, producing a level ground surface today.

leveling in recent times. Some of these channels appeared to have been filled naturally as water washed in sediment from higher ground, and others were likely filled by the city to level the surface for the modern beach and park.

Of primary interest in the search for graves was to locate areas where the bedrock was deep enough that graves could have been dug there in the past, but which still retained enough sediment and soil for easy digging. It was assumed that gravediggers would have chosen these locations for burials because of the minimal amount of labor required. Areas that had very recently been scoured by storms and filled in by the city could be discounted as locations of historic burials, as any graves that might have been located there would have been destroyed. We know this type of destruction happened often in the past, as there are historical accounts of bones and partially decomposed bodies being exposed at South Beach after storms.

An area that met these conditions was located in one of ten GPR grids collected, and hundreds of burials were found in an area of about 50 × 50 m (which is discussed in Chapter 8). The key to locating this area, and then the graves in it, was mapping bedrock depth and the recent channels that had been cut into bedrock. Using that analysis, it was relatively simple to find the area where there was a bedrock depression that had accumulated sediments and had not eroded during storms over the last century or two.

In some areas of the world, bedrock surfaces were used as natural work areas where prehistoric people performed many activities, including drying food and manufacturing and sharpening stone tools. In Queensland, Australia, a number of bedrock surfaces of this sort, of unknown age (Figure 4-9), were discovered near the Gledswood Rock Shelter in the Outback (Conyers

Figure 4-9: Bedrock slab covered with stone tool sharpening grooves. This feature is near the Gledswood Rock Shelter, Queensland, Australia.

Figure 4-10: Excavations within the drip line at the Gledswood Rock Shelter, Queensland, Australia. GPR profiles were collected from this excavation into a small sediment basin that is in front of this shelter, to the right in this photo. *Photo courtesy of Lynley Wallis.*

2011b). The shelter, situated under a small sandstone overhang, appears to have had periodic occupation for many millennia, but the covered portion within the drip line is only a few square meters in extent. Excavations by Lynley Wallis (Figure 4-10) uncovered sediments many meters thick along the shelter's back wall which contained charcoal and stone tool waste debris.

An area of about 50 × 70 m in front of the rock shelter where GPR data were acquired is surrounded by low bedrock outcrops forming what appears to be a small natural basin. It was hypothesized that this area might have trapped sediment over many thousands of years and preserved other features related to the documented occupation in the nearby shelter.

The general location of this rock shelter is on a prominent ridge, so any sediment that might have filled the small basin in front of the shelter would likely have been windblown sand or a minor amount of slope wash or possible bedrock slump blocks of local origin. The GPR survey conducted in front of the shelter was to determine depth of bedrock and to map bedrock features that might have been exposed in the past and used as work areas.

The large grid of GPR reflection profiles collected was used to identify the bedrock reflection, generated from the same Cretaceous-age sandstone formation that forms the rock shelter. It could be quickly identified in reflection profiles by its continuous high-amplitude, undulating planar reflection (Figure 4-11). Two areas where this highly reflective surface came close to the ground surface were chosen for probing with an iron bar driven into the sandy sediment until it encountered the hard sandstone. In both cases, the GPR profiles, which were corrected for estimated velocity, predicted the bedrock depth (found by probing) within a few centimeters, giving a high degree of confidence to the bedrock interpretation.

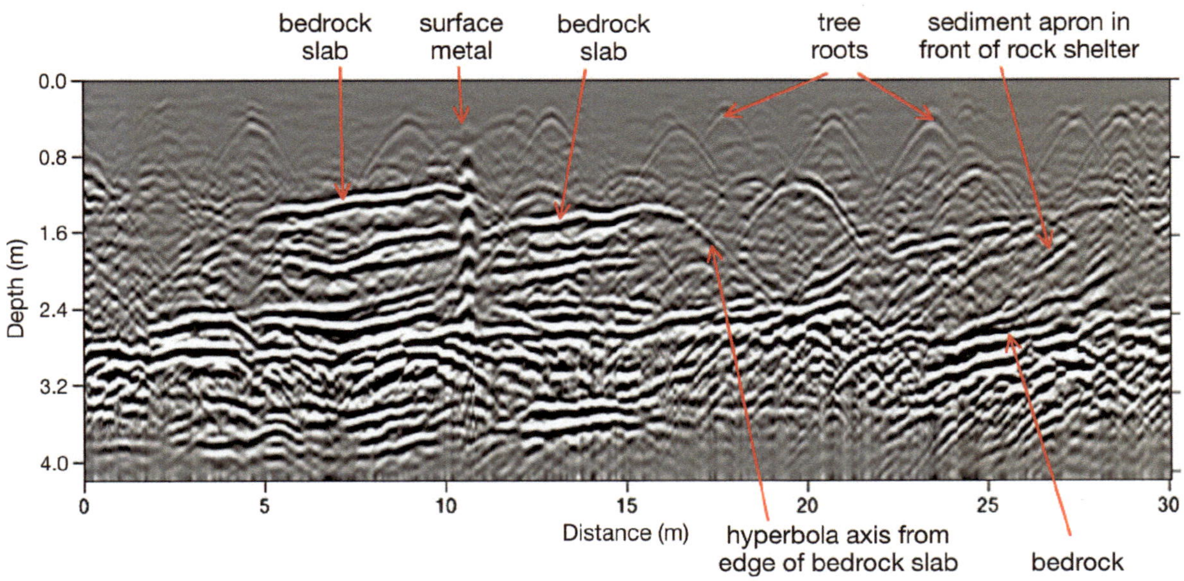

Figure 4-11: GPR reflection profile in front of the Gledswood Rock Shelter, Queensland, Australia. This profile shows dipping sediments in front of the shelter and bedrock slabs that could be prehistoric work areas. Data collected with 400 MHz antennas.

Layers of sediment that dip from the shelter into the basin were visible in many profiles (Figure 4-11), draped onto the bedrock surface. Those layers likely contain artifacts that were deposited in or directly in front of the shelter and were then redeposited in the small basin, perhaps during the initial filling of this small catchment. Within the basin itself are a number of bedrock slabs that would have been exposed during ancient occupation episodes. They are flat surfaces, much like those exposed nearby (Figure 4-9), but determining whether they contain any use wear or other modifications produced by the ancient inhabitants must await excavation. The sediment apron that likely contains artifacts in front of the shelter, and the bedrock slabs, have been completely covered with sediment and preserved more than a meter below the present ground surface. The GPR study at this site located not only possible occupational sediments associated with the rock shelter, but also potential work areas in front of it, for future excavation.

A similar bedrock mapping exercise was performed in Central Park, New York, where the purpose of the GPR study was to find artifact-rich areas for excavation. The study area had been inhabited in the early 19th century by freed slaves and Irish immigrants and was a semirural area called Seneca Village. This village was destroyed during the construction of the park beginning in 1857, and little is known of it other than some information from historic documents and maps. Bedrock in this area of Manhattan Island is Precambrian gneiss, which was scoured and polished during the last ice age, producing promontories of black rock, which are still visible in much of the park today. Prior to building the park, this area was much more rugged than it is today, with little soil and significant exposures of bedrock. Bedrock depth and sediment and soil thickness in small basins between bedrock promontories were considered important targets of the GPR research, as these areas would have been construction sites in the past and perhaps still contain archaeological materials of interest. It is known that a great deal of soil was imported into Central Park during its construction, some from as far away as New Jersey or Long Island. This imported soil was used to fill in some of the natural basins between bedrock exposures to make the ground surface more pleasing to the eye than the natural rocky and somewhat barren landscape that existed. This historic information suggested there were likely features of archaeological interest buried below the present ground surface, but these could be mapped only with extensive excavations (which were not allowed in the park) or with GPR. Discovering the bedrock reflection and then using that surface to map the historic landscape was a key to identifying areas for a few limited excavations conducted in 2011.

The bedrock reflection could be identified in GPR profiles by collecting test profiles where the antennas were placed directly on bedrock and then moved onto the adjacent grass-covered soil. In these test profiles, the bedrock reflection could be easily detected dipping away from the outcrop. A number of GPR grids were then collected in the small sediment-filled basins that were located between what had been bedrock promontories in the past, and bedrock depth was measured.

One small basin, bounded by bedrock on all sides, proved to contain a huge number of early 19th-century artifacts (Figure 4-12). During data collection, the basin was immediately visible on the GPR system computer screen by following the bedrock reflection; we could also see that the middle of the basin contained surface vegetation that was greener and more lush than that elsewhere in the grid. The bedrock surface is high in amplitude, and the reflection profile also shows a reflection from a very well-developed soil layer near the ground surface. Directly below this soil layer are clusters of high-amplitude point-source reflection hyperbolas, generated from small reflective objects. Excavations proved that this small bedrock basin contained a historic house, around which a large amount of trash had been dumped during occupation, and the disposed organic matter likely acted as a fertilizer for the overlying grass. Since the park was constructed, a much richer soil had developed in this basin; and that soil, developed on top of the midden, holds water in a way that makes it contrast with other, less nutrient-rich adjacent soils (Figure 4-12). Excavations were conducted on this feature in 2011, and while details on the finds have not yet been published, preliminary reports indicate that this small basin is a "treasure trove" of early 19th-century materials associated with this house, which tells much about the lives of people in the early years of New York City.

Similar bedrock surfaces and associated features were mapped on a Mediterranean beach in southern Israel, just below a sea cliff at the site of Ashkelon (Figure 4-13). This survey was conducted in an attempt to find the harbor of this ancient city, which has always been a mystery, as no natural harbor exists despite abundant evidence of maritime trade.

The reflection profiles collected on the beach, from the sea cliff toward the water, show a good deal of

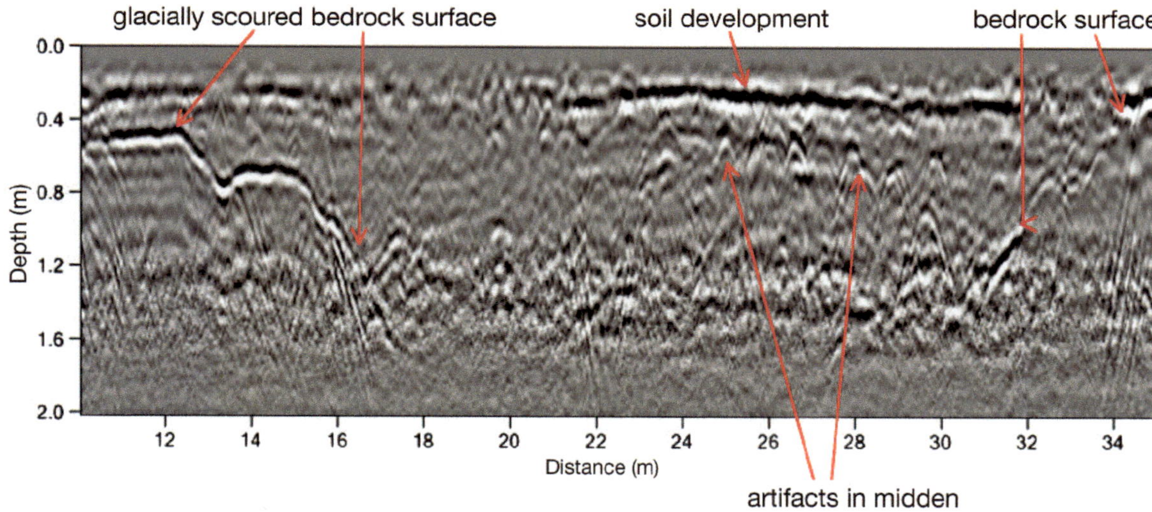

Figure 4-12: Reflection profile over a small bedrock basin in Central Park, New York. The bedrock reflection shows a small depression within which artifacts were discarded prior to 1857. Data collected with 400 MHz antennas.

Figure 4-13: Collecting GPR data on the beach at Ashkelon, Israel. Here 400 MHz antennas were used to map large bedrock slump blocks, some of which can be seen on the surface, having fallen from the sea cliff that is located just outside the frame (to the left) of this picture.

bedrock and sediment complexity, with many blocks of slumped sea cliff, now buried in beach sand. Each individual block produced a high-amplitude reflection hyperbola (Figure 4-14). The beach deposits that cover the slump blocks are only faintly laminated and dip shoreward. They are composed of sand that varies very little in composition and is well mixed by waves, as no radar reflections were generated from any interfaces of different materials. There are a few interbeds of higher-amplitude layers of beach rock within the beach sand, which are likely derived from eroded portions of the slump block or perhaps layers of clay, all of which also dip seaward. Within about 5–6 m of the water, saltwater has intruded into the beach sand and produces an environment that attenuates all propagating radar waves very near the ground surface.

The GPR results did not delineate any location that could have been a formal harbor for Ashkelon. It is likely that ships pulled up on the beach to load and unload goods during fair weather, and then perhaps avoided the Ashkelon area when waves were too high for a beach landing.

These bedrock features visible with GPR only begin to touch on the potential complexity of studying these reflective surfaces in three dimensions. Their identification in reflection profiles is always complicated, as they can be confused with other sedimentary layers. To overcome that confusion, it is necessary to take a core or auger sample or some kind of probe, or to make a direct correlation of reflections to bedrock outcrops. The best way to identify bedrock in reflection profiles is, if possible, to place the antennas directly on a bedrock surface and then collect a transect away from that outcrop as the bedrock surface becomes buried progressively deeper by soil or sediment. Amplitude maps of complex bedrock surfaces at various depths are almost always confusing and for the most part useless. Slicing up complexly buried bedrock (which is in itself complex with respect to its reflectivity) into multiple

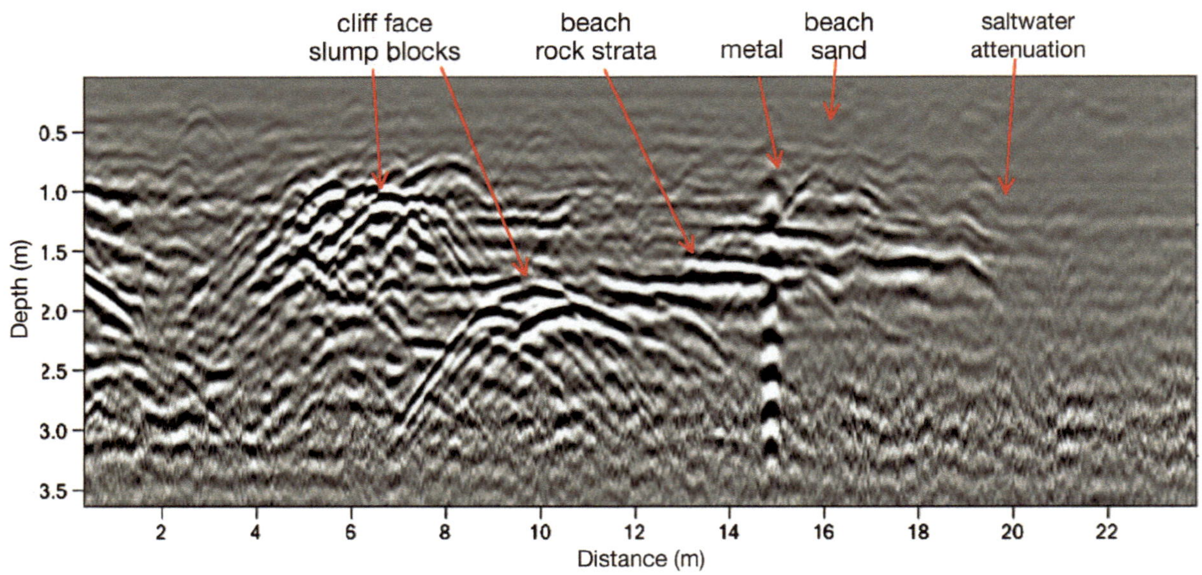

Figure 4-14: Reflection profile collected on the beach at Ashkelon, Israel. The 400 MHz reflection profile shows slump blocks from the adjacent sea cliff, which are visible as high-amplitude hyperbolas with beach strata dipping seaward. All reflections are attenuated about 5–6 m from the water's edge due to saltwater intrusion in the sand.

horizontal layers usually produces a jumble of uninterpretable reflection features. This is particularly true when the reflections from that surface area are located in many different depth maps, as this will produce an array of "anomalies" of various orientations and amplitudes. For that reason, bedrock mapping usually must be done by hand using reflection profiles after identifying the reflections generated from that surface.

BEACHES

On an uplifted beach terrace on the north shore of Oahu, Hawaii, GPR data were collected in an area hypothesized to contain prehistoric village remains. In one profile, the local bedrock, consisting of well-cemented coral sand, had been truncated during the construction of a pit feature (Figure 4-15), which has a very distinct high-amplitude base. It is known that features of this sort were commonly used as baking ovens during ancient times in Hawaii, and the high-amplitude horizontal reflection at the base of this pit is consistent with a highly reflective burned layer. The vertical edge of the oven did not produce a radar reflection in this profile, as most propagating radar waves traveled from the surface antenna parallel to the truncation face and were therefore not reflected back to the surface. The other edge of the pit, which slopes at more of an angle, produced a low-amplitude reflection because much of the energy reflected from it traveled away from the surface antennas. The beach sand and some shallow windblown material that covers this pit feature are largely nonreflective. The only shallow reflections of any amplitude were generated from tree roots.

Most of the beach sand reflections I have seen in GPR profiles are nonreflective, or exhibit only very low-amplitude reflections. That is a function of the winnowing properties of wave action, which removes clay or other different sediment fractions prior to deposition. After beach units are uplifted and exposed to weathering elements, the previously homogeneous deposits often become more reflective. During weathering, each layer of sand that contains subtle differences in porosity and permeability will accumulate differing amounts of chemical cement or secondary clays, creating layers of varying water saturation. Those more ancient beach deposits often exhibit higher-amplitude reflections derived from their sedimentary structures as compared with their modern equivalents.

An additional complexity with beach sand that has been lithified is encountered in differentiating lithified sedimentary features from those in unconsolidated sedimentary overburden. Partially cemented sand is a very suitable medium for radar propagation and reflection, and porosity differences (and therefore water percent differences) can be similar to those in unconsolidated sediments. Along the coast of Israel near Haifa, GPR

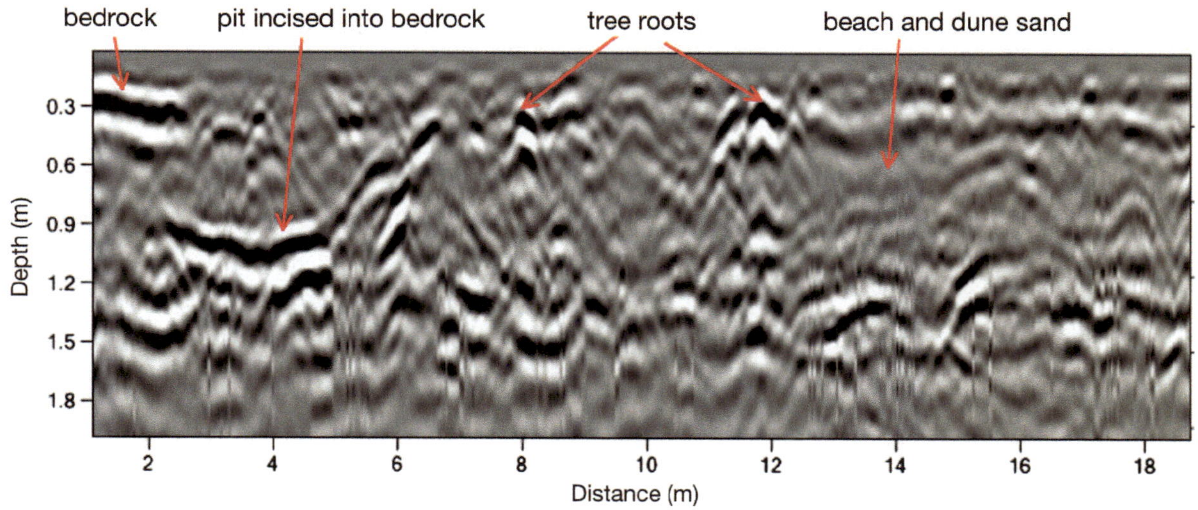

Figure 4-15: Reflection profile collected on the north shore of Oahu, Hawaii. The high-amplitude reflection at the base of a pit was likely used prehistorically as an earth oven. Bedrock was truncated during the construction of this pit feature. Data collected with 400 MHz antennas.

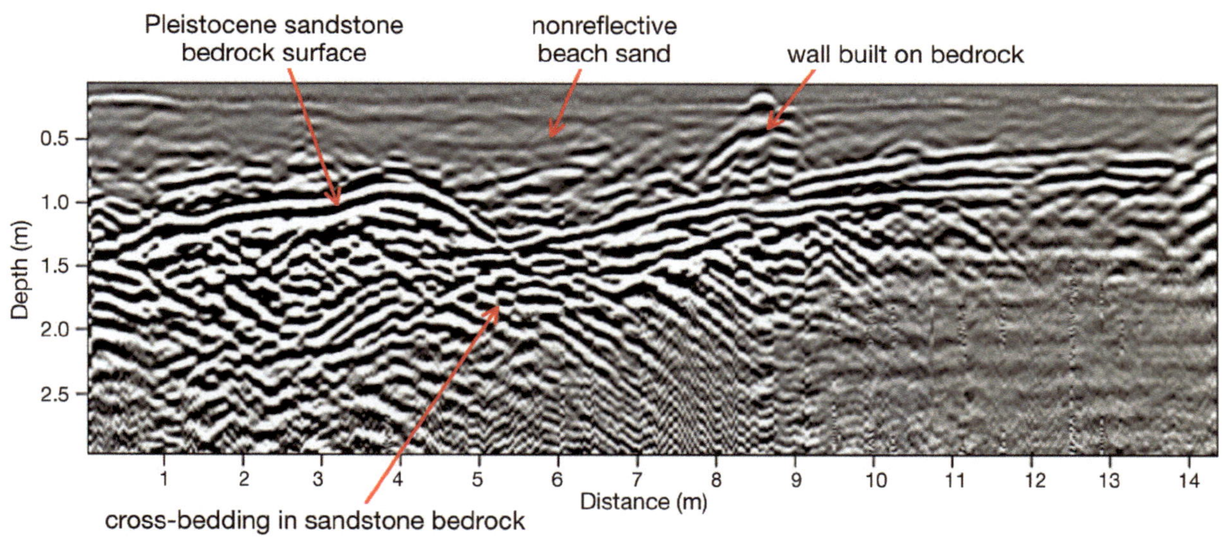

Figure 4-16: Reflection profile showing lithified sandstone bedrock, Israel. Cross bedding and two different sand dune units are visible in this profile. A wall built on this bedrock is visible, and all features are buried by very low-amplitude beach sand. Data collected with 400 MHz antennas. *Courtesy of Yossi Salmon and Jennie Sturm.*

data were collected to map an ancient harbor and other features, now buried in modern dune and beach sand. The lithified sand in this area is locally called *kurkar* and is a well-cemented Pleistocene-age sand dune and beach deposit with well-formed cross beds.

These ancient dune and beach formations make up most of the bedrock along the coast and produce spectacular GPR reflection profiles, with each of the ancient sand dune and beach units visible by their cross bedding (Figure 4-16). Built on the bedrock is a wall of unknown age; all these features are covered in nonreflective, recently deposited beach sand.

TAR PITS

An unusual GPR survey was undertaken at the La Brea Tar Pits in Los Angeles, California, in order to determine if individual oil seeps were visible with GPR mapping. The purpose of the study was to gain experience with known oil seeps that could be used as an exploration tool for discovering similar features in the Maracaibo Basin of Venezuela. The La Brea Tar Pits are famous for their excellent preservation of numerous Ice Age animals (and one human of note), who became mired in pools of crude oil that had seeped upward through vertical fractures in the ground from leaking oil reservoirs below. Excavations of these oil pools were conducted beginning in 1901 by the University of California, Berkeley, and the Los Angeles Museum of Natural History. They continue today; millions of bones and hundreds of complete skeletons of Pleistocene fauna and many plants have been discovered. The tar pits tend to be conical in shape, with the widest portion of the oil pools at the surface, probably due to their banks slumping into the viscous oil, but also perhaps widened over time as wallowing creatures struggled along the edges to escape the sticky tar (Figure 4-17).

Figure 4-17: Schematic image of an oil seep. Seeps of this conical shape are common at the La Brea Tar Pits in Los Angeles, California, where oil seeps upward along fractures from depth; the cone shape is wider at the top due to bank slumping and struggling animals mired in the oil.

One known tar pit (seep #10), which had been excavated and has subsequently filled back up with oil, was chosen as a test for GPR. The seep is about 2–3 m in diameter at the partially hardened ground surface, and was covered by a plastic sheet to keep GPR scientists from becoming encased like animals in the partially hardened tar. Reflection profiles over the seep showed that the interior of the tar pit, which is composed of only oil, produces no reflections, as there are no interfaces within that material to reflect radar energy. The surrounding layers of sediment produce high-amplitude reflections. Most of

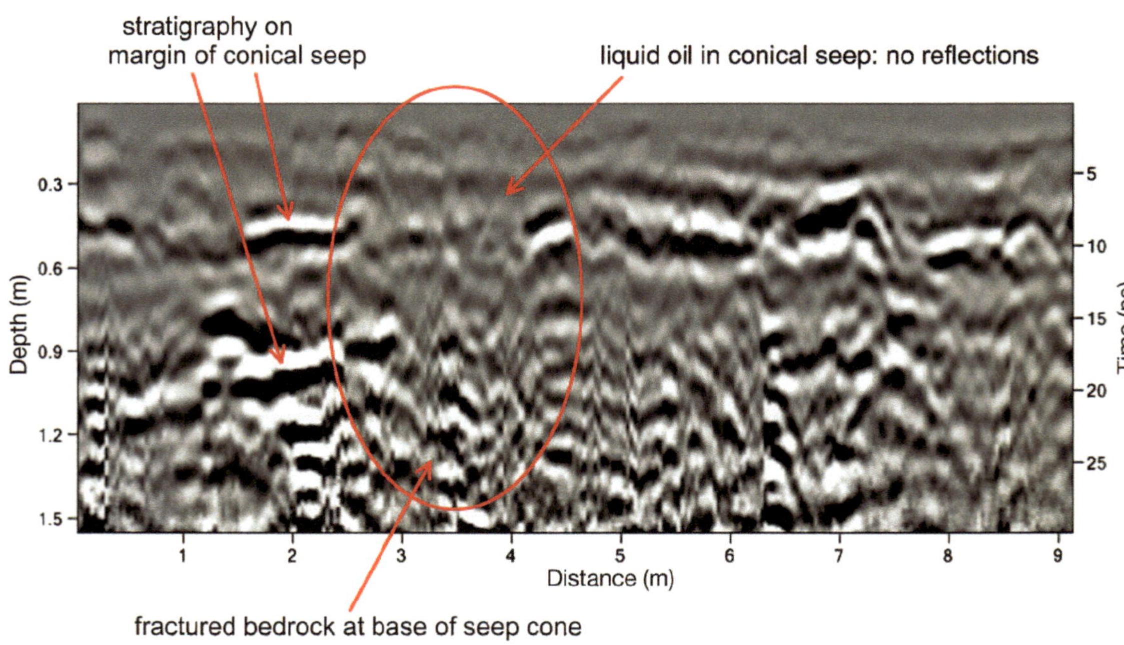

Figure 4-18: Reflection profile over an oil seep. This is oil seep #10 at the La Brea Tar Pits in Los Angeles, California, where the conical oil seep shows no reflection, as it contains homogeneous crude oil, while bedrock layers on its edge reflect high-amplitude waves. The fractured bedrock at the base of the conical seep, along which the oil migrated upward, is visible as jumbled reflections and many small point-source reflection hyperbolas. Data collected with 400 MHz antennas.

the GPR reflection profiles collected at seep #10 show in cross section what is known from analyses of many of the other nearby oil seeps (Figure 4-18): a narrowing of the oil cone with depth. At the base of the cone, the GPR profiles show many small hyperbolic reflections that were likely generated along irregular contacts of bedrock fractures, along which the oil migrated upward to fill this pit. Amplitude slice-maps of the grid collected over this tar pit reveal the extent of the nonreflective oil in the seep-filled cone and reflective layers of bedrock bounding the seep on all sides (Figure 4-19).

If there are any remaining fossils of ancient creatures preserved in oil seep #10, they are not visible with GPR, as the bones are too small to reflect energy using the 400 MHz antennas. In addition, excavations showed that most of the bones of animals that died in the seeps were impregnated with hydrocarbons, and their physical properties are now almost the same as the surrounding oil and would therefore not produce a radar reflection. To my knowledge, this information about the use of GPR to find other oil seeps has never been used either in Venezuela or elsewhere.

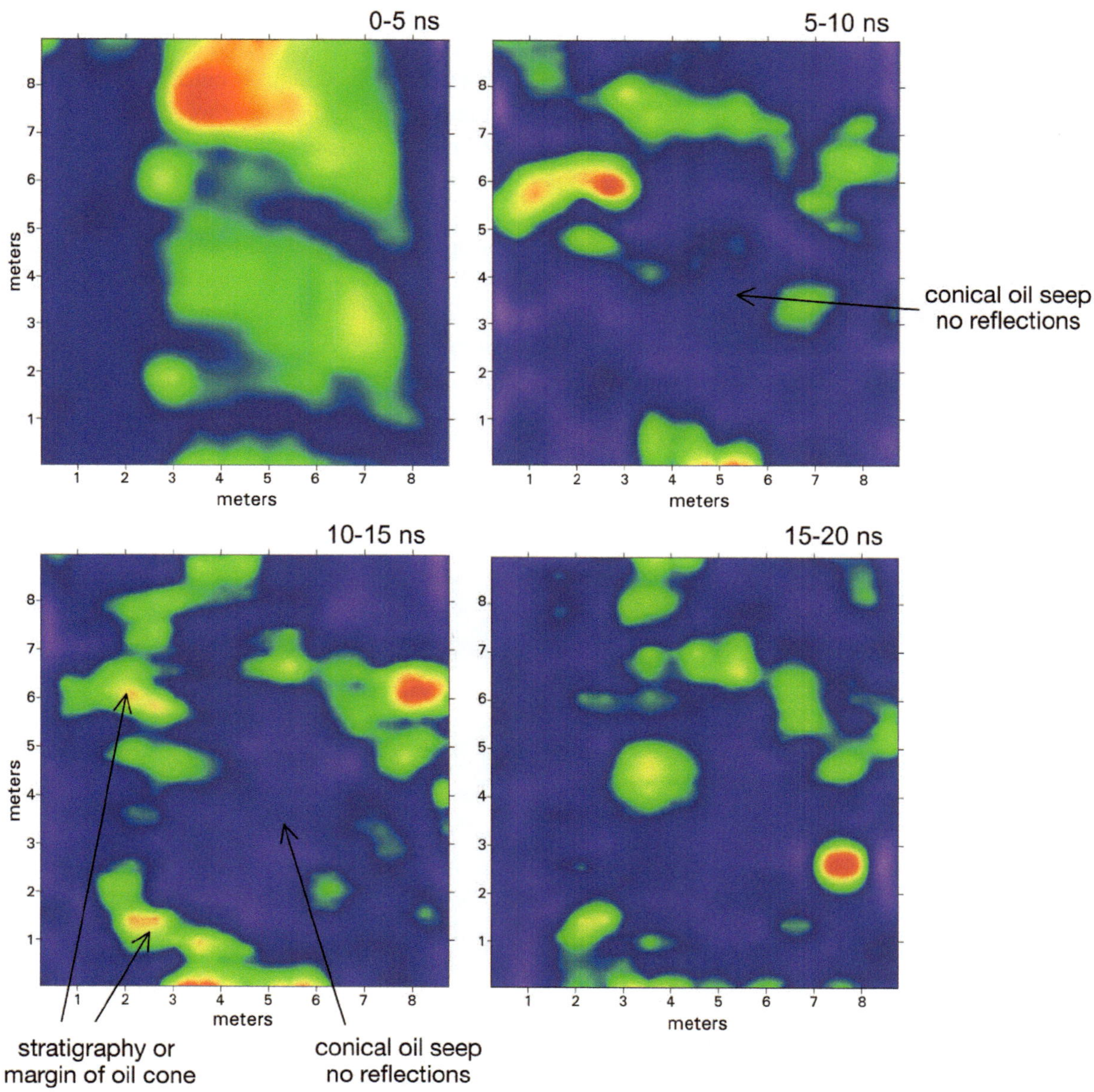

Figure 4-19: Amplitude maps of an oil seep. Seep #10 at the La Brea Tar Pits, Los Angeles, contains reflective bedrock layers on the margin of the seep and areas with no visible reflections in the interior of the seep where there is homogeneous crude oil.

In this unusual environment, the key to identifying these features is a comparison of the reflections in the surrounding sediment with those in the tar. As the tar is almost completely homogeneous, it produces no reflections but still allows radar energy to penetrate to depth. The surrounding horizontal layers contain different amounts of both water and tar in their pore spaces, producing high-amplitude radar wave reflections. As most GPR users will probably never have a chance (or inclination) to collect data in tar pits, this example shows how an analysis of the types of materials in any complex geological environment can lead to an understanding of the amplitude and distribution of radar reflections. In other places, the nonreflective features of interest similar to these features might be "sand boils" produced from soft sediment liquefaction, landslide units of homogeneous debris, or a multitude of varying reflective and nonreflective units. Both the reflection profiles and the amplitude maps are useful tools for mapping these types of features in three dimensions.

LAKES

Freshwater lakes are an excellent medium for GPR transmission and reflection, as radar waves pass through the water with little attenuation into underlying sediments. The same is also true with ice, and GPR methods have been used for decades in glaciers and for permafrost studies. While I have not collected a great deal of GPR on lakes, my limited experience has produced what I consider to be some spectacular results that demonstrate interesting geological features preserved below the lake surface. The lake profiles I have collected do not have a specific archaeological component and were collected in order to choose locations for sediment coring, with the goal of understanding past climate changes.

Prior to GPR mapping of lakes, many lake scientists had assumed that coring in the deepest part of the lake would yield the most complete sediment sequence and therefore the oldest sediments and most comprehensive record. The GPR profiles presented here demonstrate that some lakes are actually much more complicated than assumed, containing many sub-basins and packages of sediments that have very different depositional histories. Reflection data have been collected from frozen lake ice (Figure 4-20) and from fiberglass canoes (Figure 4-21), both of which allow radar energy to effectively couple with the water and propagate downward.

In high-altitude lakes in Colorado, large bedrock slump blocks likely fell into a glacially scoured basin as the ice melted. This created many sub-basins, which

Figure 4-20: Ice collection on a frozen lake. Collection was done on Palmer Lake, Colorado, using 400 MHz antennas.

Figure 4-21: Collecting GPR data from a canoe. A fiberglass canoe was an excellent means for collecting data at South Mesa Lake, Colorado.

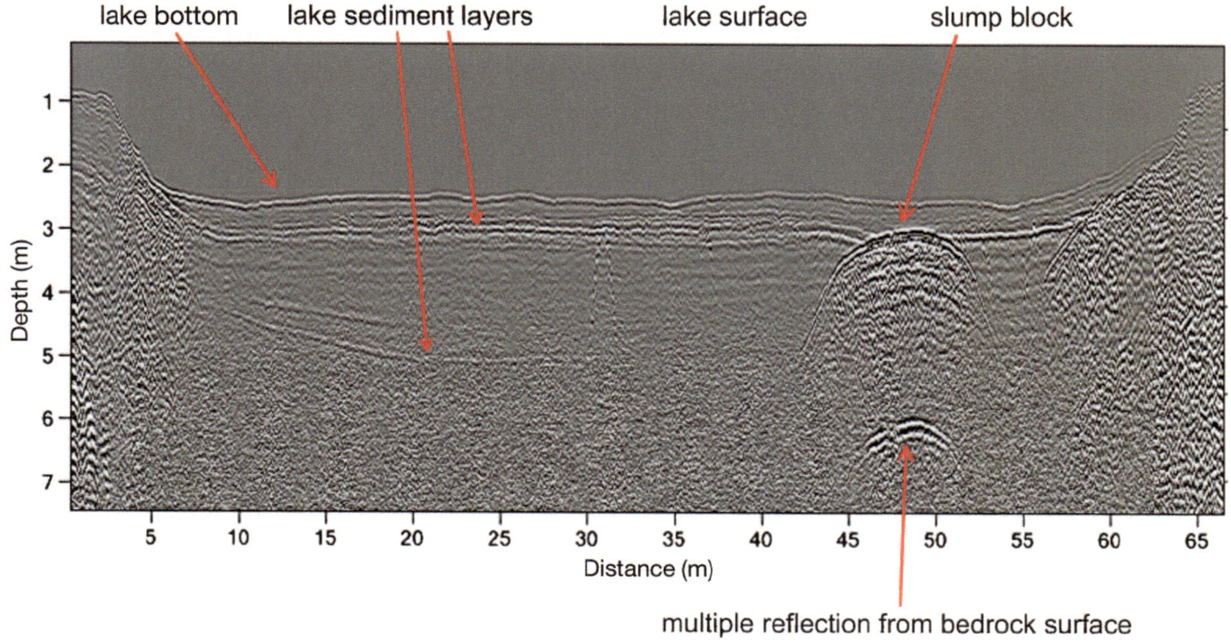

Figure 4-22: Reflection profile of data collected from a canoe. The lake bottom and sediment layers produced high-amplitude reflections above an attenuation depth of about 5 m at Lake Edith, Colorado. Slump blocks are clearly visible, as are multiple reflections from the top of a bedrock interface. Data collected using 400 MHz antennas.

then filled with sediment (Figure 4-22). Many meters of sediment then accumulated in these basins as the lake slowly filled. Today there is no visible evidence of these small intra-lake basins, as these features are covered with more recently deposited sediment. Cores taken within the sub-basins show that deposition was not constant in some of these basins. We found that some of the high-amplitude reflections visible in GPR profiles were generated at unconformities, where soil horizons and desiccation surfaces developed during warmer and drier periods when the lakes dried out. The drying episodes document periods of dramatic climate changes that were previously unknown. Other lower-amplitude reflections were produced at contacts between sandier layers, in a mostly silt and clay environment.

Some radar energy in Figure 4-22 appears to have been reflected once from the bedrock-sediment interface, traveled back to the lake surface, there to be re-reflected back into the water, perhaps from the lake-air interface (or the canoe bottom). This energy then was reflected a second time from the same bedrock interface to be recorded a second time at the receiving antenna. These "multiple" stacked reflections are visible toward the bottom of the reflection profile and can be identified as multiples because they were recorded at exactly twice the travel time as the upper, shallower reflections.

All natural lakes are formed from geological processes that produced some type of barrier to create a dam and impound water. In the Rocky Mountains of Colorado, these dams are often rock rubble that was deposited as glacial end moraines, accumulated in ridges at the toes of ice sheets when glaciers began to melt at the end of the last ice age. Two distinct packages of end moraine deposits can be seen on a GPR reflection profile from one high-altitude lake (Figure 4-23) with complex bedrock features and channels that cut through the moraine sediments sometime in the past. It is not known if these two moraine deposits are evidence of ice changes during a short period of time at the end of the last ice age, or perhaps are deposits derived from two very different glacier retreats separated by more than a hundred thousand years. Small packages of lake sediments were deposited adjacent to these moraine deposits as the lakes filled after the ice had melted. Individual rocks that make up the moraine units each produce complex high-amplitude reflection hyperbolas that give a somewhat cluttered appearance when viewed in profile (Figure 4-23).

I have not yet seen any examples in the GPR literature of archaeological features visible in reflection data collected on lakes. But it is only a matter of time before someone collects data over shipwrecks buried in lake

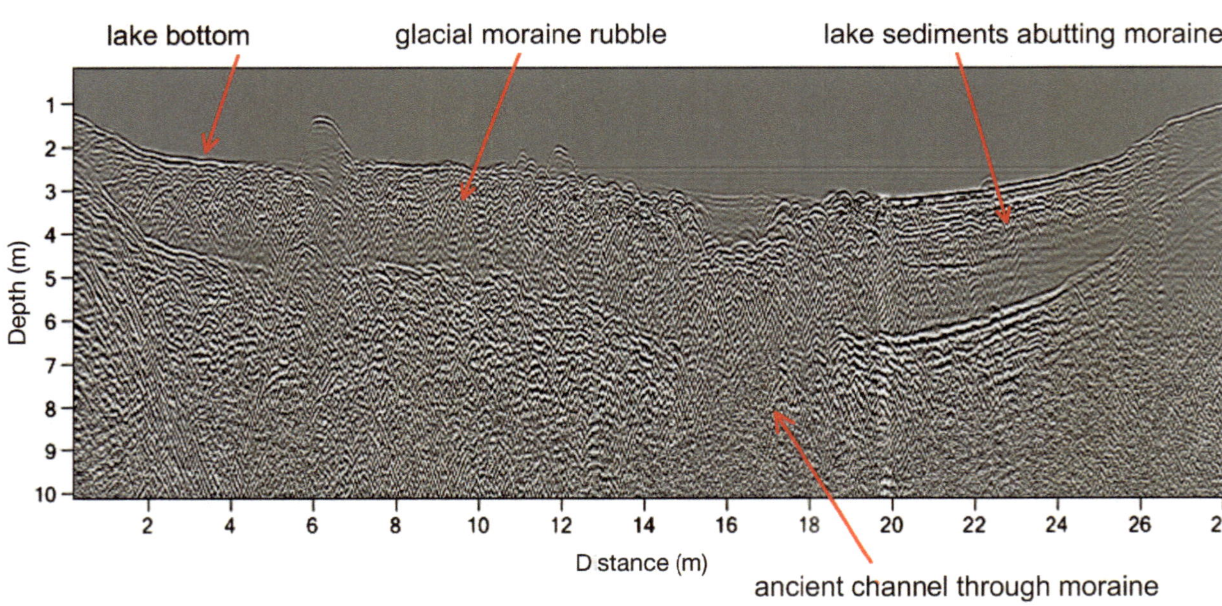

Figure 4-23: Reflection profile showing two layers of glacial moraine rubble with abutting lake sediments at Lake Edith, Colorado. The two distinct packages of moraine rubble and associated lake sediments may be the record of two glacial retreats in this small basin. Data collected with 400 MHz antennas.

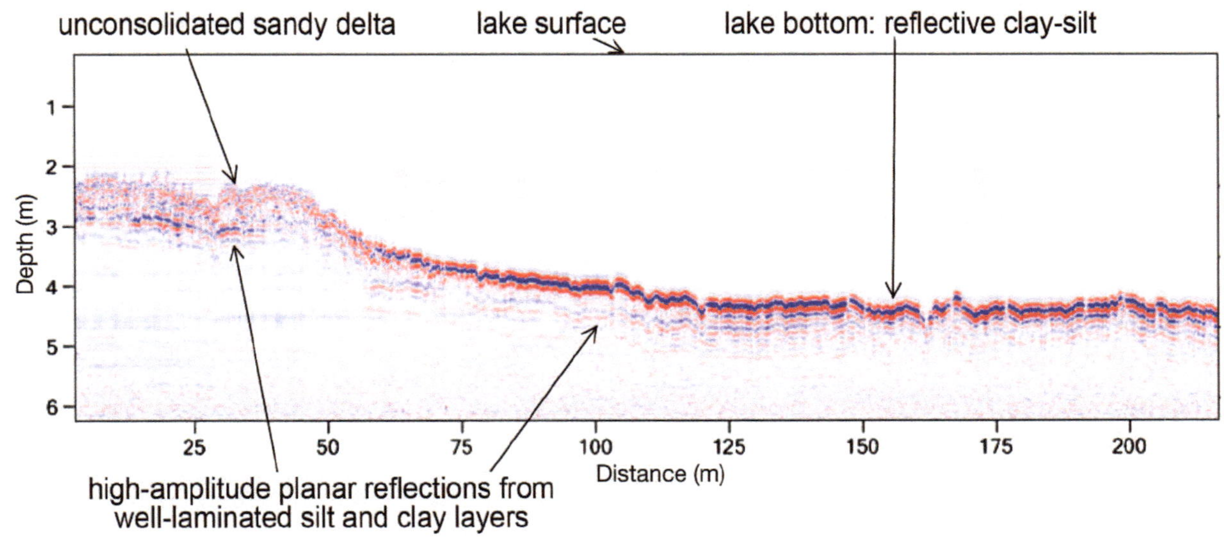

Figure 4-24: Reflection profile showing lake sediments. Collected on Lago Saladilla, Dominican Republic, this profile shows clay-rich sediment in the deeper portion of the lake, which produced a high-amplitude reflection but attenuated energy in less than a meter. The sandy delta sediments on the margin of the lake produced lower-amplitude reflections but contain less electrically conductive clay and therefore allowed radar wave transmission to a somewhat greater depth. Data collected with 400 MHz antennas.

sediments or perhaps lake dwellings built out into the water, such as those discovered in Switzerland and Scotland. It is easy to collect GPR data on ice in the winter, or in canoes or other watercraft, and the resulting reflection profiles are often spectacular, as most of the energy generated from transmitting antennas travels into the water and is available to be reflected from interfaces of interest. However, if lake sediments are just a little bit electrically conductive, an attenuating environment can be created that destroys energy at a shallow depth (Figure 4-24). One lake of this sort was studied in the Dominican Republic, where the sediment appears to have attenuated much of the radar energy within about a meter of the lake bottom. The reflection profiles are quite good at differentiating the type of sediment that has most recently filled the lake. The sandy delta units, which appear to document a recent migration of a small delta into one area of the lake, produced only random jumbled reflections, as there are no good interfaces from which to reflect radar energy. The clay and silt layers under those recently deposited delta sands are well laminated, and interfaces between layers of different sediment type produced high-amplitude planar reflections. These same types of distinct high-amplitude reflections were found in the deeper portion of the basin that did not receive any of the coarse sand sediment.

Any lake that contains freshwater is an excellent medium for GPR data collection. This is especially true if the sediments are not composed of electrically conductive clay. Collection from a boat, or on ice in the winter, are both very effective ways to conduct research with GPR. It is more difficult to place GPR data into space when they are collected from a boat, but GPS technology is beginning to make this much easier. When the antennas are placed on ice, there is excellent energy coupling, and very strong waves are transmitted into the water below, to be reflected from the lake bottom and its sediments. The same is true when antennas are placed in a fiberglass or wooden boat bottom, where that bottom is directly in the water with no air pockets or other material in the boat to interfere with propagating radar waves.

VOLCANOES

Volcanoes often bury archaeological sites, preserving them deep below the surface. In El Salvador, an eruption in about A.D. 600 covered the Mayan village of Cerén with more than 7 m of volcanic ash flows and coarser airfall units in a matter of a few days at most (Conyers and Spetzler 2002). The clay floors, and sometimes walls and columns supporting roof beams, are encased in stratigraphically complex volcaniclastic material (Figure 4-25). The clay floors of the buried structures attenuate radar

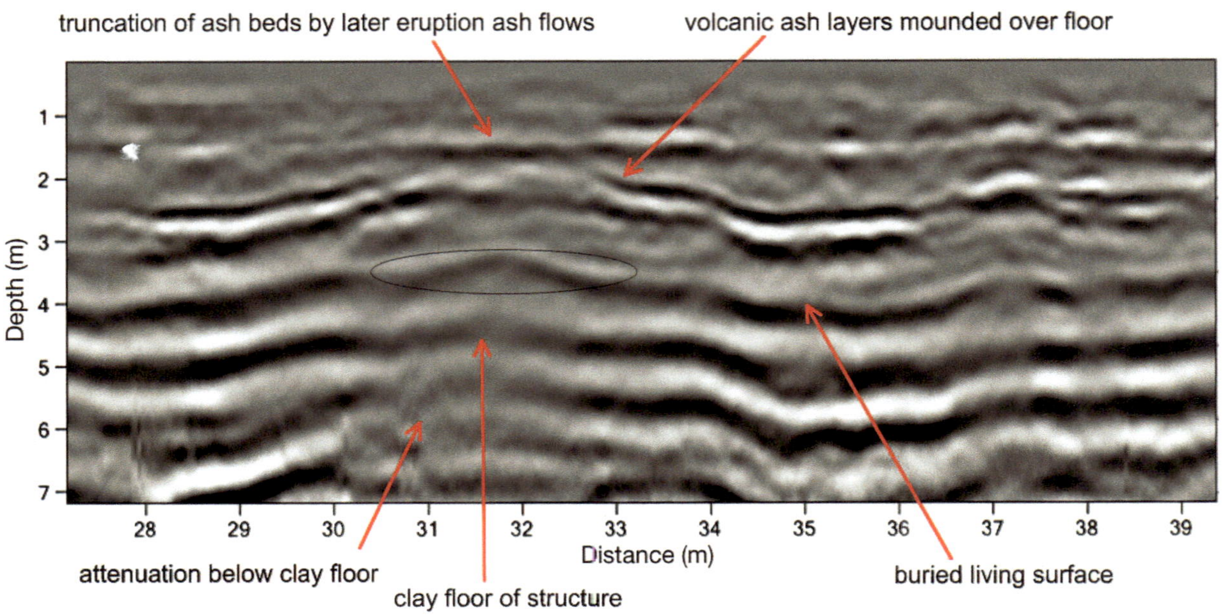

Figure 4-25: Reflection profile showing a clay house floor built on a weathered volcanic ash living surface. The clay floor attenuates radar waves that pass through it, producing only low-amplitude reflections below it. Ash layers derived from a volcanic eruption that buried the site are mounded over the structure, and a later volcanic event's ash layers rest on an angular unconformity with the partially truncated lower beds. Data collected with 270 MHz antennas at Cerén, El Salvador.

energy that passes through them, and any reflections from layers below them are very low in amplitude. Ash layers that covered the raised clay platforms are often mounded over these structures. In one GPR reflection profile from this site, the uppermost pyroclastic layers were truncated after their deposition, and subsequent ash layers were deposited horizontally over this erosion surface.

In this example, almost all that is visible in the GPR profile are reflections from the geological layers themselves. The data used to produce the profile in Figure 4-25 were collected using 270 MHz antennas, which produce a longer-wavelength radar wave than the 400 MHz, which is good for resolving some of the thicker volcanic units but can barely resolve the archaeological features encased within them. The only clue that a clay floored feature exists within this complexly bedded volcanic matrix is the attenuation of energy below that surface and the mounding of ash beds above.

Volcanoes produce a large variety of ejected materials and flow deposits, all of varying porosity, thickness, and size. When these units are newly deposited, they are almost always good mediums for GPR energy propagation and reflection. This is true for all chemical constituents of volcanic material, from the high iron and magnesium in basalt to the higher quartz and feldspar composition of andesite and granite. However, as soon as these units begin to weather, clays will form and, depending on the clay types, they will have a different electrical conductivity. Basalts tend to weather quickly in humid environments and produce more electrically conductive clays that are poor for GPR energy propagation (Conyers and Connell 2007). In Hawaii, we found that basalt cinders and flows were very poor for GPR propagation after having been weathered for about 1,000 years, if they were found on the windward (wetter) side of the islands. In those humid locations, basalt weathered quickly and produced a small amount of conductive clay, which, when moist, attenuated most radar energy within about 50 cm of the surface. The same composition basalt of the same age deposited on the leeward (drier) side of the islands was an excellent medium for radar propagation and reflection. Weathering there has been minimal and what clays have formed are dry, and therefore the medium is electrically resistive, allowing many meters of radar energy penetration. Variations in energy propagation in basalts or weathered basaltic tephra can also occur seasonally with water saturation variations. In El Salvador, the basalt-andesite composition ash attenuated radar energy in the upper meter or two during the wet season when the ground was saturated. When the same ground dried out in the dry season, radar waves had excellent penetration to about 4–6 m (Conyers 1995).

CONCLUSIONS

They key to all the geologically complex examples presented here is a differentiation between geological unit reflections and archaeological features. This is not usually possible without spending a long time looking at the landscape surrounding a research site and studying the geological units that outcrop in gullies or other exposures. An analysis of exposures and other information about buried units allows an interpreter to generate some hypotheses at the beginning of a project about what types of GPR reflections might be obtained. Soil unit thickness, clay content, and the distribution of soils, both surface and relict buried soils, can tell much about the geological history of a site. Archaeological features can then be placed within this framework, and both the geological and archaeological reflections can be identified, differentiated, and mapped.

If at all possible, reflection profiles need to be collected where the reflections visible in profile can be directly correlated to these geological exposures. If velocity of radar travel times is computed correctly, the reflections can be directly compared with those visible units, and some degree of confidence about "picking" reflections and overall interpretation results can occur. Many geophysicists spend too much time looking at their computer screens and not enough time looking around the site. Much can be learned by identifying geological units nearby that can be seen immediately within a GPR data set, so that when reflections appear on the system screen during collection, some initial interpretations can begin. The inability to interpret GPR results within the broader context of the surrounding landscape is analogous to what happens to excavation archaeologists who spend all their time in a few holes in the ground and fail to look around to get the bigger picture.

An identification of bedrock or other environments of deposition, and how those were situated in the past, can be used to understand landscape changes and identify buried landscapes that are no longer visible on the surface. This analysis of "the bigger picture," which GPR interpretations can define in buried landscapes, can be hugely important in the process of placing archaeological features into ancient environments (Conyers 2009). Most significantly, it can then place humans within those landscapes and environments, which is very beneficial to any interpretation of past human behavior.

5
Cultural Complexity

It is always a challenge to distinguish anthropological features in the ground from those of recent origin, and to deal effectively with the complexities of modern overprints and other modifications that affect GPR reflections. These issues must be confronted in almost every survey in a city, or almost anywhere that people are active or have been active in the past. When the goal is to explore for and map archaeological features, and a GPR survey area is filled with modern disturbances, the first step in any interpretive task is to determine the origin of the reflections recorded and displayed in your processed images. Only after important reflections from the ground are differentiated from general "noise" or recent cultural overprints can an effective interpretation begin.

Pipes, Foundations, and Metal

Often a determination of what is modern and what is archaeological can be simple and straightforward—for example, during a mapping project where the target was a Mogollon-period village from about A.D. 1100 within a pipeline right-of-way (Figure 5-1). In Alamogordo, New Mexico, the local water department was considering placing a water pipe in a narrow strip of land under its control. Local archaeologists had identified ancient ceramic artifacts on the ground surface in the general area, which had been brought up from an unknown depth by burrowing animals, and they speculated on whether an ancient village existed in one area of the proposed pipeline construction. Employees of the city of Alamogordo notified the local archaeologists and me that there were no archaeological sites in the area of the construction and suggested that my survey would be a waste of time. As they had done previous construction in the area and had placed many pipes in this right-of-way, I was somewhat pessimistic as to whether we would be capable of finding anything with GPR. In addition, the area to be surveyed was cluttered with modern buildings, vehicles, boats, trash, and vegetation, which made it difficult to lay out grids and move antennas. However, within the first few minutes of collecting reflection data, the floors of the pit houses could be seen in reflection profiles, and the ancient village could be easily mapped (Figure 5-2). Three large-diameter pipes were also visible in profiles, and the vertical trenches excavated to emplace them could be seen truncating some of the ancient house floors. Both the modern and ancient materials buried in the ground were easy to differentiate, and the partial destruction of the cultural remains by modern trench digging was readily documented. The city employees and the construction firm tasked with placing a new water line in this area were not happy with the ability of GPR to so quickly map these buried features. In contrast, the archaeological community was overjoyed to have found an ancient village to study, before it was destroyed further by construction.

Metal within paving surfaces can greatly affect radar energy penetration, and GPR can therefore be an effective tool for mapping reinforcing metal and other materials in concrete roads, runways, and floors. But for archaeological mapping, metal interferes with radar wave penetration to such an extent that it can make some profiles unusable, depending on the frequency of the antenna. A survey was conducted on a street in downtown Reno, Nevada, in an attempt to find historic buildings associated with the construction of the Transcontinental Railroad in the 1860s. Both 200 and 500 MHz antennas were used for this survey, where antennas had to be moved across the street in transects only when the pedestrian "walk" sign was illuminated, and collection had to be finished before buses and cars sped by (Figure 5-3). One benefit to this survey was that we had a man dressed up as a leprechaun (who was supposed to lure gamblers into a nearby casino) help us move the antenna cables to keep them from being run over by buses.

Figure 5-1: Collecting GPR data within a pipeline right-of-way. This was a cluttered location with an abundance of modern cultural complexity in Alamogordo, New Mexico.

The reflection profiles generated from the 500 MHz data showed distinct reflection hyperbolas from the iron rods in the wire mesh embedded within the concrete roadway (Figure 5-4: top). However, a good deal of radar energy still penetrated between the metal bars and traveled to the layers of interest below the roadway, producing very good reflections from important historic surfaces covered by the road. The 500 MHz antenna was much better at transmitting between the wire mesh than the 200 MHz (Figure 5-4: bottom) because it produces a more focused transmitting beam of energy. Very little of the more widely spread-out 200 MHz energy passed through the mesh, and the energy that did get through produced only very blurry, poor-quality reflections of the same sub-road reflections.

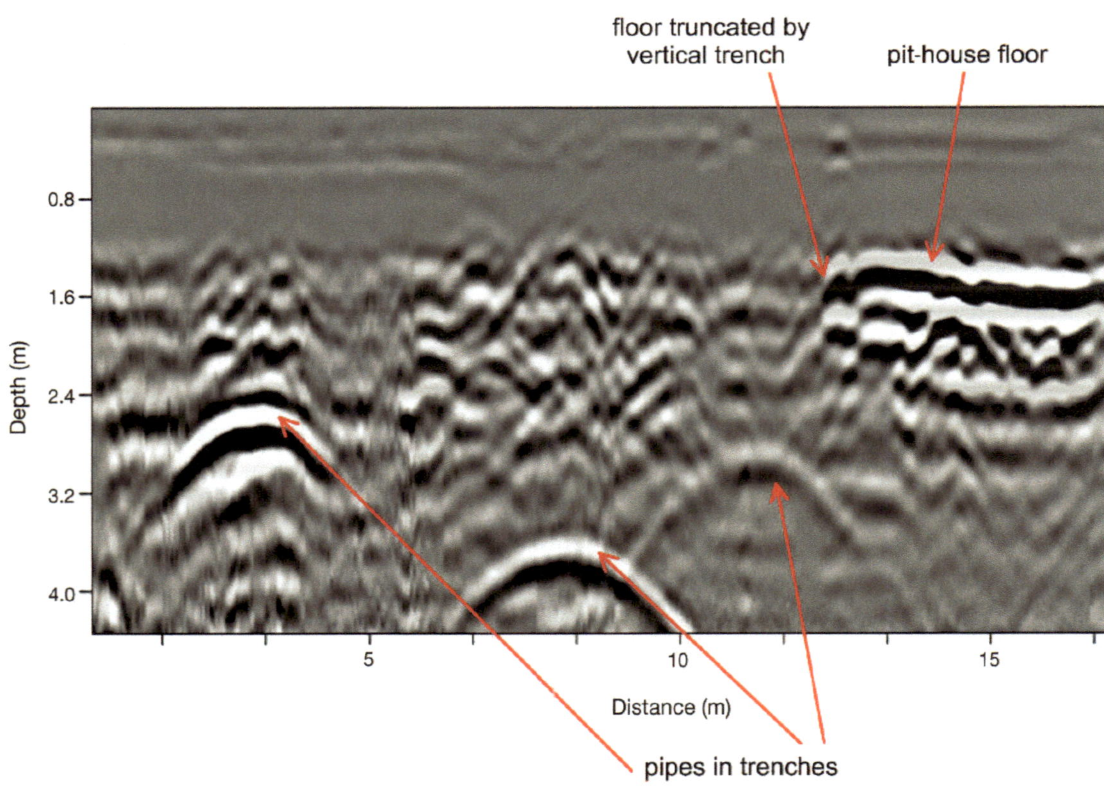

Figure 5-2: Reflection profile showing a truncated pit-house floor and three water pipes, one of whose vertical trenches has truncated the floor of the pit house, which dates to about A.D. 1100, in Alamogordo, New Mexico. Data collected with 400 MHz antennas.

Figure 5-3: GPR profile collection over a concrete road surface in Reno, Nevada. The road surface had metal reinforcing mesh embedded in it, which interfered with energy penetration. Data collected with 200 MHz antennas.

What this study demonstrates is that near-surface metal need not preclude a GPR survey in areas where most other geophysical methods would be impossible. The more focused, higher-frequency antennas can propagate enough energy through the near-surface metal to reflect from buried interfaces of interest. Lower-frequency energy, which tends to spread rapidly after leaving the transmitting antenna, produces reflection hyperbolas that obscure what little energy moves to depth, and is less likely to produce useful results.

In a parking lot in Santa Fe, New Mexico, we were asked to find the location of possible human burials and other historic features near where a high-rise parking structure was to be built. This area, owned by the Roman Catholic church, sits in the oldest part of town which has seen a great deal of construction and change since it was first occupied by Pueblo people in about the 11th century. The Spanish built a church in this general location in the late 17th century, and there has been a good amount of development and redevelopment ever

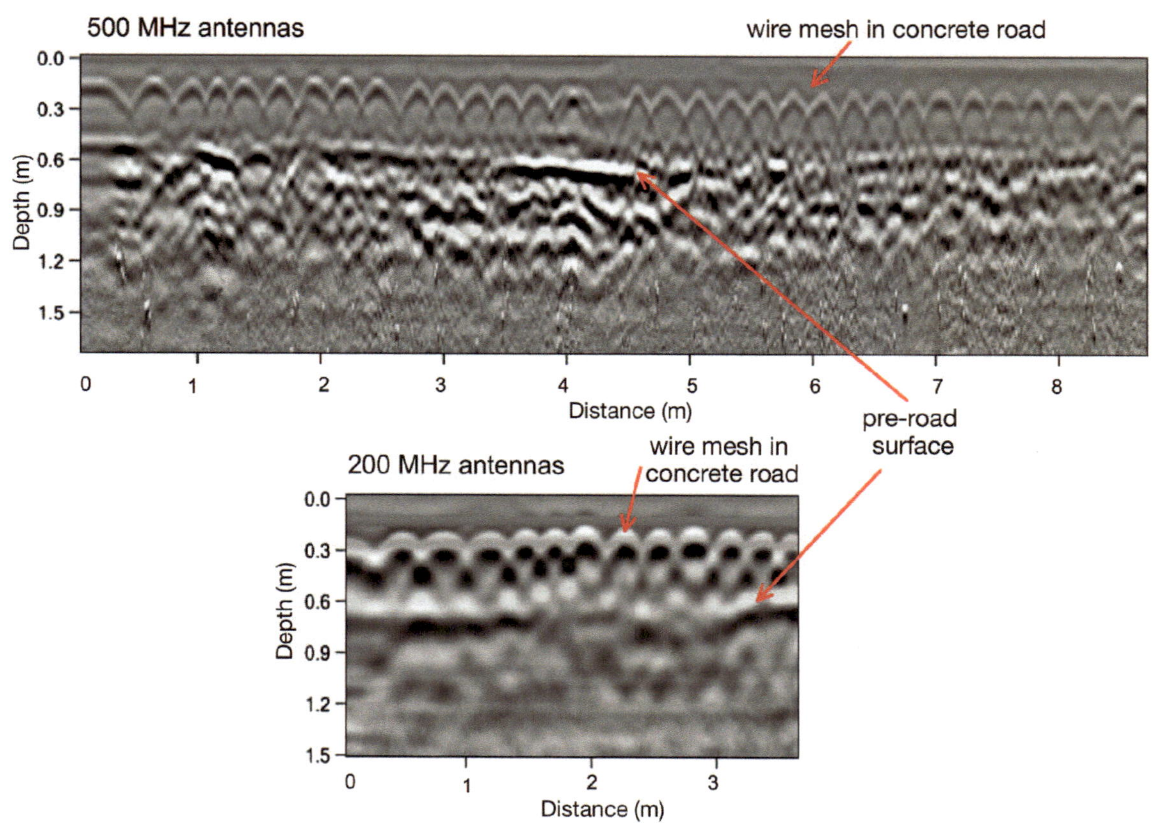

Figure 5-4: Reflection profiles collected on a road surface, the same one shown in Figure 5-3. The 500 MHz antennas have a narrow transmitted beam of energy, and therefore more energy makes it through the iron mesh to reflect from pre-road surfaces below. The 200 MHz energy spreads more widely as it leaves the transmitting antenna, and the mesh hyperbola reflections obscure much of the energy that makes it to surfaces below the road.

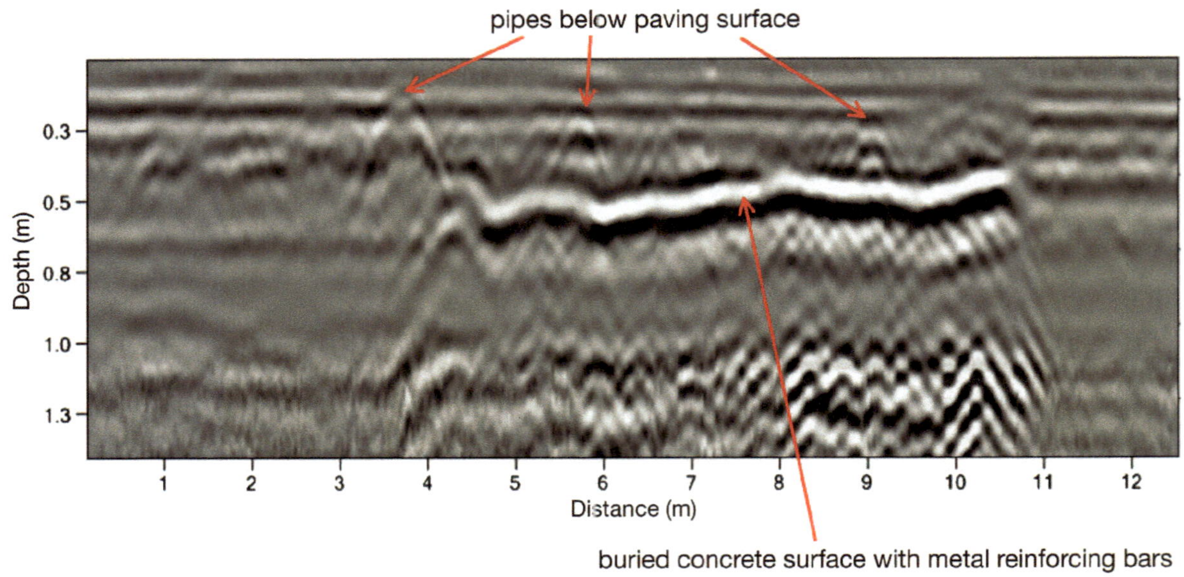

Figure 5-5: Buried concrete surface of a road or historic house floor in a parking lot. This surface is a possible road or historic house floor constructed with reinforcing metal below a parking lot in Santa Fe, New Mexico. Data collected with 400 MHz antennas.

since. Our many GPR grids located a number of historic houses and features of marginal interest, and much to the chagrin of the church, a graveyard was located in the middle of a parking lot.

Because of the complexity of cultural features in the ground in this area of Santa Fe, determining what was of historic interest and what might be only the remains of buildings demolished during the last 50 or 60 years was difficult to determine. Some reflection profiles were quite good at displaying buried house floors made of concrete with reinforcing bars, a construction technique used in this area only in recent times (Figure 5-5). This method of construction, easily determined by examining the GPR reflection profiles, could be used to define some buried features as potentially important historically, and others less so. The large number of buried utility pipes was also easily mapped using GPR (Figure 5-6).

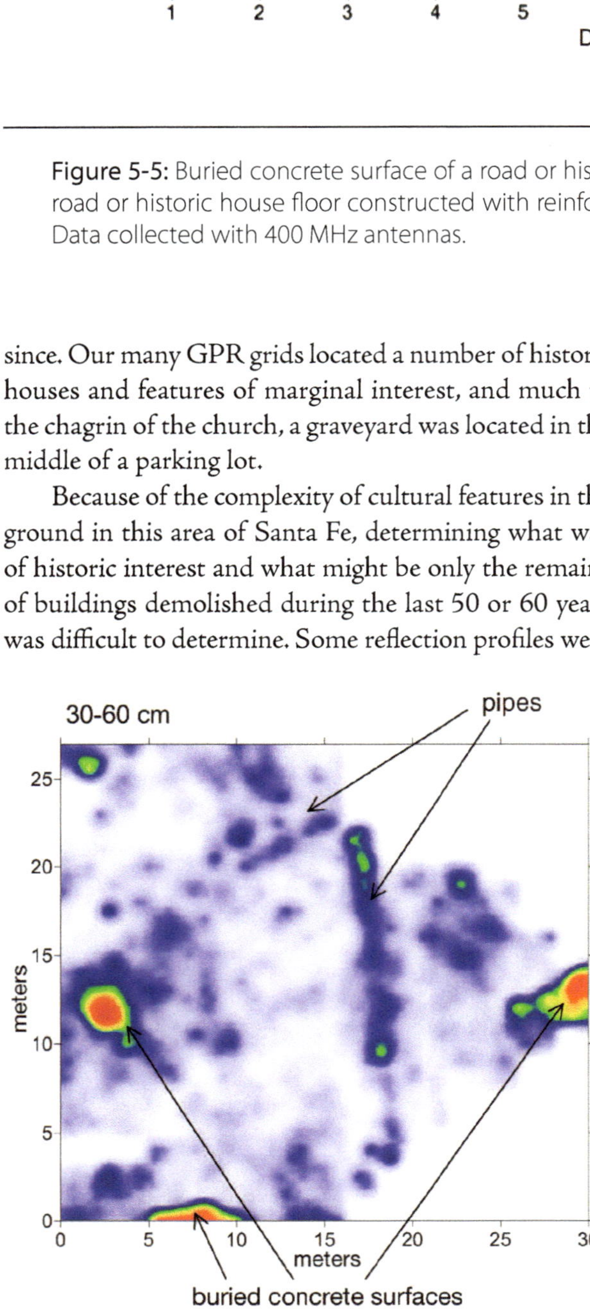

Figure 5-6: Amplitude map from a parking lot with buried modern materials in Santa Fe, New Mexico. The map shows that the buried concrete surfaces are visible as high-amplitude reflections and the pipes as linear reflection features.

In this parking lot, the abundance of modern disturbance from pipes and from the buried modern building foundations produced a complex array of features that were difficult to "interpret around." If modern materials of this sort can be identified with GPR, they must be annotated in maps as modern, and the important reflections from other older features in between can then be concentrated on in interpretations.

AIR WAVES GENERATED FROM SURFACE OBJECTS

Often, subtle reflection features visible in profiles can be produced by radar waves moving in air from the trans-

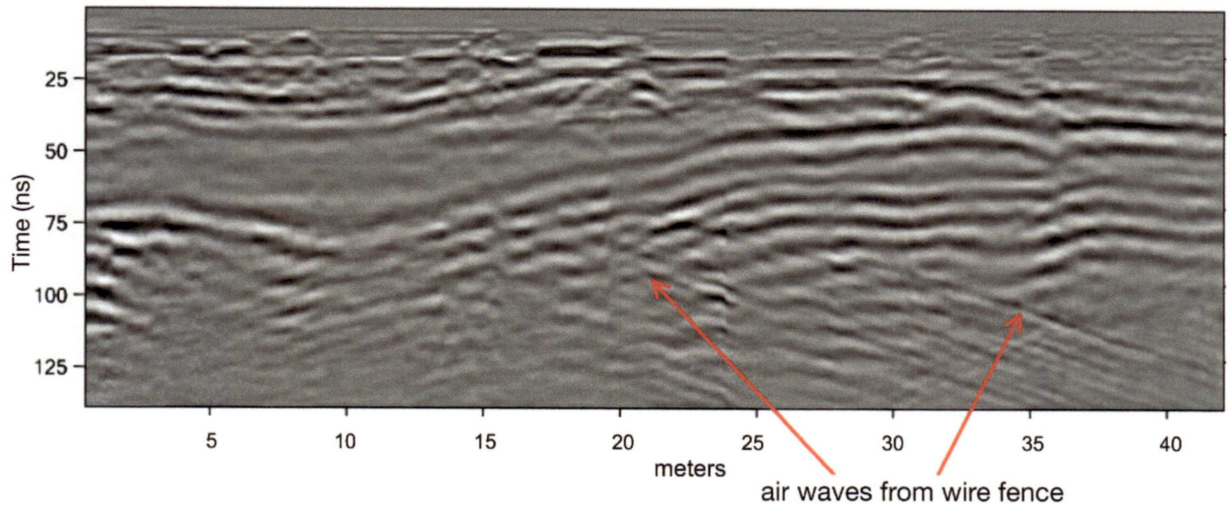

Figure 5-7: Air waves generated from a wire fence. These subtle air waves were produced from a very thin wire fence about 3 m away from the antenna transect. Air waves are always visible as straight reflections in profiles, as the radar energy moves at a constant velocity (the speed of light) to and from the reflection surface. Data collected with 300 MHz antennas in El Salvador.

mitting antennas and then reflecting from surface objects, commonly referred to as *air waves*. These recorded waves (Figure 5-7) are always straight, when viewed in reflection profiles, and thus can be readily identified. They are produced as energy moves from the surface antenna to some above-ground reflection surface and then back to the receiving antenna, at the speed of light. As the velocity of the radar waves never varies in air, but the distance between the antennas and the reflective surface changes as antennas are moved along a transect, the reflections are recorded in ever increasing time, making them appear to be dipping in the ground. When the amplitudes are strong enough, these air waves might be confused with stratigraphic horizons. However, as they are always straight, and dipping beds are rarely that regular, they can be easily differentiated.

Lower-frequency antennas are often more susceptible to transmitting waves that move in air and are not propagated directly into the ground, as their energy tends to spread out more after it is generated at the antenna. In addition, lower-frequency antennas are more difficult to shield and therefore "leak" energy that reflects off any reflective surface nearby. I have noticed air waves generated from trees, buildings, cars, and even passing people or livestock nearby from antennas that are 400 MHz or lower in frequency. When collecting GPR data along a street in Portugal with the 400 MHz antennas, energy moved in the air to a curb along the edge of a sidewalk, which produces very high-amplitude air waves, obscuring the important reflections from the bedrock below (Figure 5-8). While it is possible to remove air waves in some processing programs, they often overwhelm the reflections generated from within the ground and can drastically distort the reflection images produced.

I have seen reflection profiles collected in caves and rock shelters that have distinct air-wave reflections generated from the stone ceilings. I have also seen amplitude maps where interpreters sliced up profiles into horizontal layers and interpreted air waves in progressively deeper slices as "steps" leading down into the ground. In this "step" interpretation (unfortunately presented in public, to the chagrin of the interpreters later on when these air waves were identified), the air waves had been generated from a standing wall along one edge of the GPR grid, and the profiles were collected moving perpendicular to that wall. In the progressively shallower slice-maps, the "steps" appeared to lead directly up to the wall, an unusual location for steps! A quick look at the profiles showed that these so-called step features were nothing more than air waves produced by the wall, which were collected in greater time as the antennas were moved away from the wall. When sliced into horizontal amplitude maps, these high-amplitude waves looked much like evenly spaced steps, a huge mistake in interpretation.

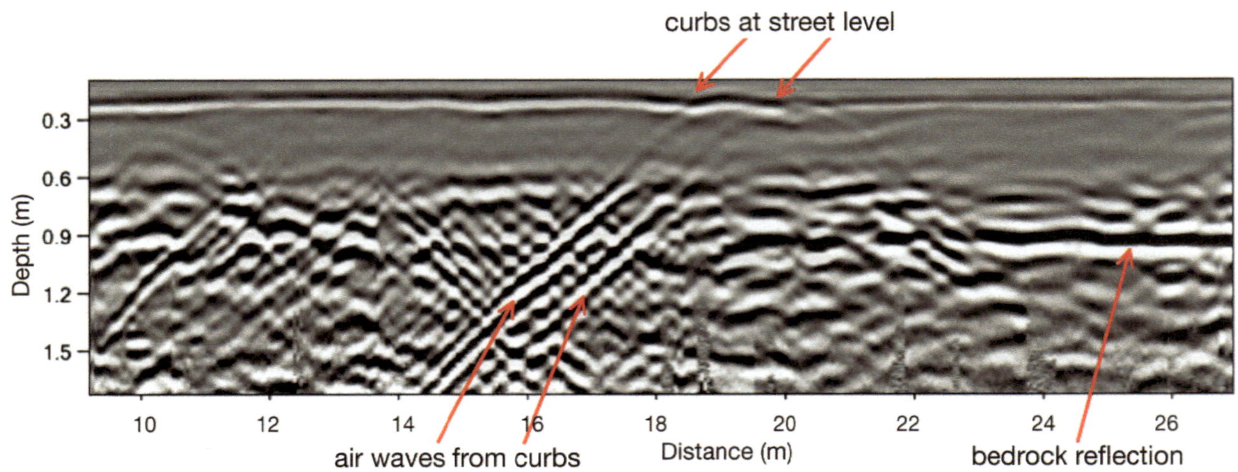

Figure 5-8: High-amplitude air-wave reflections from raised curbs along the edge of a sidewalk in Castro Marim, Portugal. The curbs generated reflections from radar energy that moved parallel to the road surface in the air. These reflections have overwhelmed many of the reflections that might have come from within the ground.

GROUND-SURFACE VARIATIONS

A common type of modern modification on the ground that can be a great detriment to GPR data quality is a bumpy ground surface, such as that produced by agricultural plowing. When the ground surface is not flat and the antennas move over the uneven surface, they will sometimes be in contact with the ground and other times in the air. Antennas in contact with the ground couple energy more effectively, and it will be transmitted efficiently downward. When the radar energy must travel through air, and then into the ground as it leaves the transmitting antenna, it couples differently with the ground, and its strength, and the amplitude of resulting reflections generated from it, is different from that generated when the antennas are more ground-coupled. Many such bumps along any one reflection profile can produce an amazingly complex series of reflections that can be very difficult to interpret, and few of them are likely to be from units of interest buried in the ground.

In highland Ethiopia, it seemed that every time we set out to collect some GPR data, the night before fieldwork was to begin someone had plowed the field we were to work in. We wondered what people were doing plowing their fields at night and why their work seemed always to be in the fields we were just about to survey. Apparently, word of our plans to survey with GPR got out among the local farmers, and they went to great lengths to "claim"—by plowing—the land we were interested in. Later we found out that because of the very weak land tenure laws in Ethiopia, the only way to have any effective control over land is to actively work it. It appears that the local farmers had been watching us late in the day when we chose locations for GPR surveys in preparation for our next day's work. To them, we were just strangers with unusual equipment who, they believed, were connected in some way to people of authority; so they would stay up all through the night to plow their land prior to our arrival the next morning, as a way to gain more effective control over it (Figure 5-9). Of course, we had no intention of appropriating their land,

Figure 5-9: Recently plowed fields in highland Ethiopia. This recent disturbance produced a very difficult surface on which to collect GPR data, as the antennas were always changing their coupling characteristics.

but the local people had good reason to be suspicious of us, as they have been subjected to decades of arbitrary and cruel edicts by the authorities. They were quite effective at ruining the areas we wished to study with GPR, and our reflection profiles were all but unusable when collected on the newly plowed ground (Figure 5-10). Even land that had been plowed a few weeks in advance produced very uneven energy coupling and poor reflection data.

To overcome the severe coupling problems encountered in these plowed fields, we attempted to employ a number of collection and processing techniques. But the only method that was at all productive was collecting profiles using a running average stack of traces (horizontal smoothing), which took the average of every five traces to produce one average trace (two in front and two behind each trace recorded), as can be seen in Figure 5-10 in the lower profile. Even this method produced very difficult-to-interpret reflection profiles and also effectively removed all point-source reflection hyperbolas. Point-source hyperbolas from buried stone architecture, which might have been a clue to the location of buried stone features of interest, became invisible in profiles. A sampling of all the averaged traces and spatial mapping of all the high-amplitude reflections to produce amplitude maps did show us the location of house rings and buried walls in plan view. But the individual profiles were mostly uninterpretable because of the nature of the ground surface.

Other types of ground-coupling changes that affect energy penetration and the resulting reflections recorded can be the product of differences in the type of material

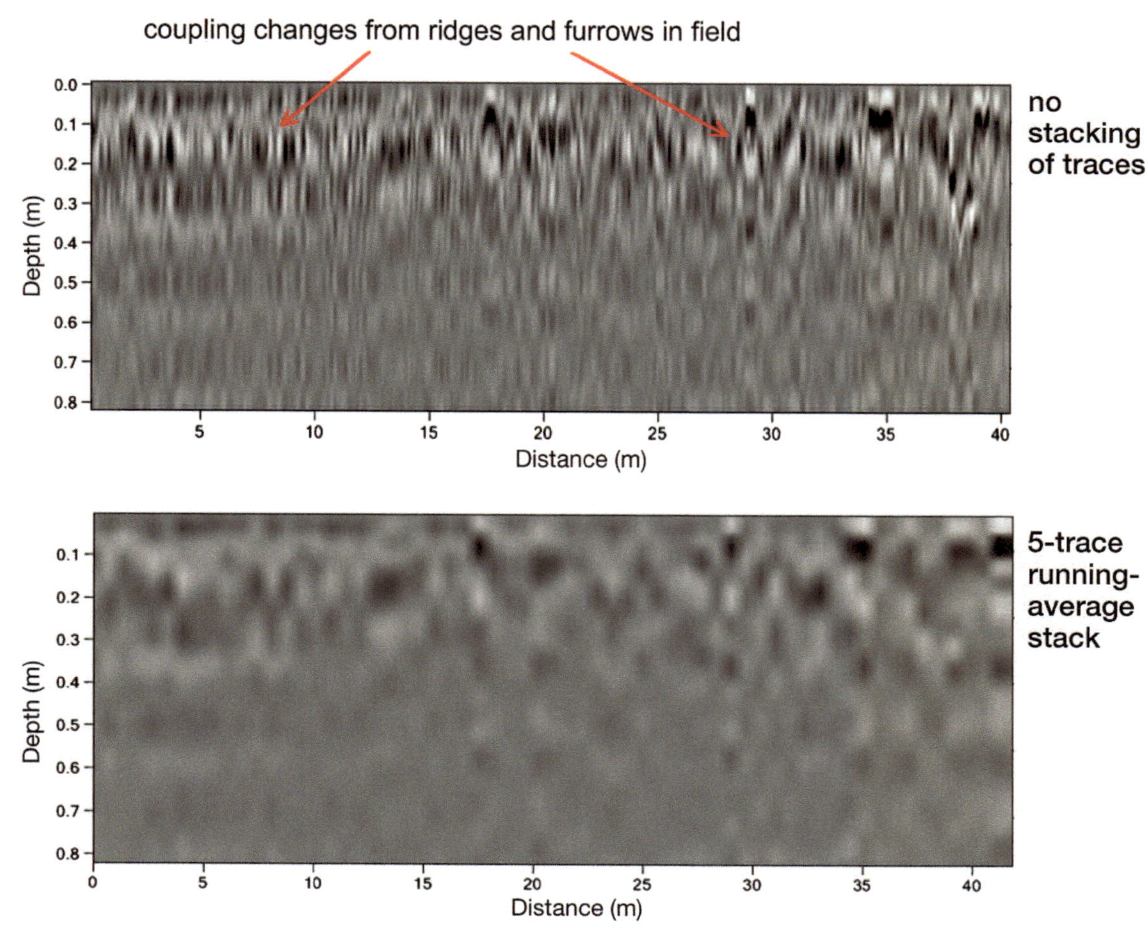

Figure 5-10: Reflection profiles collected in recently plowed fields. The top reflection profile shows an abundance of vertical changes in reflection characteristics due to the antenna coupling changes on recently plowed ground. Some stacking of traces in a running average during collection, or later during data processing (lower reflection profile), will remove some of the worst of the coupling changes, but this profile is still of very poor quality.

that covers the ground. The very best surface for collecting GPR is ice, or frozen ground. One survey we conducted in Colorado in the middle of winter was begun in the early morning while the ground was still frozen from the very low nighttime temperatures. Very good reflections were recorded in those conditions. As the sun came up and the ground surface thawed, I noticed the amplitude, depth of penetration, and overall data quality started to decrease. By midday, the energy penetration depth and the amplitude of reflections had decreased dramatically because of the lack of ice directly in contact with the transmitting antenna on the ground surface. This phenomenon has long been understood, and I was told that a GPR colleague made an attempt at imitating winter conditions by placing frozen gel packs (made for drink coolers) under his antenna during collection to improve energy coupling with the ground. Unfortunately, as those gel packs thawed during the day, energy coupling also decreased and the data quality changed accordingly. That grid of data was unusable for spatial amplitude mapping, as there was no way to compare profiles in a grid collected with such varying energy coupling and the resulting differences in reflection amplitudes.

In southern Portugal, a number of GPR profiles were collected along a street in an attempt to map the bedrock as it dipped into what was thought to be an ancient harbor at the bottom of a hill. This seemingly simple task was complicated not only by the air waves from curbs (Figure 5-8), but also by the coupling changes due to the types of paving surfaces. When the antennas were moved along limestone cobblestones (Figure 5-11), there was energy penetration to about 1.5 m or so, but as soon as they were moved onto the asphalt paving, wave penetration depth increased to about 2.5–3.0 m (Figure 5-12). I am not absolutely sure what it was about the asphalt that allowed the increase in energy depth penetration, but it could have been a number of factors. The small limestone cobbles might be somewhat more electrically conductive and therefore attenuated energy more than the asphalt. Or perhaps there is more water in the ground under the porous cobble pavement than in the impermeable asphalt, which produced a more electrically conductive medium and therefore wave attenuation and less energy penetration and reflection. I like the latter hypothesis least, as more water would likely make energy move at a slower velocity. If that were the case, then the bedrock reflection, which can still be identified under the asphalt (with higher amplitude), would likely be recorded at a longer two-way travel time

Figure 5-11: Collecting a GPR profile on limestone cobbles and an asphalt surfaced road. This profile was started on limestone cobbles, where the antenna is located in this photo, and then moved onto the asphalt-surfaced road in the background using 400 MHz antennas, in Castro Marim, Portugal. The profile collected here is shown in Figure 5-12, with the curb that produced the air waves in Figure 5-8 in the background.

beneath the cobbles and plotted at a deeper depth. That is definitely not the case (Figure 5-12). Or perhaps there is something about the limestone cobbles, and the way they are placed next to each other, that diffracts and reflects energy at angles where it is effectively lost and any reflections from those waves are not recorded. Whatever the case, this example shows how dramatically different reflection data can be depending on the type of material at the ground surface.

I often worry that many of my interpretations about variations in reflections that appear to be located at some depth are really a function of changes in surface antenna coupling differences. It is always important to pay attention to the roughness of the ground and to changes in material types on the surface during collection. As the antennas move over these surface features, the reflections obtained from depth can then be qualitatively analyzed to determine how much they are being affected. It is not

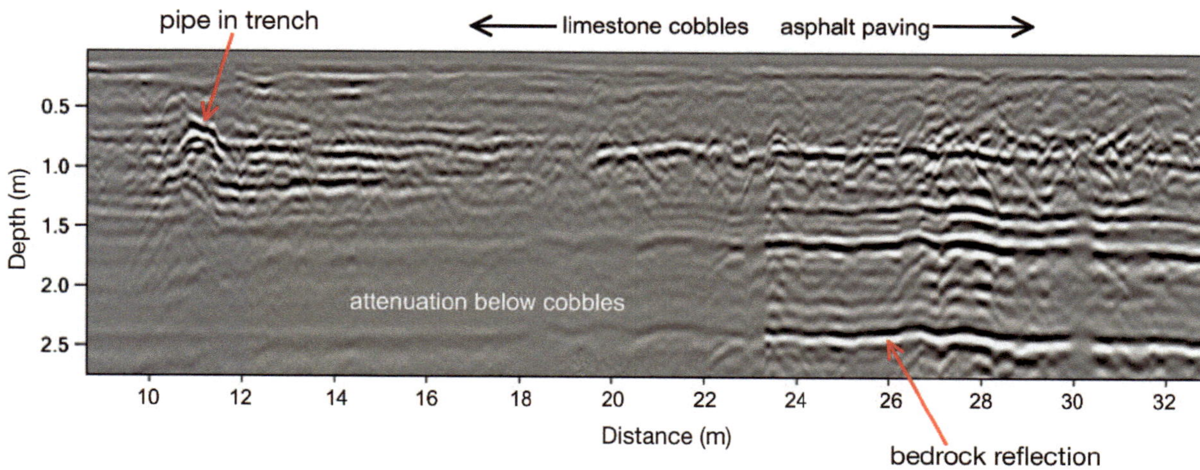

Figure 5-12: Reflection profile collected on limestone cobbles and an asphalt-paved surface. This profile was collected when antennas were moved from limestone cobbles on the surface onto the asphalt-paved surface. There was energy attenuation below the cobbles at about 1.5 m in depth, but as soon as the antennas were moved onto the asphalt surface, the energy penetration depth doubled. Data collected with 400 MHz antennas in Castro Marim, Portugal, shown in Figure 5-11.

often that we every get to collect GPR data in parking lots, well-manicured lawns, or football fields where there is no variation at all in the ground surface. While plowing ridges and furrows can be the most limiting of ground conditions, even small bumps caused by tufts of grass, or aerially restricted pools of water, can drastically change the nature of energy propagation into the ground and reflection from features of interest.

EXCAVATION DISTURBANCES

On the outskirts of the modern city of Pueblo, Colorado, sits the locally famous Goodnight Barn, built by the pioneer cattle driver Charles Goodnight in 1868. This historic entrepreneur was important in opening up the large-scale cattle drives between Wyoming and Texas in the middle of the 19th century, and this barn was built as a stopping point for cattle drives and stagecoaches for many years. In an attempt to locate archaeological features in the vicinity of the barn, GPR grids were collected widely around it. A quick view of the aerial photos of this area should have given us all the clues necessary for interpreting the GPR data, as the barn sits as a lone island in a disturbed landscape of gravel-extraction operations and artificial lakes produced by the removal of huge amounts of construction aggregate (Figure 5-13).

Many large grids of GPR profiles were collected around the historic barn (Figure 5-14) and processed into reflection profiles and amplitude maps. The profiles all show a very complex array of heavy-equipment scars that were part of the gravel-removal operations. Some of these large trenches appear to have been backfilled with layers of sediment that differ from the gravel matrix of the site, producing high-amplitude planar reflections (Figure 5-15). The gravel is largely nonreflective, as it is homogeneous and the individual small clasts do not produce

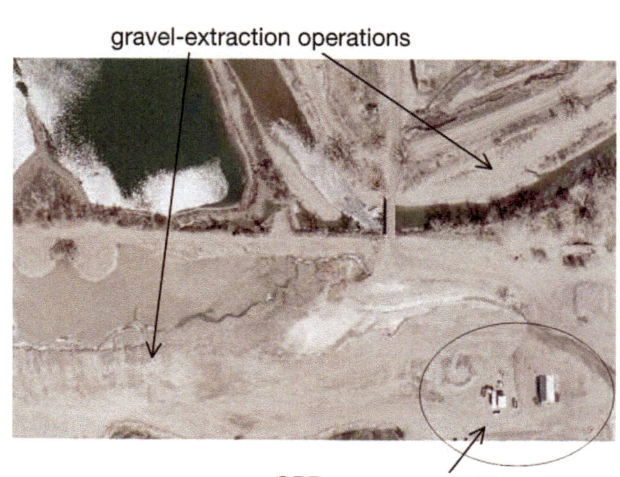

Figure 5-13: An aerial photo of the Goodnight Barn site, Pueblo, Colorado. This photo shows gravel-extraction operations nearby.

Figure 5-14: Collecting GPR data at the Goodnight Barn site. The 400 MHz antennas were used to collect data around a flat surface created by leveling during gravel-extraction operations near the site in Pueblo, Colorado.

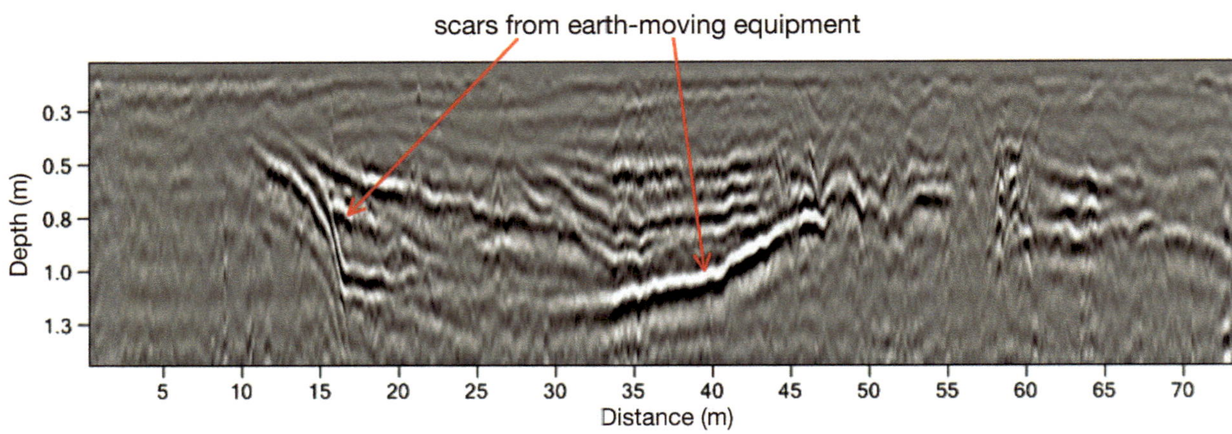

Figure 5-15: Reflection profiles collected around the Goodnight Barn site showing compaction and leveling scars created by heavy earth-moving equipment. Any cultural features that might have existed in this area have been destroyed. Data collected using 400 MHz antennas near Pueblo, Colorado.

distinct reflection hyperbolas. Linear high-amplitude reflections are visible in all maps from various depths, and the large-scale excavation operations could be mapped, which were not of any great interest to this project. No historical features of any sort were discovered in this highly disturbed ground. We could have saved a great deal of time and money by having paid more attention to the gravel operations and where they had been conducted.

Large earth-moving scars of this sort have been mapped at various locations around the world, and they almost always mean that all near-surface archaeological materials have been destroyed. I have seen them at Roman sites in Israel and Italy, cemeteries in the United States, and in locations where local farmers were just burying trash (Figure 5-16).

Large scars on the ground caused by heavy equipment are more common than we would like to believe in many locations. Landowners and managers are often hesitant to admit that they have been burying their trash in certain areas or illegally obtaining gravel or sand for construction purposes. It is often these types of activities that lead to archaeological discoveries, if the construction workers are paying attention to what they dig up and are also motivated to tell archaeologists about interesting buried finds. More often than not, their activities go unrecorded and then can lead to a good deal of destruction of the buried cultural resources. The scars of these operations are often easy to identify with GPR; fortunately, sometimes archaeological materials are somehow preserved between or below the disturbed areas.

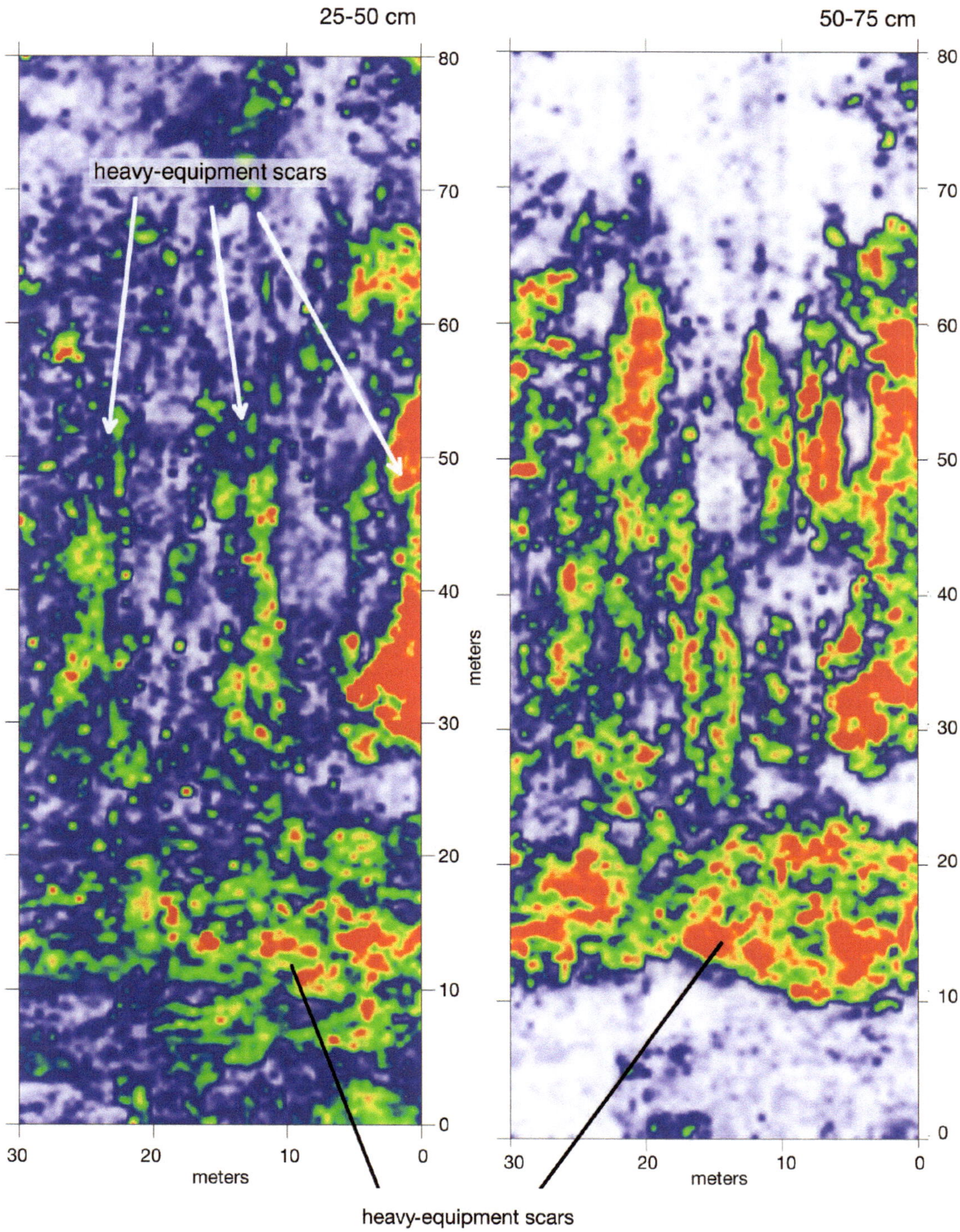

Figure 5-16: Amplitude slice-maps showing the heavy-equipment compaction and leveling scars. The heavy equipment produced scars created by large-scale excavation at Goodnight Barn, Pueblo, Colorado. Data collected with 400 MHz antennas.

Background Noise

Another common aspect of GPR collection is interference from external radio sources, which often transmits energy at about the same frequency as GPR antennas. As most GPR antennas are broadband, they are capable of receiving and displaying many radio transmissions, which can often disrupt, interfere with, or totally destroy radar waves coming from reflection surfaces within the

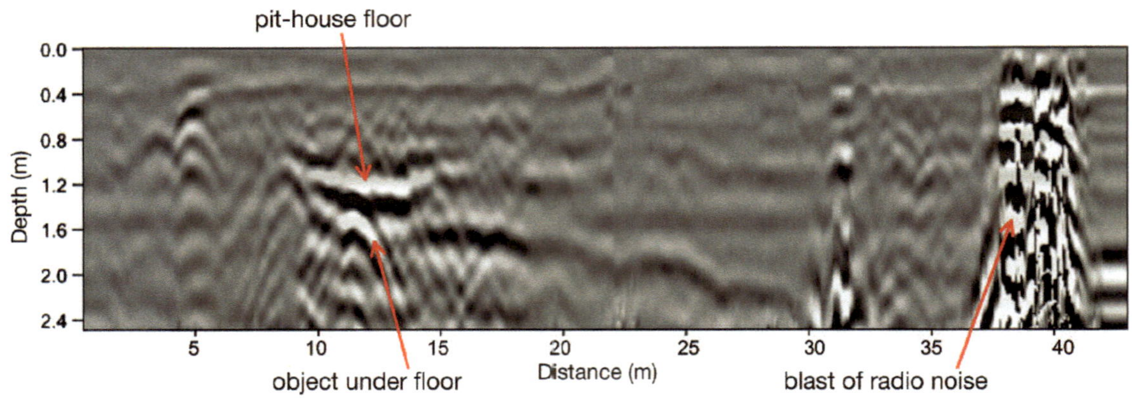

Figure 5-17: A short burst of radio noise interfering w th reflections from the ground. This radio noise has totally overwhelmed the receiving antenna during collecticn of this reflection profile, destroying reflections from the ground between 37 and 41 m. Fortunately, there was no external radio noise at the beginning of collection, where a pit-house floor and a subfloor feature are visible. Data collected using 400 MHz antennas near Alamogordo, New Mexico.

ground. While some GPR scientists claim to be able to define the specific frequencies of interfering energy, I have never had any luck doing so but plenty of experience seeing its effect on radar reflection quality, which is dramatic (Figure 5-17). When radio transmissions of particularly high intensity are recorded, they can totally overwhelm any of the weaker waves coming from within the ground. During collection of GPR profiles in Alamogordo, New Mexico (Figure 5-1), we were periodically blasted with short-duration pulses of radio waves from an unknown source. This area is near the White Sands Missile Range, where the U.S. military is always testing new equipment, and those operations are likely the source of these GPR-wave-destroying transmissions. I have seen similar disruptions at the Los Alamos National Laboratory, also in New Mexico, and almost everywhere in Israel, which is the most radio-noisy place I have ever collected GPR data in. Even a small cell phone placed near a GPR antenna can disrupt energy reception, but not usually to the extent shown in Figure 5-17. Sometimes blasts of radio noise have lasted many minutes, and all one can do when this happens is to stop collecting and begin collecting again when the transmissions cease.

I am occasionally confronted with such extraordinary background energy noise that all GPR data collection seems to be impossible. Usually this occurs when collecting in an area where the ground is attenuating energy anyway, and the range gains have been increased in an attempt to boost the reflection amplitudes from deeper in the ground than is possible. This happened to me in Tucson, Arizona, with 500 MHz antennas, where I had the time-window opened to 60 ns in the hope of collecting reflections from deeper than turned out to be possible. The gains were set quite high during collection, especially for the lower half of the time-window, and the only amplitudes that were being enhanced were those of background radio transmissions. Also, the survey area was directly next to the airport and in a very busy urban location, where there must have been a great deal of radio noise to begin with. The profiles of the unprocessed reflections showed high-amplitude banding from this radio noise in the lower half of all profiles (Figure 5-18). In these raw profiles, some archaeological features were just barely visible above the zone of total noise.

These noisy GPR profiles were collected in an area where pre-Classic Hohokam people had built a large pit-house village along the eastern bank of the Santa Cruz River in southern Arizona. The village was abandoned in about A.D. 1150 and is covered by alluvial sediment. Many GPR profiles showed indications of reflections from the floors of these buried pit houses, and there were many other features that were barely visible in the profiles. To obtain usable reflections in the profiles, each profile was frequency-filtered to remove any waves recorded above 600 MHz. From the remaining waves, I removed a composite background trace of all horizontal bands recorded in all traces in each profile (Figure 5-18: middle). All remaining traces were then stacked, with each three adjoining traces averaged into one trace, in a running average, to produce a horizontally

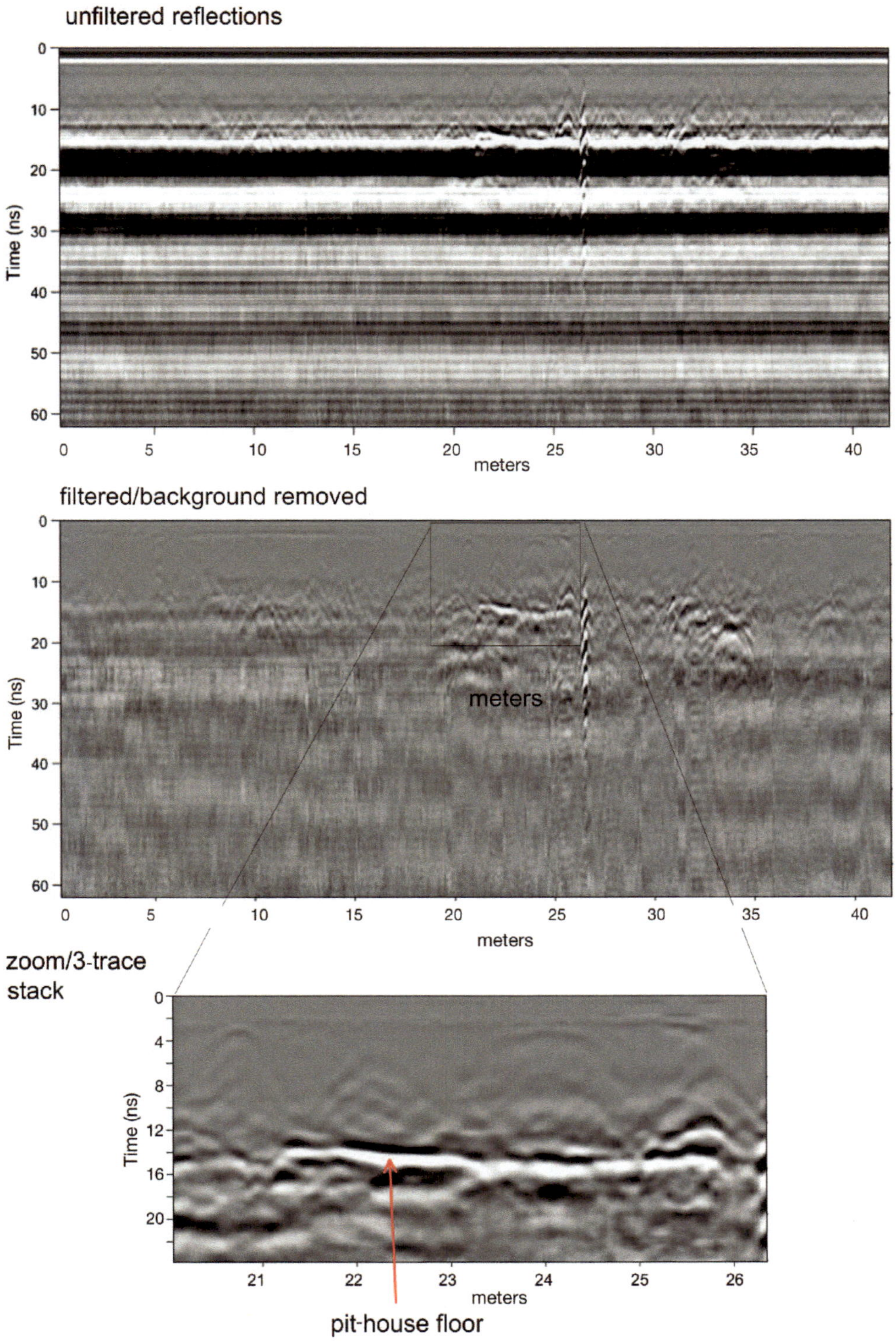

Figure 5-18: Raw reflection profile processed to remove background radio interference. One reflection profile collected with 500 MHz antennas shows how it appeared with no processing. These data were collected near the Tucson International Airport in an area with a large amount of background radio noise of many types. The pit-house floor is barely visible in the upper, nonprocessed profile, but becomes visible when the frequencies above 600 MHz are removed and then all background traces are removed. When the portion of the profile is zoomed in on and the traces are stacked with a three-trace running-average stacking, the pit-house floor becomes distinctly visible.

averaged reflection profile. When each reflection feature of interest was viewed singularly (Figure 5-18: bottom), each pit-house floor could be viewed with good clarity in profile, and the resulting profiles were sliced into amplitude maps for display. Without conducting these processing steps, this data set would have been largely unusable due to the high degree of background radio interference.

In general, the less electrically conductive the ground, the less of a problem normal background radio will be. When the ground is electrically resistive and is conducive to radar wave propagation, whatever background noise there is will likely be insignificant. When a majority of radar energy is attenuated at a shallow depth, the background noise from deeper in the time-window is the only data available to display in reflection profiles. It is then very apparent but can be removed by simple background-removal filters, or just ignored during interpretation. Sudden bursts of radio noise is another problem altogether, and when this occurs there is nothing that can be done but repeat the data collection when that background radio transmission ceases.

CONCLUSIONS

It is impossible to avoid these kinds of cultural complexity in the crowded and complicated world we live in. Usually the better the GPR data, the less one is bothered with complexity when viewing reflection profiles or amplitude maps. The useful high-amplitude reflections from within the ground are just more visible than the unwanted background noise or other complexities. Or perhaps with good data (those with distinct high-amplitude reflections), interpreters can simply concentrate more on the plethora of useful reflections from the ground (those that are visible and easier to interpret) and can ignore the others. Whatever our interpretation foibles and habits, it is always important to be on the lookout for features that are a product of modern overprints or noise (perhaps here we can define noise in a similar way that gardeners define weeds: something that we don't want or interferes with what we care about). There can be nothing worse than interpreting features as reflections of importance in the ground, only to find out later that the actual origin is modern cultural complexity. The reflection features discussed here are also very difficult to identify after many reflection profiles in a grid are resampled and amplitudes are mapped over a broad area. Only by looking at the profiles in some detail first and identifying the origins of reflections can many of these potential problems be discovered and possibly removed before amplitude mapping.

6

Attenuation and Depth of Penetration

Once radar energy has left the transmitting antenna and coupled with the ground, depth of penetration is mostly governed by the amount of attenuation that energy is subjected to as it propagates downward. The greater the attenuation of the medium through which the waves are propagating, the greater the loss and the less depth of penetration. In almost all cases, electrically conductive ground is the greatest factor in attenuation. Very conductive ground usually attenuates all energy, no matter what its frequency. In less conductive ground, lower-frequency antennas are often capable of transmitting longer wavelengths of energy, which can penetrate deeper in the ground than the higher-frequency energy (Conyers 2004a: 49).

I have heard various highly educated physicists try to explain why short-wavelength (high-frequency) radar energy tends to attenuate at a shallower depth in the ground but have never completely understood their answers. There are equations that explain this, but I tend to be "equation-phobic." In desperation, I asked a more layperson-friendly physicist this question and finally got an answer I could understand. She related energy loss of a propagating wave in the ground to the energy loss by a person taking a walk. If a walker uses a very short stride (analogous to a short wavelength of propagating energy), more energy is expended in a shorter distance of walking (analogous to propagating energy moving into the ground), and the walker becomes tired faster and finally has to stop (propagating energy finally attenuates). While this explanation is no doubt quite troublesome for some readers with a more detailed understanding of wave propagation physics, it works just fine for me and most archaeological geophysicists.

One of the primary issues concerning attenuation and depth of radar energy penetration into the ground is that they are integrally related. Attenuation causes energy loss, which causes depth of penetration to decrease. Sometimes attenuation is gradual, and sometimes abrupt, depending on the types of materials the propagating radar waves encounter during their propagation. Another factor that causes energy loss with depth is how the transmitted "beam" of energy from a surface antenna spreads with depth. Spreading out in a conical pattern from the antenna causes radiating waves of radar to spread out over a larger mass of ground, and when some waves are finally reflected from a buried interface, there is less energy available to travel back to the surface to be recorded. In almost all cases, attenuation continues to occur progressively with depth, due to greater water saturation and more weathered materials being encountered, which contain more clay from being subjected longer to weathering processes. When some clay types are wet, they become more electrically conductive and therefore act as an attenuating agent. It is conceivably possible that the ground could become less attenuating with depth, but I can think of only a few ways this could happen. If there was a layer of extremely resistive quartz sand under a more conductive surface material, energy passing into that sand would undergo little attenuation. Another situation could be a void space in the ground, within which there is no attenuation of propagating waves whatever (discussed in Chapter 10). In almost all other areas of the world, attenuation of radar energy in the ground occurs with propagation distance downward, and radar energy is completely lost at some measurable depth.

DETERMINING DEPTH OF ENERGY PENETRATION

One of the first questions always asked in the field by onlookers or collaborators when conducting a GPR survey is "how deep are you seeing?" or something to that effect. It is often difficult to determine this precisely during data collection, but an answer can be obtained if velocity estimates have been made and two-way travel times are converted to depth (Conyers 2004a). Once depth is calculated from measured wave travel time, it is usually quite easy to determine the maximum depth penetration of transmitted radar energy by looking for the point in a trace where coherent reflected waves coming from the ground change completely to only recorded background noise. The depth at which there is nothing but noise being collected is the depth at which all transmitted radar energy that could be reflected back to the surface has been attenuated. Once that depth is reached, the question of how deep one might be "seeing" can be answered.

Noise is usually identifiable in individual traces as high-frequency waves of variable amplitudes that change their character constantly during collection. This is what common background radio noise—which always surrounds us—looks like on a GPR system screen (Figure 6-1). Some areas of the world are "noisier" than others, but I have yet to encounter a location where there isn't at least some noise of this sort. Usually the "deeper" in the time-window, the greater the effect of background noise, because at some depth most transmitted energy has been attenuated and, as noted above, there is nothing left to record but that background noise. However, the noise will be visible in individual traces only if the range gains are set high enough that the variable high-frequency waves can be viewed on the computer screen. Most of the GPR systems I have used allow the internal system software to adjust the gains throughout the time-window prior to collection so that all received waves are "on-scale" and visible while reflection profiles are being collected. When those automatic range-gain programs are applied, the software will enhance the waves throughout the time-window and automatically increase the amplitude of background noise that is being collected in the deeper portions of the time-window. When viewed in profiles, that noise is visible as horizontal banding (Figure 6-1), which can be readily removed by most software programs when data are displayed for interpretation.

All GPR collection systems allow the user to view traces during setup routines so that depth of energy penetration can be determined and the time-window adjusted accordingly. However, some manufacturers have internal software controls that save the recorded reflections with no gains whatever (Sensors & Software, and Mala Geoscience systems) or some percentage of the gains that are applied in the field (GSSI). In all cases, gains must then be reapplied during reflection profile processing and data interpretation.

Depth of radar energy penetration can be more precisely determined once reflection profiles are processed for analysis. With the *y* axis plotted as approximate depth in the ground, deeper reflections can be gained and maximum depth of resolution determined (Figure 6-1). When background noise is removed in the processing of profiles for visualization, the depth at which there are no more reflections is the attenuation depth. Most profile processing software programs also allow individual traces in profiles to be viewed in order to determine if there are variations in depth penetration along profiles. As reflections from deeper within the ground are lower in amplitude, they can be gained to be visible; and if they are reflections in the ground from horizons of interest, they will usually have the same wavelength as those higher in the profile (Figure 6-1).

In some cases, attenuation depth is abrupt (Figure 6-1), and in others it occurs gradually with depth (Figure 6-2). These differences are a function of water saturation and the changes in the chemistry of units in the ground, and they are often difficult to predict or explain without detailed analysis of the material through which the radar energy is passing. Radar energy passing into the ground will always lose amplitude as it spreads out with depth and is absorbed by electrically conductive materials (and to some extent by magnetic components) that make up the matrix of a study area.

ENERGY ATTENUATION AS A FUNCTION OF GROUND CONDITIONS

In the early days of GPR, much was written about the environments where GPR could be employed, and there were many discussions about the conditions thought to be unsuitable. I heard many people say that GPR would not work when the ground was wet, or in clay (or worst of all, in wet clay). For a while I actually believed these

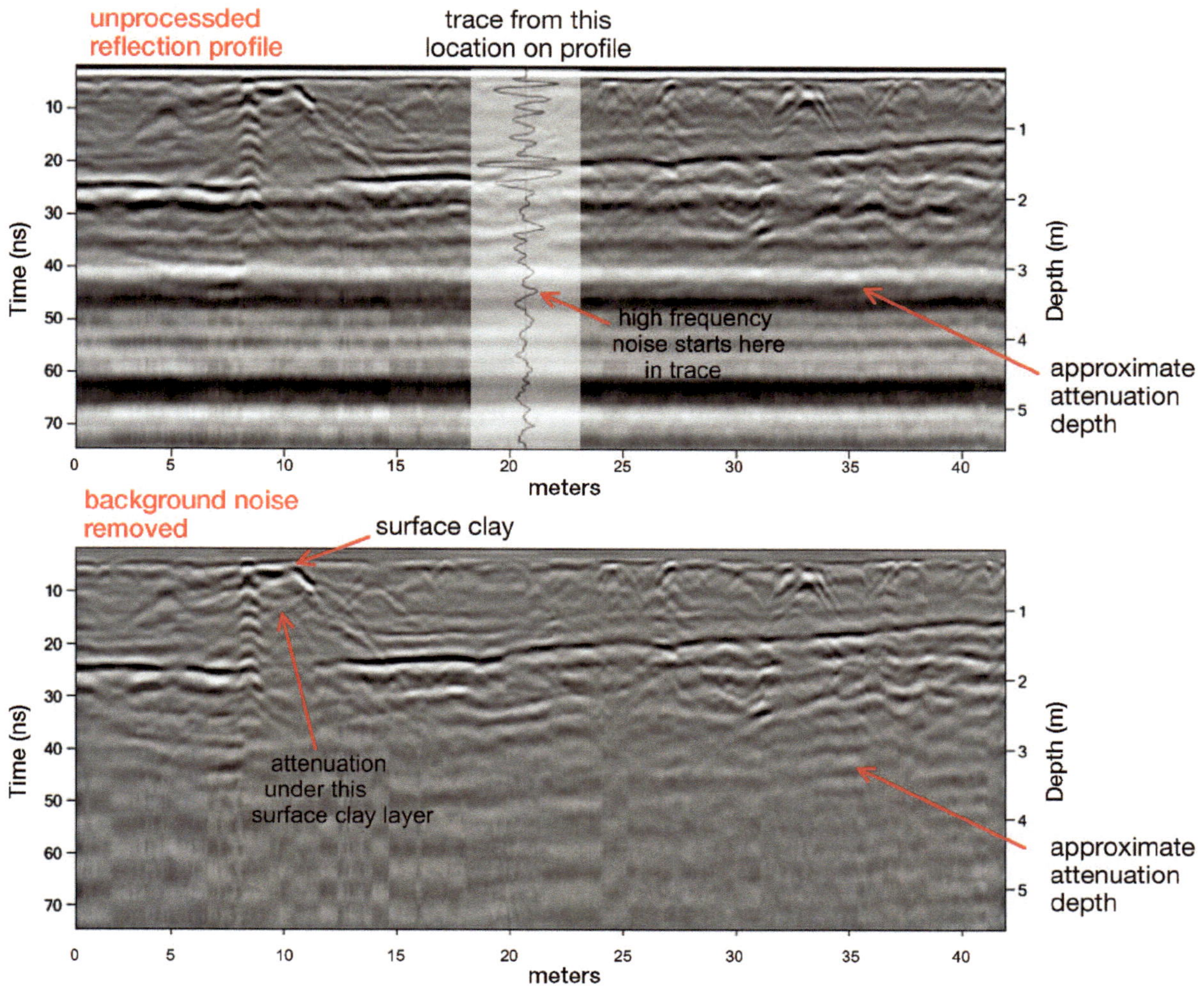

Figure 6-1: Abrupt attenuation with depth in a reflection profile. This example shows an abrupt attenuation when waves encountered a very electrically conductive layer. In the upper profile, background energy remains in the traces and only high-frequency noise is being collected below about 45 ns (about 3 m in depth). In profile, that noise appears as high-amplitude horizontal bands of recorded waves, which can be removed in a simple background-removal filter during processing, seen in the lower profile. Data collected with 300 MHz antennas at the Cerén site, El Salvador.

analyses, figuring that those making such statements must have had a reason for arriving at their conclusions. Interestingly, archaeological geophysicists in the early 1990s had totally written off the British Isles as an area where GPR would work. I was told this was because an early paper that reported on GPR results in England (which will remain uncited here) was given initial credibility and then was later found to be either wrong, flawed in some way, or just faked. While I heard this from a number of reputable colleagues, it is difficult to understand how one published paper could have managed to spoil the GPR method for the whole United Kingdom (and Ireland too, I wonder?). But for at least a decade or so, it was "common knowledge" that GPR would not work there. I suspect that some of these initial GPR attempts were found to be less than adequate because the ground is so wet there most of the year, or perhaps archaeological geophysicists were having such good luck with magnetic and earth-resistance methods that they didn't have time for GPR. Not having had a great deal of experience myself at that time, I accepted the "prevailing wisdom" about wet ground and clay soils and always planned my fieldwork during the dry seasons, or even refused to try GPR in ground that had clay in it. That turned out to be a very big mistake, as I missed out on many opportunities to do GPR in warm tropical areas during the winter, places where I now know it would have worked well.

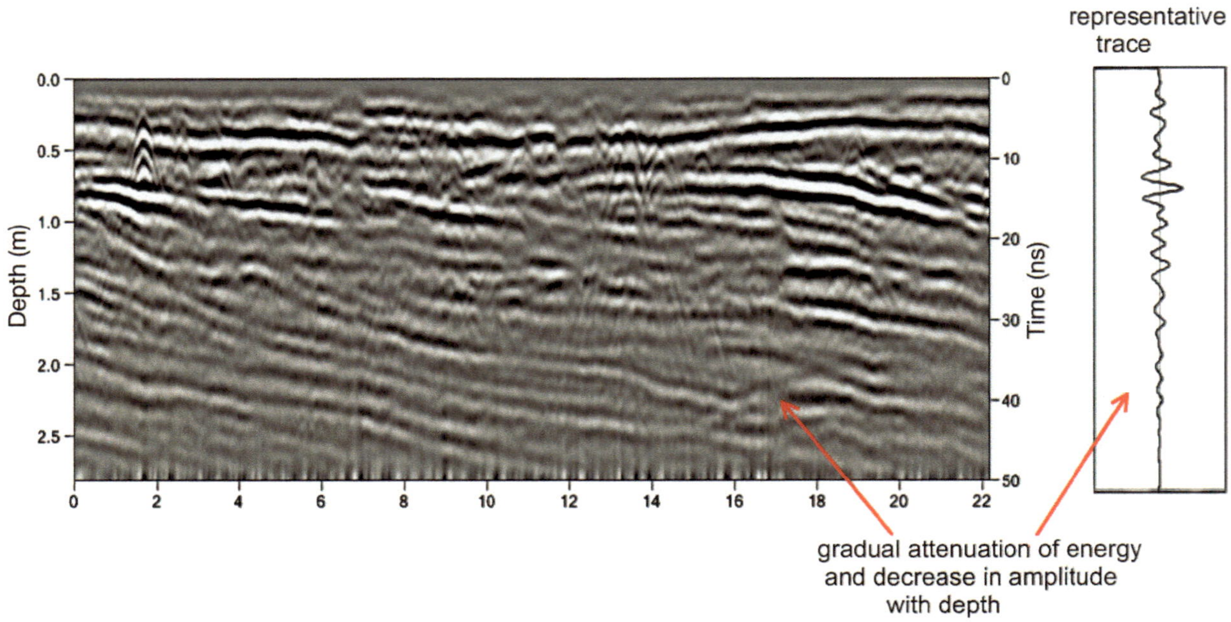

Figure 6-2: Reflection profile showing gradual attenuation of energy with depth and therefore decreased reflected wave amplitude, as can be seen in the representative trace from this profile. Collected with 400 MHz antennas in sand dune layers at Batemans Bay, Australia.

As time went on and I found myself talked into doing GPR in many areas of the world, it became inevitable that sooner or later I would collect some reflection data in wet and clayey conditions. One such study was done in the middle of a very rainy winter on the banks of the Columbia River in Oregon. The targets were known to exist about 3 m below the ground surface in a mostly clay levee of the river. As I wanted this study to have at least some chance of success (and still believing the "wet clay is bad" dogma), I told my collaborator that the survey area must be stripped of about a meter of sediment before my arrival. I also estimated I would collect a 30 × 30 m grid of data, and that grid size was planned in advance. A week prior to my arrival, the local archaeologists took heavy equipment and dug a hole exactly 30 × 30 m to a 1 m depth, which then proceeded to fill with rainwater and create a "mud swimming pool" (Figure 6-3). Needless to say, I was shocked when I arrived, but as I had made a commitment, I proceeded to collect the data anyway. To my great surprise, the transmitted radar energy passed easily through the mud and to about 2 m in depth, where a basalt layer composed of cobbles was visible (Figure 6-4). Later laboratory analysis of the clay at this site showed it to be mostly alluvial in nature, with the diagenetic clay component composed of mostly kaolonite. Both these clay types have a very low electrical

Figure 6-3: Collecting GPR data in totally saturated mud. This mudhole, a difficult place to collect data, was along the banks of the Columbia River, Oregon, using 200 MHz antennas.

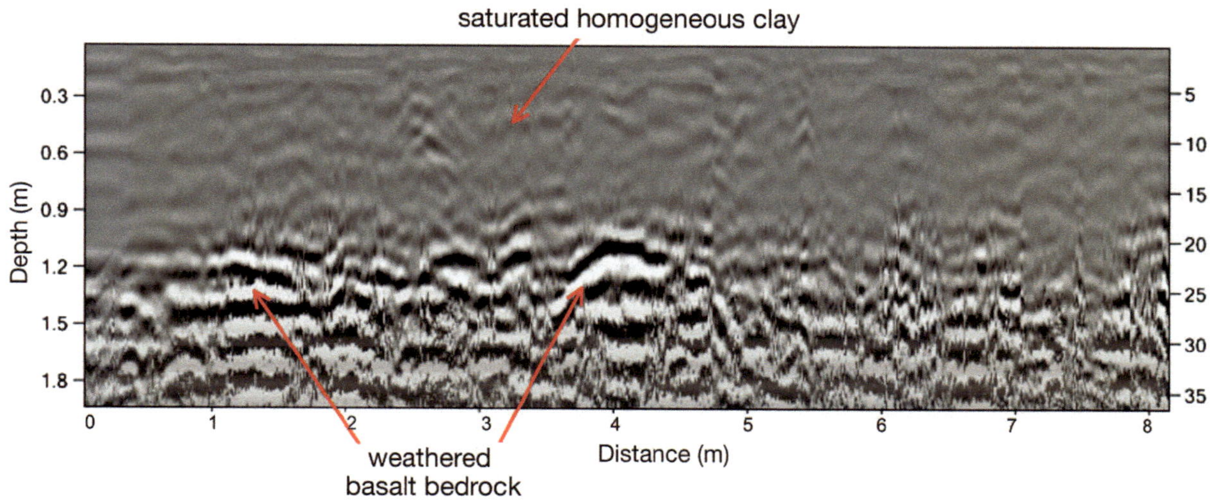

Figure 6-4: Reflection profile collected in totally saturated mud (the mudhole shown in Figure 6-3). Energy penetrated the low-conductivity mud, and good reflections were received from weathered basalt about 1.2 m below the ground surface. Data collected with 200 MHz antennas.

conductivity even when wet, which explains why radar energy readily traveled to that depth. This mudhole site, and others I conducted like it later in Oregon, are at one end of the spectrum with regard to moisture and clay content, and it soon dawned on me from those studies that the "prevailing wisdom" at the time was flawed.

About the same time that I was rethinking the wet clay constraint of GPR efficacy, I traveled to the north coast of Peru, where it rains only about every century or so, and the sandy ground was exceedingly dry (Figure 6-5). As I was still somewhat under the spell of the prevailing idea that dry sand is "good" for GPR, I expected very deep energy penetration. To my surprise, the deepest reflections I was capable of collecting using 200, 270, and 400 MHz antennas were from only about 50 cm in the ground (Figure 6-6). In these studies, I had good stratigraphic information that showed there were geological and archaeological surfaces well below that depth, which should have reflected radar waves. However, despite trying every trick we could think of in data collection and processing, the radar energy (irrespective of frequency) that was transmitted was not capable of traveling deep enough to reflect from the surfaces of interest. I did not perform any analysis on this sediment, but I have heard of other studies conducted in the general area that found hydrous clays in the sandy sediment, which also retained salt. This bound water and salt was likely incorporated into the sand from the marine fog, which covers the landscape for much of the year. The sandy ground appeared to be dry but had apparently still retained enough moisture and salt within the clay constituents to produce a very electrically conductive medium. For me, this study was the opposite end of the "perception spectrum" with respect to moisture and sediment type from the Oregon mud tests. At that point, I was dubious of all foregone conclusions with respect to GPR's usability in any environment and initiated a study to see if we could predict how GPR would work in different ground types, provided we knew the constituents

Figure 6-5: Collecting GPR data in very dry sand. On the north coast of Peru it has not rained in decades, and the sand is very dry.

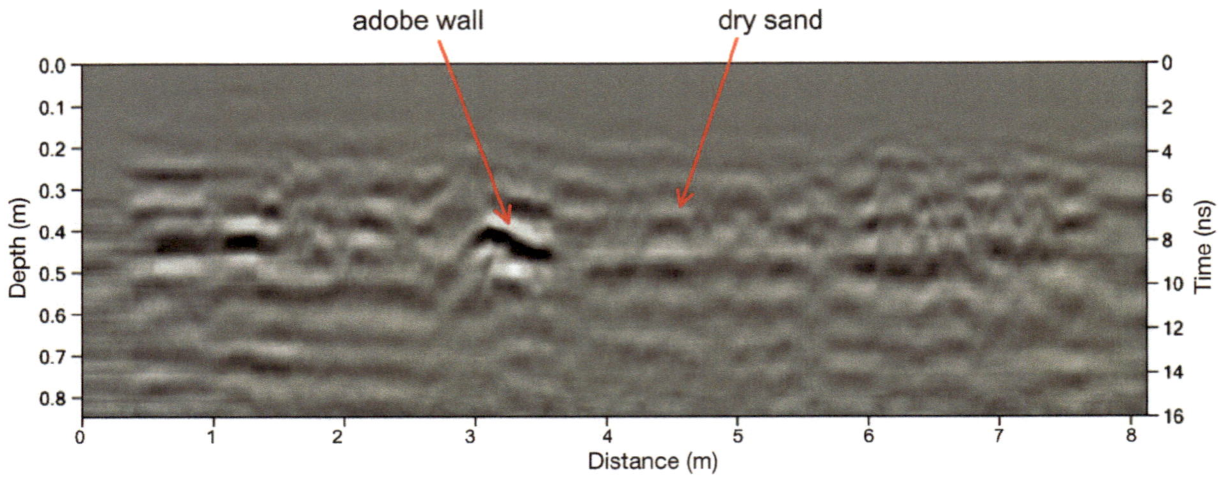

Figure 6-6: Reflection profile collected in totally dry sand on the north coast of Peru, where it had not rained for many decades prior to this study. All energy was attenuated at about 50 cm in depth in this electrically conductive sediment. Data collected using 400 MHz antennas.

of that ground in advance (Conyers 2004b). Similar work has been done on this topic by soil scientists who use GPR, and maps have been prepared on the suitability of GPR in every one of the United States except Alaska (Doolittle et al. 2007).

Our experiences in Peru and Oregon forced me to rethink all my preconceptions about which ground conditions are "good" for GPR and which are "bad." However, confusing the matter was evidence that others had collected GPR data in western Oregon and had very poor energy penetration, whereas some data collected in active coastal dunes on the coast of Peru showed many meters of energy penetration. Without a doubt, there are other factors that affect GPR wave penetration and reflection that do not fit the old models of "good" and "bad," which I will discuss below. The detailed work that Doolittle and his colleagues (2007) have conducted throughout the United States suggests that radar penetration depth is almost totally a function of the ground's electrical conductivity, which can be partially, but not totally, adjusted for by the frequency of the transmitted energy.

One of the very best areas for GPR collection that I have ever had the pleasure to work in was central Florida, where the ground is composed primarily of quartz sand with only a shallow surface soil layer. In a study conducted there to map the burrows of gopher tortoises, an endangered species in this area of the USA, I was able to use 900 MHz antennas (Figure 6-7) to transmit radar waves to more than 4 m in the ground. While I am not sure if this is a world record for depth penetration with that frequency antenna, it is surely in contention. In this case, the dry sand was an almost perfect medium for radar transmission, as it is almost all quartz, which is as electrically resistive a material as can be found in nature. I was so flabbergasted with the depth penetration and resolution in this area, which was much deeper than my target burrows (Figure 6-8), I neglected to test the limits of energy penetration by opening up the time-window even more. I suspect energy could have been transmitted even deeper than 4 m. The ground there has almost no electrically conductive clay, as most

Figure 6-7: Collecting GPR profiles in central Florida within quartz sand. This ground is composed of sediment with almost no clay and a minor amount of organic matter. The blue hose demarcates the subsurface location of a large tortoise burrow that was mapped with 900 MHz antennas.

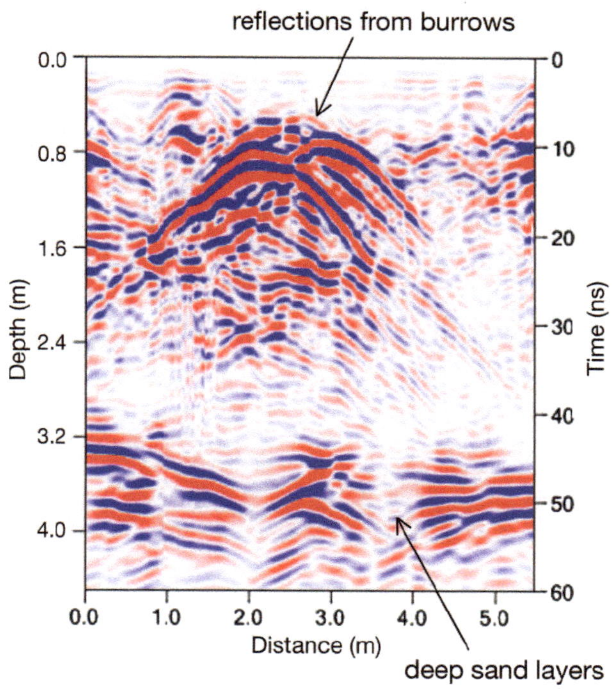

Figure 6-8: Reflection profile in quartz sand. Reflections were generated from a tortoise burrow in central Florida, showing good reflections from deep sand layers to more than 4 m in depth using 900 MHz antennas.

Figure 6-9: Attempting to collect GPR data in saturated cow manure, using 400 MHz antennas near Salinas, California. There was absolutely no energy penetration whatever in this ground, due to the very high conductivity of the medium.

has been leached out of the sand during the exceptionally heavy summer rainstorms that occur in this area. What clay remains has been chemically altered to kaolonite, which is a very electrically resistive material.

The very worst GPR energy penetration I have ever experienced was in wet cow manure at a dairy farm on the coast of California (Figure 6-9) during a very wet winter. It was apparent to me during the first few minutes of fieldwork that there was no radar energy penetration into the ground at all, as the only thing being recorded on the receiving antenna was background noise. My colleagues, who were very interested in finding the walls of a historic adobe ranch house thought to have been preserved somewhere in the area, kept encouraging me to keep collecting all day in the hope of finding something different (the GPR definition of insanity). I ended up collecting 12 grids of data while wallowing all day in cow manure, and I ate one of the most disgusting lunches ever, there being no place to dry off or get clean prior to eating. Not only was it the most terrible GPR survey I have ever conducted, but it also produced totally useless results. The reflection profiles showed attenuation at the ground surface, with nothing but noise throughout the time-window (Figure 6-10). I neglected to take a sample of the ground and perform any chemical analyses on it, but I assume that the high bovine urine and fecal content in the saturated ground produced a very electrically conductive medium, attenuating all energy as soon as it left the transmitting antenna.

Variations in ground covering materials that are electrically conductive can also affect energy penetration. In Key West, Florida, the city workers in a park had imported sand, mixed with some kind of binding agent, to use as a surface for bike and pedestrian trails. They could not tell me exactly what the binding agent was, but it was exceptionally electrically conductive, as all radar energy that entered this thin sand was immediately attenuated. As soon as 400 MHz antennas moved off this surface layer, waves traveled to about 40 ns, which is about 2.5 m deep in this area (Figure 6-11).

Similar variations in surface material that affect radar energy penetration have been found in more usual situations, such as the variations in grass cover and sidewalks. On my university campus, we have found that the grassy ground attenuates almost all radar energy, while the energy that passes through the adjoining sidewalks

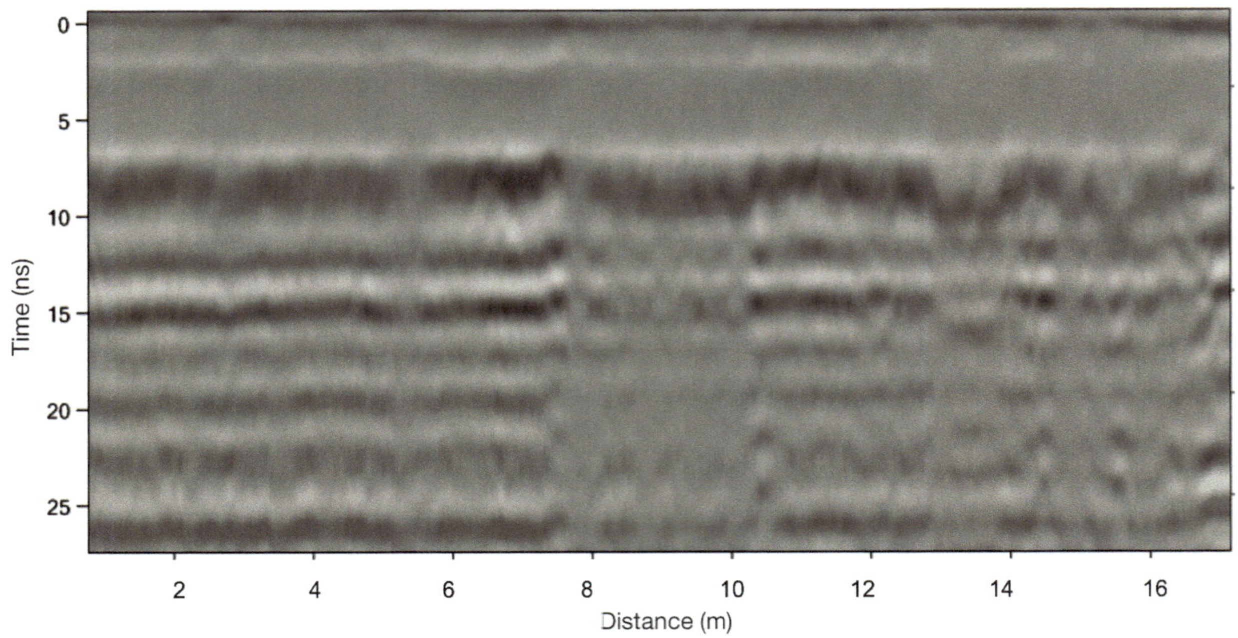

Figure 6-10: Reflection profile collected in wet manure. Nothing is displayed but noise throughout the time-window. Data collected near Salinas, California, with 400 MHz antennas.

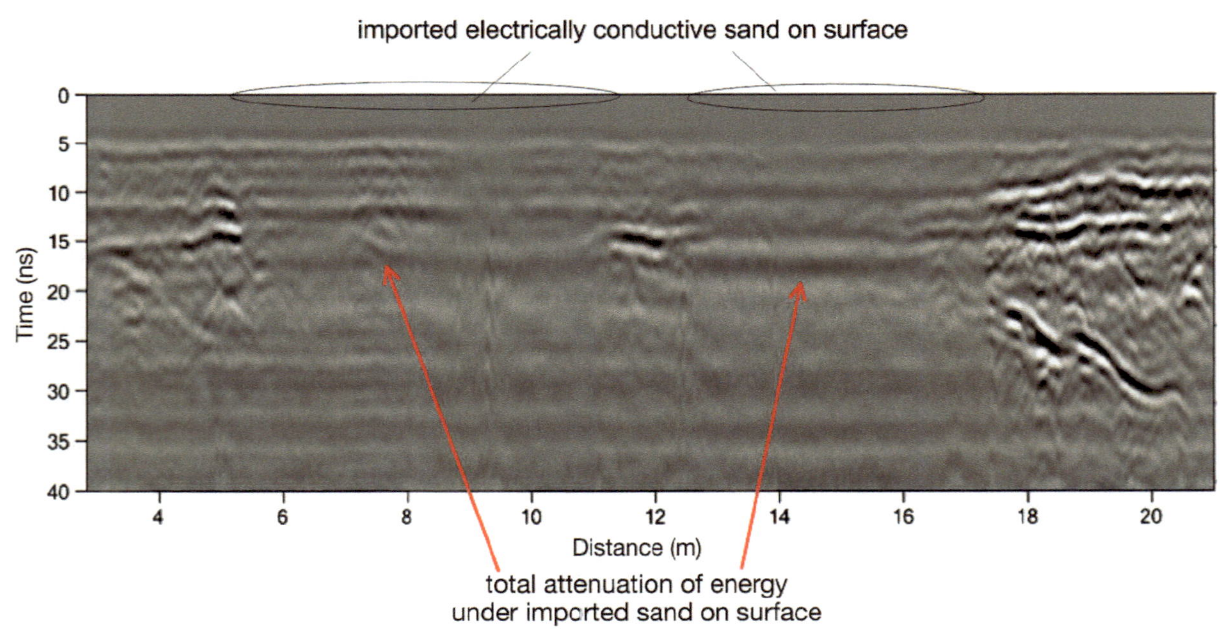

Figure 6-11: Reflection profile across imported electrically conductive sand. An electrically conductive binding agent was incorporated with the sand, and the mixture was used to produce bicycle and walking paths, with the layer creating a highly attenuating environment. All radar energy was destroyed very near the surface when the antennas were on this sand, but easily traveled to 35 ns (about 2 m deep or more) when there was no sand present. Data collected with 400 MHz antennas at Key West, Florida.

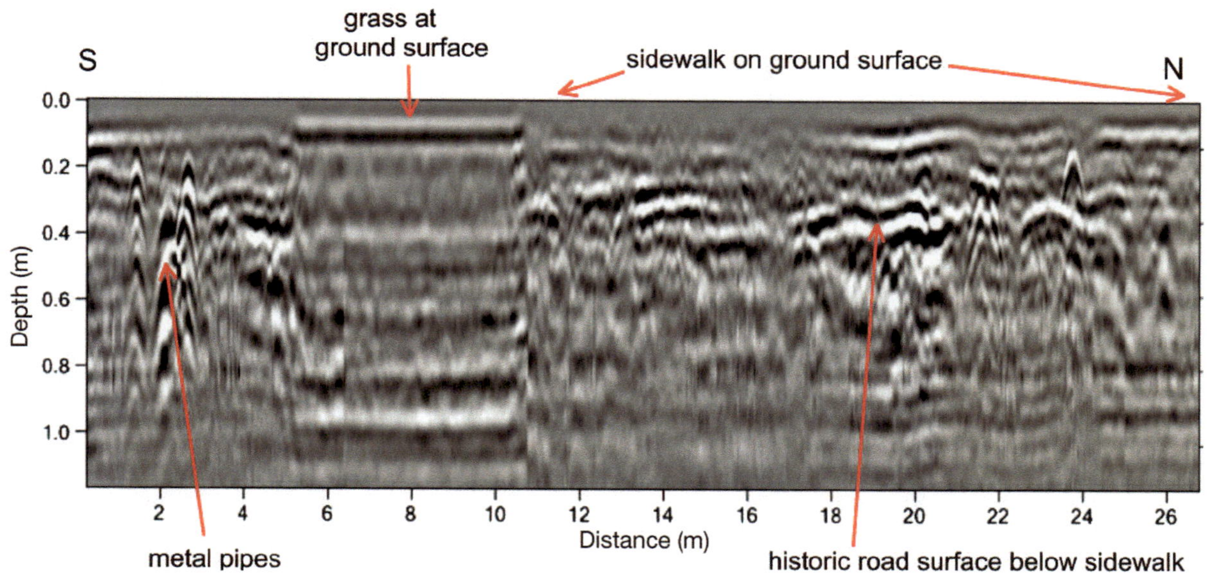

Figure 6-12: Attenuation of energy in grass. Grassy ground surface attenuates all energy when the antennas are moved over it, but when they pass over a sidewalk, the radar energy penetrates to about 60 cm. Data collected with 400 MHz antennas on the University of Denver campus, Colorado.

travels to about a meter or so in depth (Figure 6-12). In this case, an analysis of the silty clay soil shows it to be moderately electrically conductive, with montmorillonite and bentonite clay mineral constituents. I suspect another factor in the grassy area's attenuation is its watering by sprinklers, as the added moisture to the clays would produce a higher conductivity. There is no watering done on the sidewalks, and therefore they act as a protective shield for the drier sediments and soils below. The electrical conductivity is thus somewhat less under the sidewalks, allowing energy penetration deeper in the ground before being attenuated.

Amplitude maps made from grids of reflection data collected on both grassy ground and sidewalks must be interpreted carefully due to the variations in energy attenuation (Figure 6-13). The interpretable radar reflections are visible only under the sidewalks, and buried features visible in the maps there must then be projected into the grassy areas where there are no usable data. The amplitude maps produced this way on the University of Denver campus show a buried historic road under the sidewalk, which is now completely buried by landscaping.

Buried layers that are highly conductive can also complicate interpretation, as was discovered at the Cerén site in El Salvador. In this study, an unusual layer was found in one part of the site, which appeared to attenuate all transmitted radar energy below it (Figure 6-14). Our initial interpretation was that this layer might be imported clay that had been used to pave a plaza or other work surface used by the ancient Maya in this village. We knew that these people had imported clay to build the floors of their houses, and our experience showed that this clay was electrically conductive and attenuated radar energy in a similar way. However, this hypothesis soon became difficult to support, as our GPR mapping showed that this buried unit continued for many hundreds of meters to the east of the site. That information, when used to test the plaza or work surface hypothesis, was untenable, as this feature, mapped with GPR, was much larger than any known work area or plaza ever documented anywhere in Central America. The only other hypothesis we could come up with was that this attenuating clay unit, which must be very electrically conductive, was an unknown geological layer that had never been uncovered or described in the area.

To test our hypothesis about this buried unit, we used a hand auger to dig up 3 m of sediment in the area where we had mapped this unknown unit (Figure 6-15). A 20 cm thick dark brown mud layer was found interbedded with the well-known volcanic ash layers, which correlated directly to the attenuating layer visible in the GPR reflection profiles. It had the same color as the oldest soil known from the area, which we knew from other geological studies underlies all the volcanic ash beds

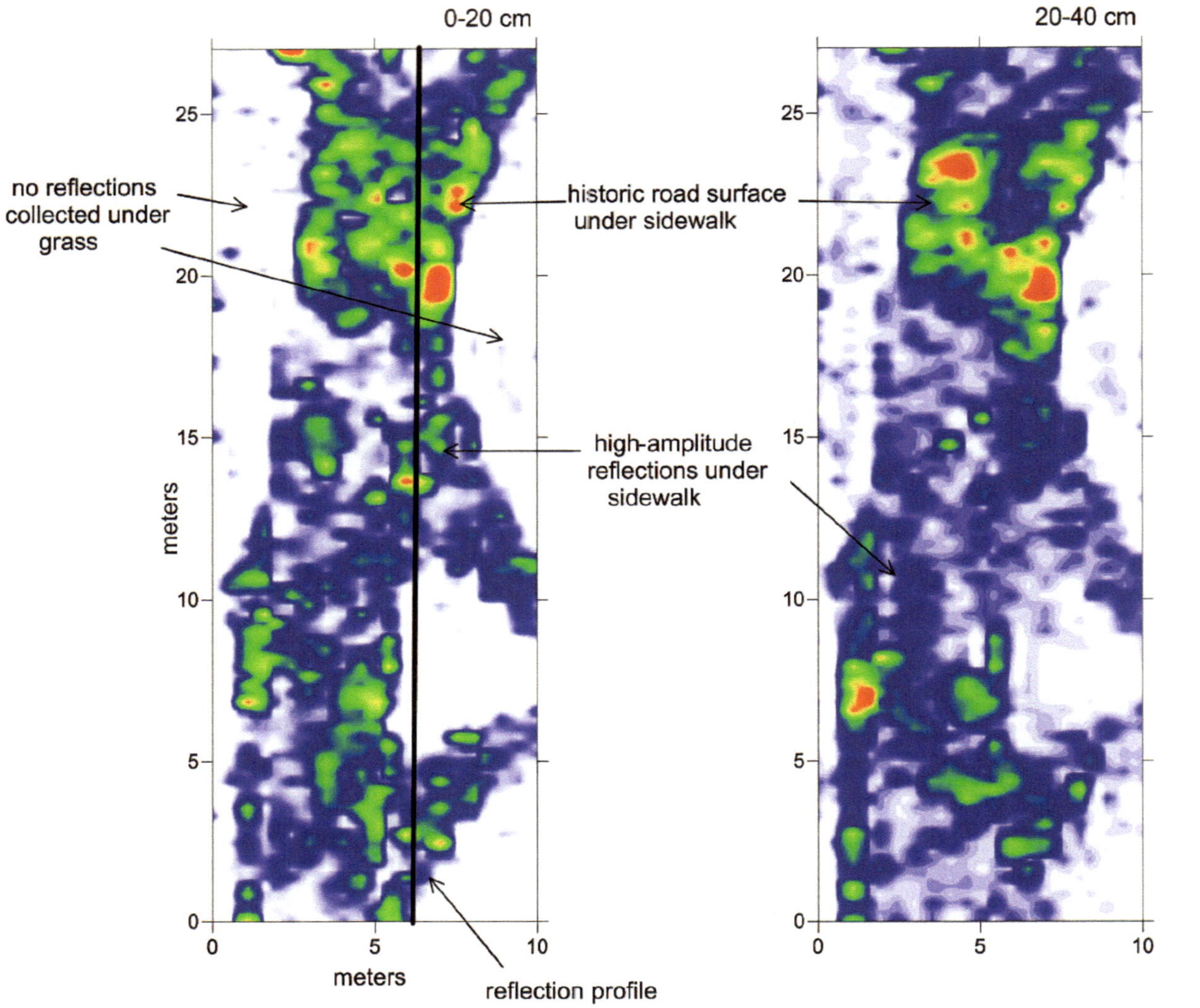

Figure 6-13: Amplitude maps showing differential attenuation in different ground. These maps show a sidewalk and adjacent grassy areas, illustrating the differences in amplitude with depth over these two mediums. The grassy areas allow almost no energy penetration and are shown as white in these maps. The reflection profile is shown in Figure 6-12. Data collected with 400 MHz antennas on the University of Denver campus, Colorado.

that covered the ancient village. This confused us for a while, as we could not figure out how this presumably very ancient soil had become encased within the volcanic ash sequence that is known to have been emplaced over only a few days, and completely covers the site. As it turned out, all that was necessary to understand this unusual stratigraphy was to explore the exposures on the riverbanks to the east of the site. There I found exposures of the stratigraphic relationship, just as we found in the auger samples, and could map in the GPR profiles. These exposures showed that the attenuating dark brown clay unit was a thin mudflow layer composed of material that had been eroded from an exposure of the very ancient soil upstream from the site during what must have been a torrential rainstorm during the volcanic eruption. It was encased in the sequence of ash from the volcano and was preserved only in the area that received deposition of the mudflow, to the east of the site. What began as a possibly interesting GPR discovery of a cultural layer previously unknown at this site turned out to be a marginally interesting unit of geological concern.

Saltwater totally attenuates all radar energy that encounters it, as it is highly electrically conductive. This is important to appreciate when working anywhere near the ocean, and I was interested to see how close to the water the GPR antennas could get before the saltwater

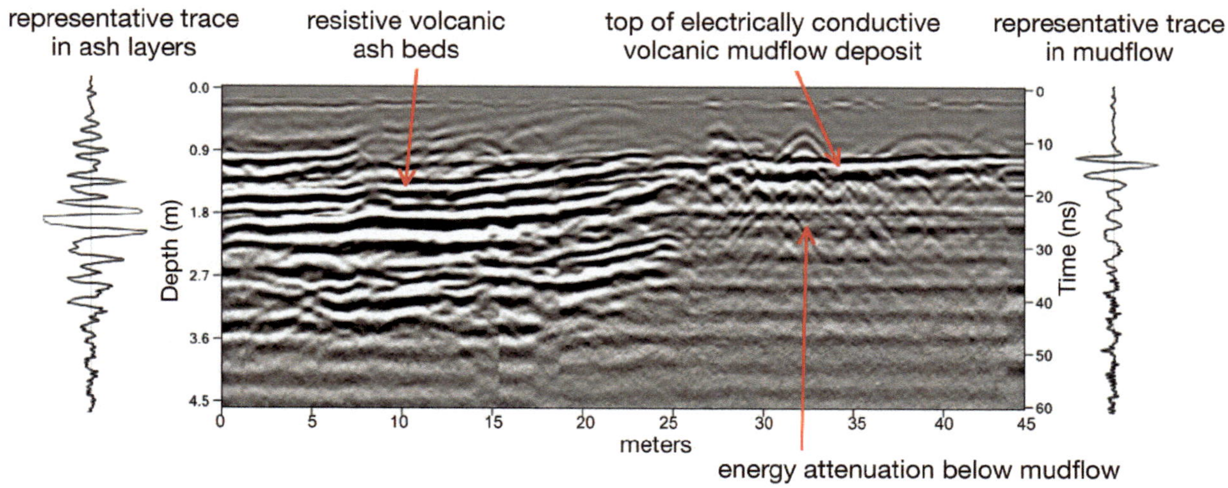

Figure 6-14: Reflections generated from the electrically resistive volcanic ash beds and an interbedded mudflow. The electrically resistive volcanic ash beds are high in amplitude, and there was good energy penetration to about 4 m in depth. A volcanic mudflow was deposited only across the area on the right portion of this profile, with an upper surface about 1 m below the ground. This unit is electrically conductive, and all radar energy below it is attenuated. Representative traces from the right and left of this profile demonstrate the depth penetration of radar in these two media. Data collected with 300 MHz antennas at the Cerén site, El Salvador.

Figure 6-15: Auger samples in volcanic ash units. These samples were collected at the Cerén site, El Salvador, to understand the attenuation shown in Figure 6-14. Payson Sheets and Jeff Lucius did the digging, with their results on the right. A brown mudflow unit was found encased in volcanic ash, which is the highly electrically conductive unit that attenuated all radar energy below.

attenuated all transmitted radar energy. When working at Pearl Harbor, Oahu, Hawaii, I collected a few reflection profiles by placing the antennas directly in the water and then moving them in a line from the beach away from the shore (Figure 6-16).

Figure 6-16: Testing saltwater attenuation on a beach. This test was done on a coral beach at Pearl Harbor, Hawaii. The antennas were placed in the saltwater and then moved up the beach to determine the distance from the saltwater-beach interface where there remained radar energy attenuation.

The results of this study showed that, as expected, all radar energy was attenuated when the antennas were placed in the saltwater. As soon as they were moved onto dry sand, there was some energy transmitted to about 10 ns or so in the ground (Figure 6-17). Within about 1.5 m of the edge of the water, there was good energy penetration and no apparent influence of saltwater attenuation in the sand (Figure 6-17). The boundary of partial attenuation with good energy penetration is the interface between saltwater and freshwater within the beach sand. I have conducted this experiment elsewhere and have found that in beach areas where there is more wave activity, the interface between freshwater and saltwater is typically farther inland, but usually not more than a few meters from the swash zone of waves. This same test was conducted two years apart on the beach at Ashkelon, Israel, and dramatically different results were recorded. In 2007 the saltwater-freshwater interface was just a few meters away from the edge of the water. When the tests were conducted again in 2009 along the same area of the beach, the interface was highly variable, with an apparent interface many tens of meters inland in some areas. I asked the local inhabitants about any unusual wave activity during the winter of 2009, and they commented that there had been very high surf on this beach just a few months before the data were collected. Perhaps saltwater had been

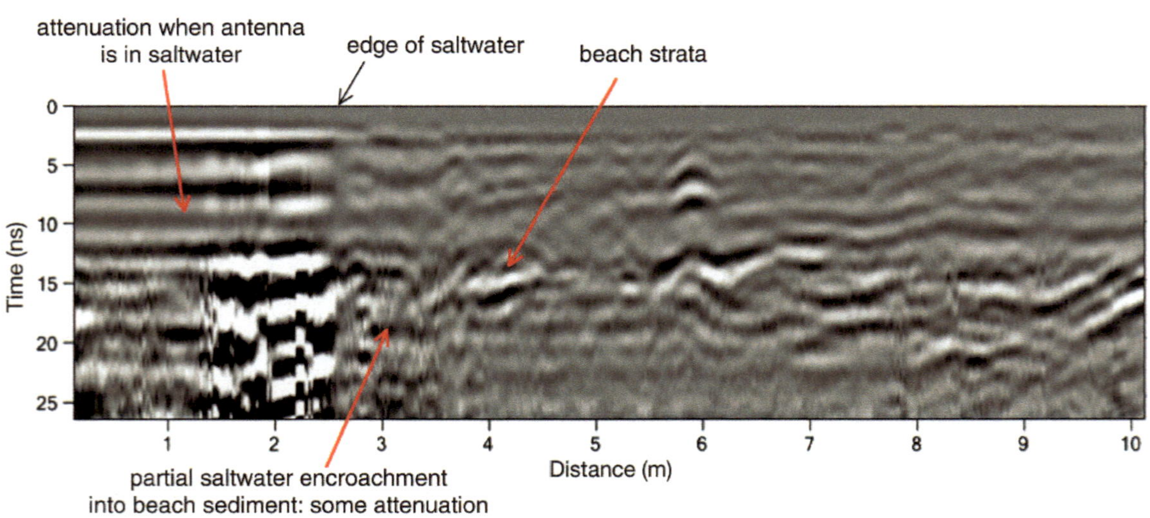

Figure 6-17: Reflection profile collected on a beach to test attenuation in saltwater. The antennas were placed in the saltwater at the beginning of the transect where there was total energy attenuation. At the edge of the saltwater, radar energy penetrated the sand but was still partially attenuated until about a meter away from the edge of the saltwater, due to partial saltwater encroachment into the beach sand. About 2 m away from the edge of the saltwater, there was no evidence of energy attenuation. Data collected with 400 MHz antennas at Pearl Harbor, Oahu, Hawaii.

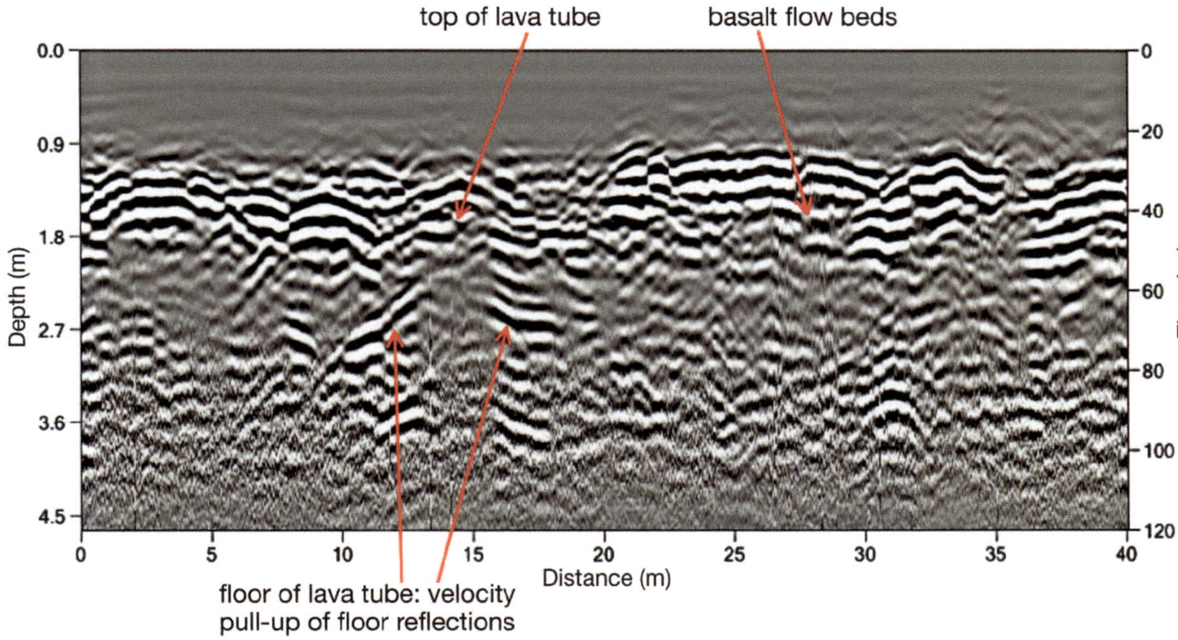

Figure 6-18: Reflection profile collected on recently deposited basalt flows. Good energy penetration was achieved to about 4 m in depth with 270 MHz antennas. The lava tube ceiling produced a high-amplitude reflection, and the floor of the lava tube was pulled upward, which is common in flat surfaces under a void space. Data collected on the leeward side of the island of Hawaii.

pushed up on only some portions of the beach during that event and was creating the variation in attenuation along this coastline seen in the GPR profiles.

Radar waves are composed of conjoined magnetic and electrical fields, which propagate in unison (Conyers 2004a: 24). When one of the fields is attenuated, the wave will no longer propagate, and attenuation of transmitted energy occurs. While much has been written about electrical conductivity as a very important variable in radar wave depth penetration, there is less research on the magnetic component of wave propagation in the ground. Theoretically, any ground that has a high magnetic permeability will also effectively remove one of the oscillating fields and create attenuation. I had read of only limited GPR work conducted in magnetic ground prior to 2004 and therefore had no way of assessing the claim made by some that magnetic ground was poor for GPR. In 2004, at the Tenth International Conference on GPR, I attended some presentations by British and Icelandic scientists working on basalt flows in Iceland, who showed examples of reflection profiles collected with low-frequency antennas where depth penetration was up to 10 m or more. That suggested to me that perhaps the highly magnetic basalt there was not actually affecting the propagating radar energy as I had thought it would. So when working in Hawaii in 2005, I decided to test GPR on recent lava flows on the south coast of the island of Hawaii. The purpose of that study was to explore for lava tubes within flow sequences, which are known to have been used for the burial of ancient Hawaiian people (Conyers and Connell 2007). We found that on flows about 2,000 years old or less, the 270 MHz antennas were capable of transmitting radar energy to about 4 m in depth, with good resolution not only of the basalt layers, but also of the void spaces in the lava tubes (Figure 6-18). On the windward side of the island of Oahu, however, we conducted similar tests on similar but much older lava flows, and all radar energy was attenuated very close to the surface. In this case, the flows had been weathered, and electrically conductive clay minerals had formed in them (Conyers and Connell 2007). These tests appear to support Doolittle's conclusion that radar depth penetration is much more affected by electrical conductivity than by magnetic permeability (Doolittle et al. 2007).

Conclusion

Penetration depth of radar waves can be quite variable, depending mostly on the electrical conductivity of the ground. It is rarely possible to have good data on the chemistry and physical properties of the ground prior to conducting a GPR survey, but perhaps in the future there will be handheld devices that can determine these properties for us. Doolittle's maps that estimate GPR effectiveness for almost every state in the United States can be a very good guide for how GPR will work in some general areas (Doolittle et al. 2007), but in my view, they should be taken only as a broad guideline. In testing his conclusions, I have found that moving only a few kilometers one direction or another from where the maps indicate a certain GPR efficacy, to soils of a different type, can produce very dramatic differences in depth penetration.

Barring the unlikely scenario of obtaining detailed soil and sediment studies in advance of a GPR survey, GPR users usually must visit a site and try out different antennas and collection options to determine depth penetration and attenuation. While this can sometimes be an unfavorable option, it is always a good idea to make informed plans regarding which antennas to bring and how long to spend in the field to perform the tasks necessary, and to have an idea about how successful you will be. That said, I have found that when making this type of plan related to GPR performance, I am frequently wrong and have to adjust to whatever conditions are confronted once data are collected.

7
Historic Sites

In the United States, we tend to categorize sites into two general classes, historic and prehistoric. Historic sites here generally start about 1492 or whenever there was contact with Europeans; also any sites that have historic documents that can be used as resources in interpretation fall into this category. This means that most of the sites prior to the arrival of Columbus in the Western Hemisphere, which are commonly studied by archaeologists, are considered prehistoric (with the exception of Mayan sites that have some written records and one or two Viking sites in Canada). I will therefore use those designations here even though this might be a foreign concept for others who work elsewhere in the world.

Many North America historic sites can be discovered and mapped quite easily using GPR, as they were often constructed of substantial materials that do not readily weather and erode and are therefore often well preserved in the ground—materials such as brick, metal, concrete, and stone. However, because there are some historic sites constructed from materials very similar to those at prehistoric sites, this basic segregation of site types by their buried materials is not at all comprehensive or definitive.

Cellars and Basements

In the Chesapeake region of Maryland, near Washington, DC, the remains of a tobacco-exporting town, aptly named Port Tobacco, are preserved. At one time a thriving trading center with shipping access to the Chesapeake Bay, this town was abandoned at the end of the 19th century, as the arm of the bay that allowed ship transport to the town silted up and was no longer navigable. Historic documents tell of a number of floods that occurred prior to that abandonment, with the sediment deposited in the channel derived from newly cleared upland forest land used for producing the high-revenue-generating tobacco, sediment that choked off access to the town by water. In an open area where some remains of pottery and building materials had been discovered during a systematic shovel test survey, a number of grids of GPR were collected to explore for and classify some of the buildings in the old town.

In one grid, the cellar of a house was preserved, filled with bricks and possible stones from collapsed portions of the foundation that fell or were pushed into the cellar. The GPR reflection profiles show a jumble of architectural rubble in that cellar, each chunk of which generated a reflection hyperbola within a distinct incision to about 2.5 m below the ground surface and a planar floor reflection (Figure 7-1). One portion of the wall that lined the cellar can be seen as a stacked series of high-amplitude reflections, consistent with even courses of brick, each of which produced its own reflection. The regular spacing of these reflections supports an interpretation of evenly spaced bricks. Some of the objects preserved in this historic cellar are likely stones or bricks from the wall that lined this subfloor room. A concentration of these high-amplitude-generating reflective objects in one area of the cellar represents the remains of the chimney that collapsed soon after abandonment. If objects within features such as this cellar are of interest, they could be classified by size and density according to the shape and density of the hyperbolic reflections. Other nearby features, such as pits of unknown origin, are visible outside the limits of the cellar feature (Figure 7-1).

In complicated features such as this cellar in Maryland, the amplitude maps tend to show the outlines of the preserved feature much better than the individual reflection profiles (Figure 7-2). The outline of the cellar is very distinct in the deeper slices, and individual objects can be picked out as distinct high-amplitude reflections. The stratigraphy within this cellar, visible in the individual profiles, can help in understanding the sequence of events that occurred in this area of the historic town soon after abandonment. This is possible because of the precision of the GPR method, where individual reflection profiles show stratigraphy, and the amplitude maps in depth slices indicate the aerial placement of objects within the feature.

In this cellar, the concentration of bricks and other stones from the chimney and perhaps the fireplace are visible in one end of the subsurface feature. The chimney materials fill up the deepest portion of the cellar (to about 200–250 cm), indicating that they likely collapsed into the newly abandoned cellar hole soon after the house was abandoned (Figure 7-2). Directly over the distinct floor reflection, which appears to be composed of compacted earth and not highly reflective stone, is a layer about 25 cm thick that is nonreflective and likely a deposit of silt. Sedimentation due to agricultural practices and other deposits from large storms hastened the town's abandonment beginning in the first half of the 19th century. In the reflection profiles, this sediment layer, from one or more depositional events, sits directly on the cellar floor and is overlain by architectural rubble that produces high-amplitude reflections from each individual piece. The architectural debris that fills the remainder of the cellar was emplaced after that flood event.

The cellar features are easy to see in both profiles and amplitude maps when they have been filled by artifacts or sediment that is different from the surrounding material. The stratigraphy and placement of artifacts in the fill materials can sometimes be used to reveal the behavior of people during abandonment and possibly activities that took place post-abandonment. With enough analysis of GPR images, combined with strategic excavations, the types of fill and artifact types could tell a good deal about these "storehouses" of historic information below the ground surface.

TRASH MIDDENS

Trash middens are especially prolific features for historical archaeology due to the abundance of artifacts and often their time-depth, which can encompass decades or even centuries of trash disposal behavior. In GPR profiles, they are visible as concentrations of many high-amplitude reflection hyperbolas, which produce concentrations of small reflections when the amplitudes are displayed in depth slices. The historic midden discussed

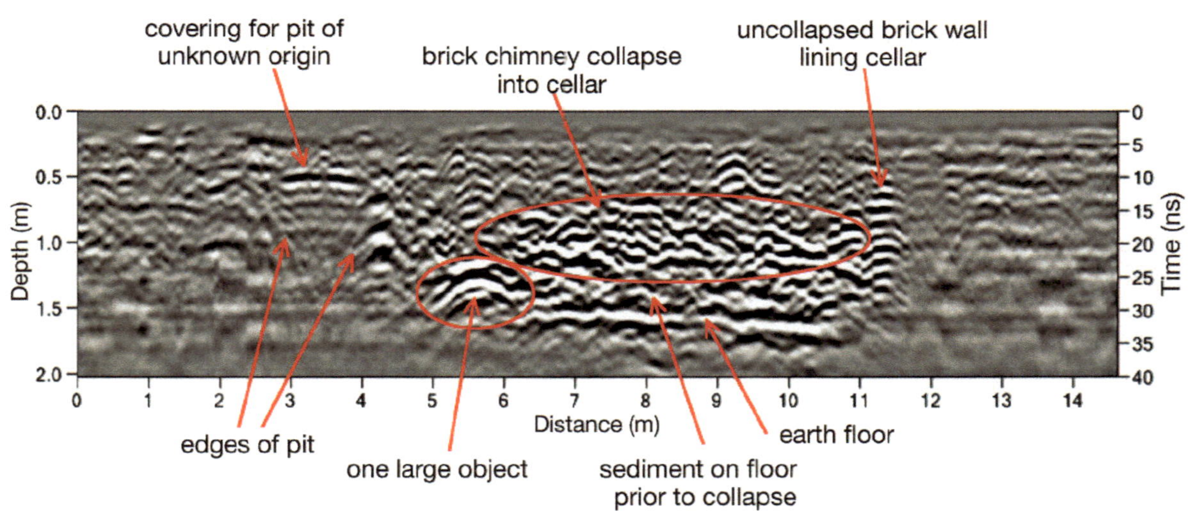

Figure 7-1: Reflection profile across a cellar in a 19th-century house at Port Tobacco, Maryland. The cellar floor is overlain with sediment, perhaps deposited from one of the many floods that finally led to the abandonment of the community. Brick and other architectural materials fill in the cellar hole. Data collected with 400 MHz antennas. *Courtesy of Peter Quantock.*

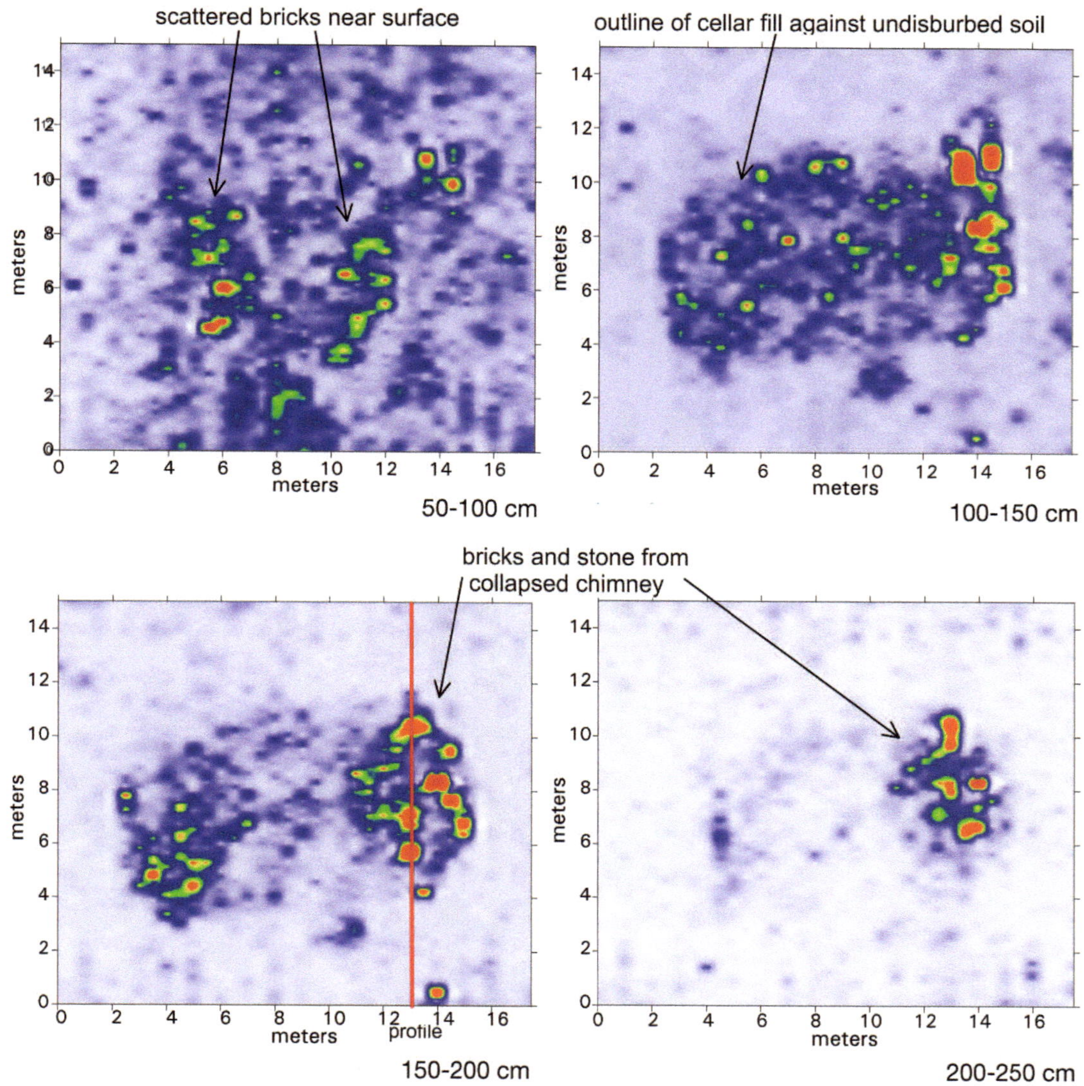

Figure 7-2: Amplitude maps of a cellar. This house's cellar in Port Tobacco, Maryland, shows chimney collapse materials. Data collected with 400 MHz antennas. *Courtesy of Peter Quantock.*

in Chapter 4 (Figure 4-4) at the Pio Pico site in California demonstrates how these artifact concentrations appear in amplitude maps (Figure 7-3). The old river channel within which the trash was discarded can be generally outlined by the concentration of artifact reflections. In this example, only the larger artifacts are displayed, as the 400 MHz antennas used for this survey do not generate high-amplitude reflection hyperbolas from objects less than about 15–20 cm in diameter. This midden was excavated and found to contain thousands of objects much smaller than this, which did not generate their own individual reflections.

Middens are hugely important in historic archaeology, and their identification for possible excavation using GPR can be quite critical. Depending on the frequency of the antennas used, resolution of individual artifacts of various sizes can be variable. A concentration of point-source hyperbolas is the reflection *signature* of these artifact deposits (Figure 4-4). With three-dimensional analysis using a high-resolution GPR grid, it is possible that total midden volume could be calculated. The volume of trash in a midden determined using GPR mapping could then be integrated with excavation information to study a number of interesting aspects of human behavior. For instance, if the artifacts in a midden found with GPR could be age-dated and analyzed with regard to trash production by a certain number of people over a given time, these data could potentially be used to

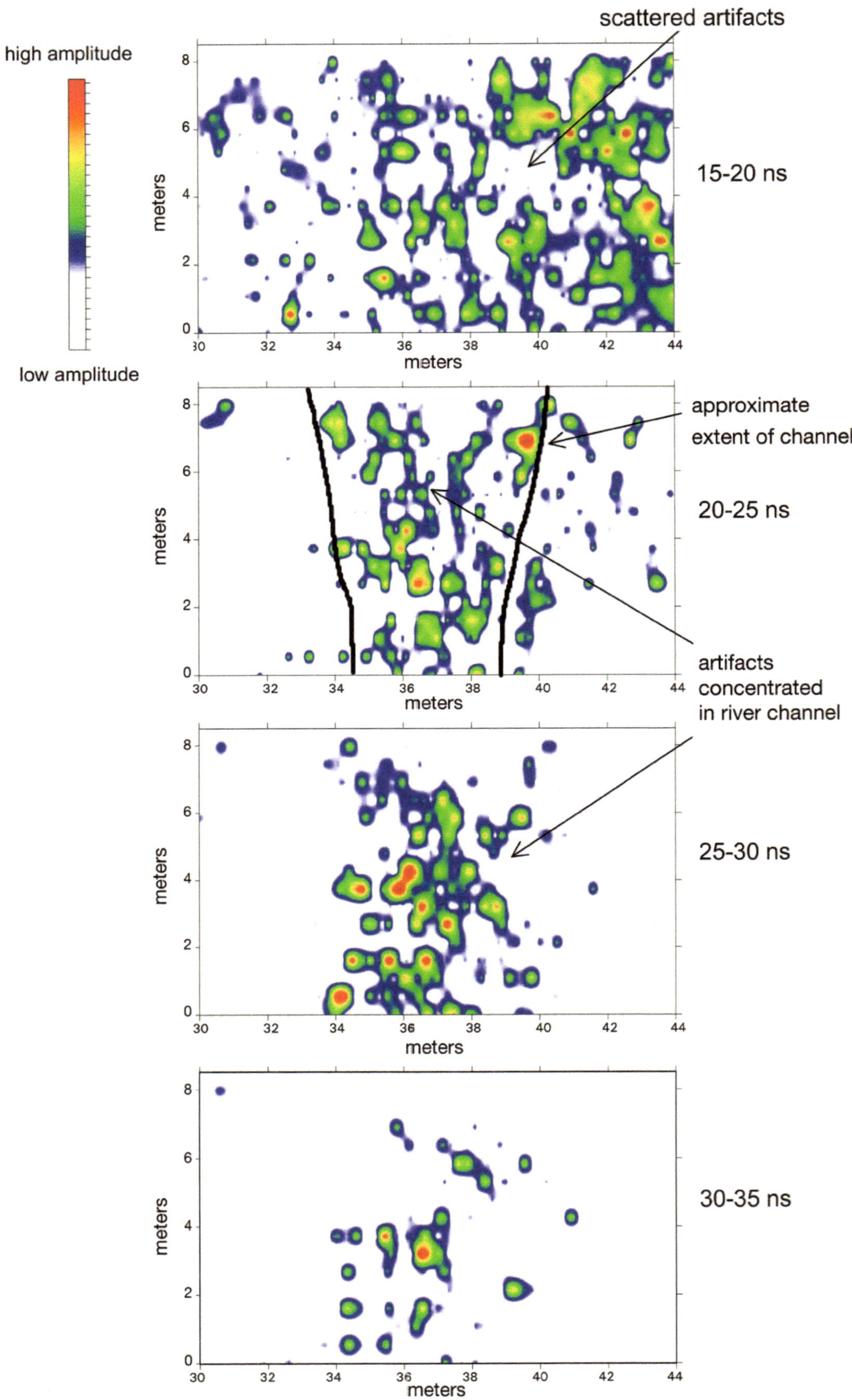

Figure 7-3: Amplitude maps showing the larger artifacts in a cellar, which appear as distinct high-amplitude reflections in a trash midden at the Pio Pico site, California. Only artifacts larger than about 15–20 cm in diameter are visible in this slice, due to the resolution of the 400 MHz antennas.

calculate population size. With enough spatial and temporal information, cultural changes and population dynamics over time could be studied using an integration of GPR mapping and standard archaeological analysis.

UNUSUAL FEATURES: PONDS AND WATERWORKS

Historic documents and maps show that there was a lily pond somewhere in the backyard of the 19th-century Governor's Mansion in Denver, Colorado. During a renovation of the garden in 2005, GPR profiles were collected on a driveway leading to the carriage house, and the pond was discovered in an unexpected location (Figure 7-4). This pond, which was excavated and found to be lined with plaster, produced a high-amplitude planar reflection from its bottom (Figure 7-5). That surface appears to be undulating in the GPR profile, but it was found to be flat when exposed to view. The undulations in the profile were caused by variations in retained moisture within the sediment that fills this depression, which brought about small changes in the velocity of radar energy across the feature. Those variations in velocity, when converted to depth in the ground, produced the undulations in the floor when viewed in profile.

A second planar reflection is visible under the bottom of the pond, which could be an earlier pond feature or perhaps the base of a layer used in the construction of an impermeable layer during its construction. Excavations that exposed the pond did not extend to this depth.

In Central Park, New York, a number of historic features were discovered with GPR and excavated in 2011. One reflection feature in particular was very distinct and appeared to have been produced from metal, due to its

Figure 7-4: Collecting GPR data over a parking lot surface. This used to be the formal garden of the Governor's Mansion, Denver, Colorado.

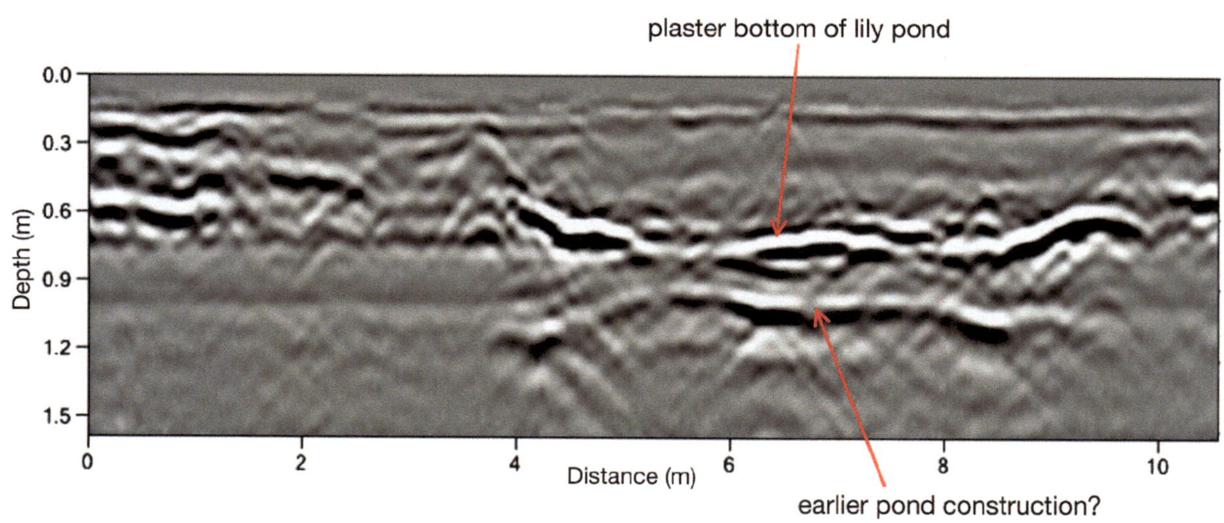

Figure 7-5: Reflection profile across a historic lily pond in the garden of the Governor's Mansion, Denver, Colorado. The plaster bottom of the pond produced a high-amplitude reflection, which undulates somewhat due to subtle velocity changes in the overlying fill material, creating velocity pull-ups and pull-downs. Data collected with 400 MHz antennas.

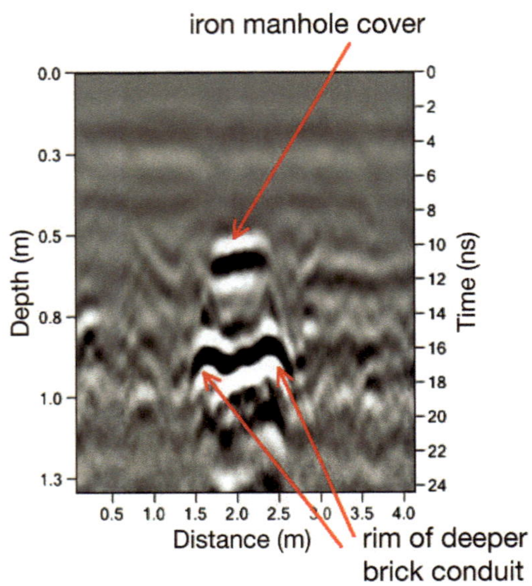

Figure 7-6: Reflection profile of the top of a buried iron manhole cover, with a lower brick conduit leading to a very deep water conduit in Central Park, New York City. Data collected with 400 MHz antennas.

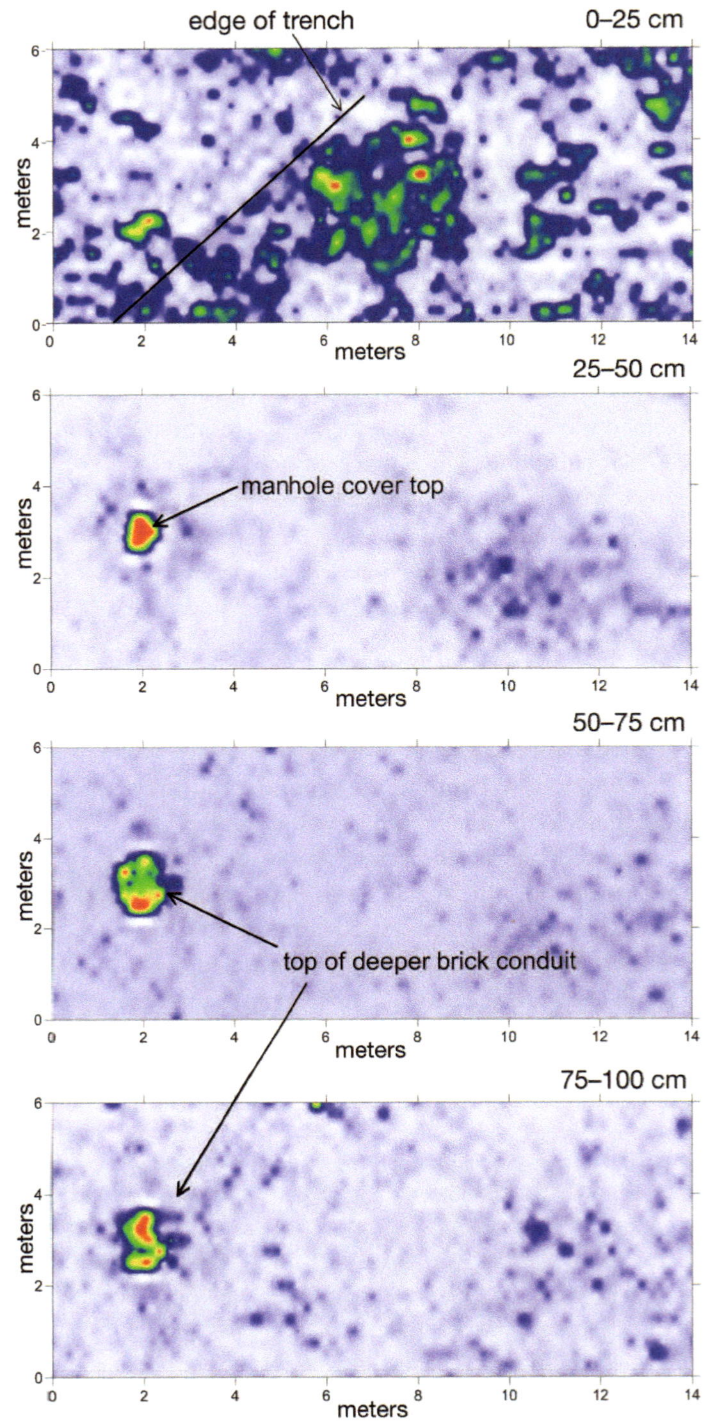

Figure 7-7: Amplitude maps of a manhole cover with a brick conduit. The shallowest feature is the iron cover with a brick conduit leading to a deep water pipe conduit below, in Central Park, New York City. Data collected with 400 MHz antennas.

high-amplitude reflection (Figure 7-6). Two layers of reflections were visible in the profiles, while the overall feature appeared small, with an unusual three-dimensional shape (Figure 7-7). We hypothesized that it might be a cased water well or perhaps pre-park sewer or drainage feature. When excavated, the upper high-amplitude reflection was found to be a flat, iron manhole cover (Figure 7-8) dating to the early part of the 20th century. This cover was opened and a brick-lined conduit was found leading to a very deep conduit containing very old water lines. The lower reflections were generated from the rim of the lower, wider brick conduit that led downward to the historic horizontal water conduit that took water to a reservoir somewhere in the park to the east. The vertical brick sides of both the upper and lower vertical conduits that led downward were not visible either in the GPR profile or the amplitude map because they are vertical and did not reflect any radar energy back to the surface. The rim of the lower conduit generated two hyperbolic reflections, which merged in the reflection profile into one undulating reflection (Figure 7-6). That deeper horizontal conduit was below the depth that radar energy could penetrate using 400 MHz antennas.

STRUCTURE WALLS AND FLOORS

Fort Garland, in Colorado's San Luis Valley, was constructed by the U.S. Army in 1853 to protect settlers and gold and silver prospectors from Ute Indian raids. Its most famous commander was Kit Carson, the so-called Indian scout who lived with a number of tribes in the West, married two Native American women, and worked as a guide for settlers in the western United States. The Utes were forcibly confined to reservations in the early 1880s, and the fort was abandoned in 1883. Today some of the fort has been reconstructed for tourists, but much of the remains of the original structures are invisible below the ground surface. Historic documents suggested that there was a building used as officers' quarters in one location, bounded by an interior plaza or parade ground.

To determine if there were any remains of the original construction, one GPR grid was collected over an area adjacent to the parade ground. We assumed that the original buildings had been constructed of adobe but, as with most adobe structures in this area of the American Southwest, the foundations for the adobe walls were made of stones. In the GPR profiles, the stone wall foundations were clearly visible as reflection hyperbolas, with compacted earth floors between them visible as moderately reflective planar surfaces (Figure 7-9). In some places, it appears the wall foundations might have been several layers of stone thick, which produced collapse features visible in the GPR profiles. Those concentrations of stones might also be the bases of chimneys used to heat the officers' rooms. The stones used as foundations for the walls are readily mapped in amplitude slice-maps (Figure 7-10), with distinct reflections from the compacted earth floors within the walls.

In downtown Denver, Colorado, under a parking lot, historic maps showed some late 19th-century row houses, which documents indicated were occupied by elderly women. Two of these structures were modified into a duplex (two structures sharing a common wall) in the 1930s, before the entire region was converted into a block of auto repair garages. When I moved to Denver in 1975, I remember getting a muffler replaced on my 1967 Ford Mustang in one of the establishments on this block. The GPR amplitude maps readily show the foundations of one of the row houses in the shallowest slice (Figure 7-11). This feature is likely one of the row houses, or part of the duplex, which was torn down and then paved over for a parking lot in the 1980s. It was excavated, and the brick foundations of these houses were exposed. In what was the backyard of these houses, a deep, small structure was also discovered (feature A in Figure 7-11), which was not excavated and whose function remains unknown.

The deeper slices show the surviving brick foundations of the basements of these houses, which were used to store coal and firewood for the furnaces and also functioned as root cellars and storage space. These were also excavated, and many artifacts from the early 19th century were found in these deep features (Figure 7-12).

In an orchard in the ground of Four Mile Historic Park near Denver, Colorado, a grid of GPR was collected in an area where it was considered probable to find historic buildings that remained undocumented. The park today contains gardens and orchards surrounding the original house and associated horse stables, which were built in 1859 as a stage stop four miles outside of Denver. Stagecoach passengers stopped here to freshen up before arriving in the city about a half-hour ride away on the stage road along the bank of Cherry Creek.

The GPR maps produced in the orchard showed a distinct metal pipe, which had been truncated, crossing the grid (Figure 7-13). At the same depth as the pipe, an almost perfectly square, high-amplitude planar feature was noticed about 4 × 4 m in dimension in the amplitude map, hypothesized to be a building floor. We first took soil and sediment samples with a coring device, both inside the square feature and outside it, and compared them. This exercise showed no noticeable difference in constituents at the depth indicated in the amplitude maps. Two 1 × 1 m excavations were then conducted, one in an area where the metal pipe was visible and the square feature could be encountered, and the other outside this feature for comparison. The stratigraphy, sediment, and soil types seemed to be exactly the same in both excavations, and the metal pipe was encountered at the correct depth as mapped in the GPR images, so we were confident that our time-to-depth conversions were correct. At that point, I began to have worries that the GPR images were producing spurious "anomalies" that just happened to be square in

Figure 7-8: Manhole cover and brick conduit. When uncovered, the manhole visible with GPR in Figures 7-6 and 7-7 turned out to be from an abandoned water pipe. The cover was opened and can be seen to lead to a brick conduit leading to a deep water main used for domestic water to New York City. *Images courtesy of Meredith Linn and Herbert Seignoret.*

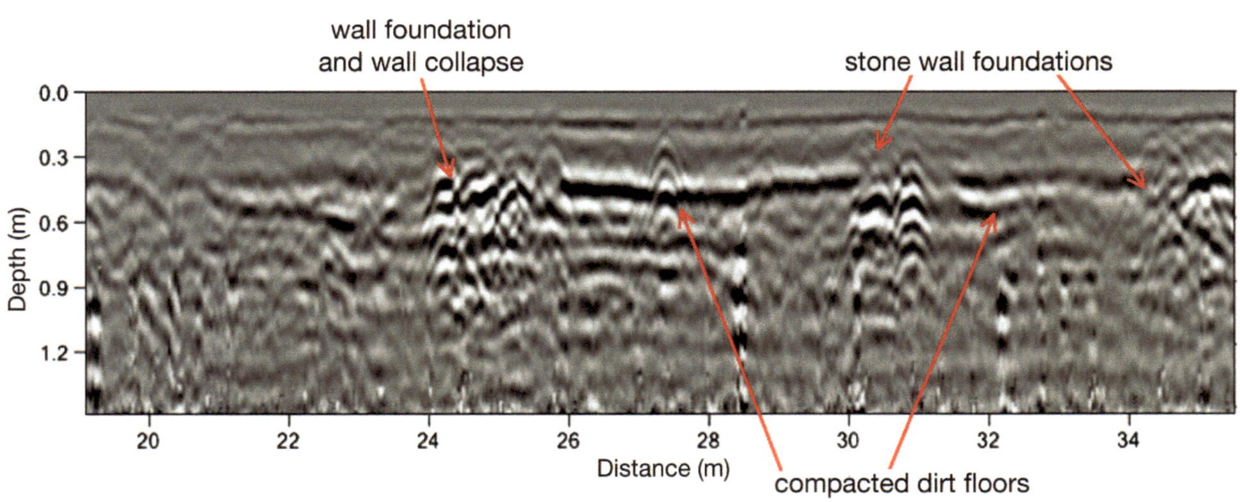

Figure 7-9: Reflection profile across a historic fort, with an earth floor and stone foundations. This profile, collected at Fort Garland, Colorado, illustrates a weak planar reflection from a compacted earth floor, and stone foundations and possible wall collapse as high-amplitude hyperbolic reflections. Data collected with 900 MHz antennas. *Courtesy of Jennie Sturm.*

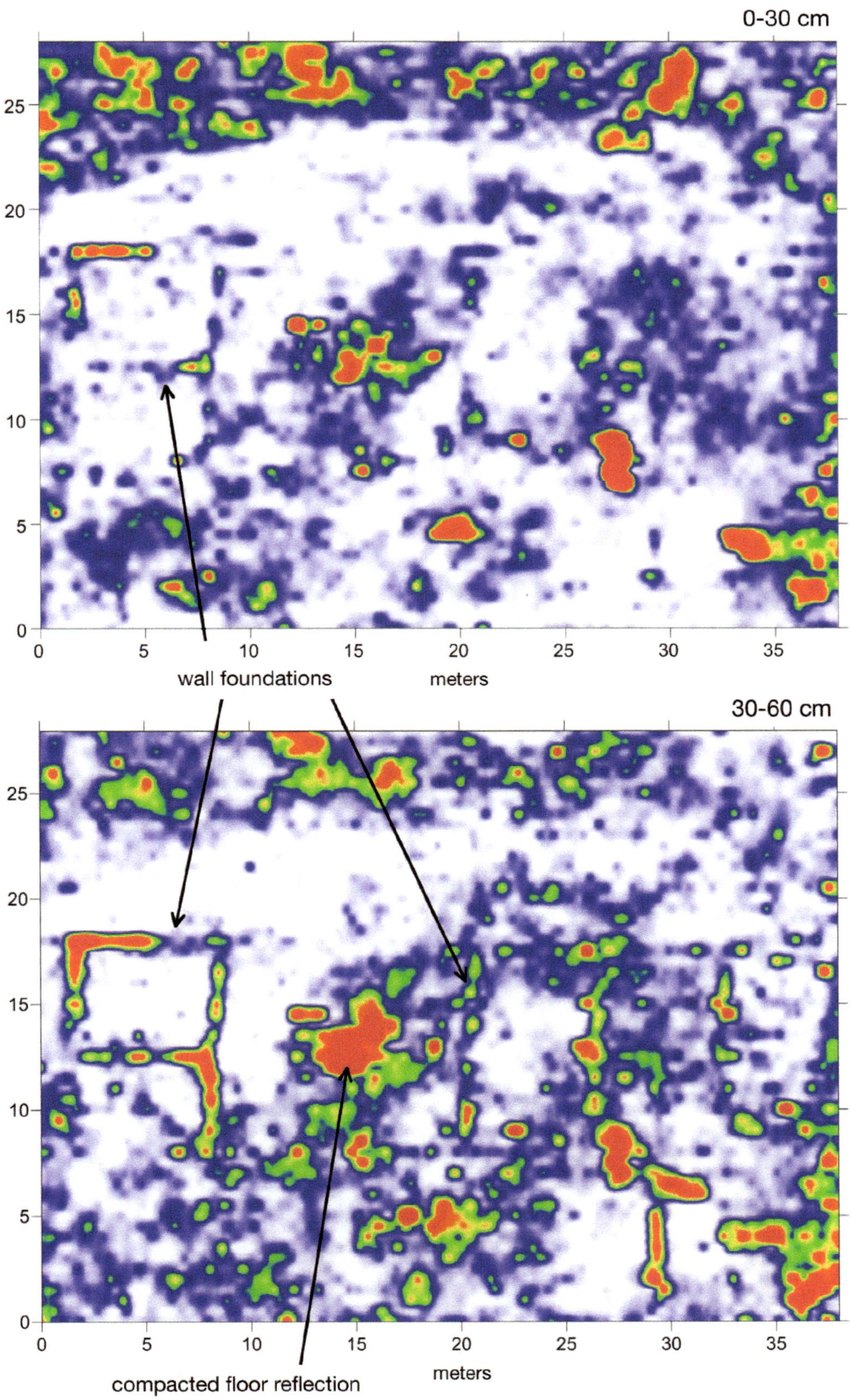

Figure 7-10: Amplitude maps of wall foundations of officers' barracks at a historic fort, Fort Garland, Colorado. The wall foundations are visible, as are portions of the compacted earth floor. Data collected by Jennie Sturm using 900 MHz antennas.

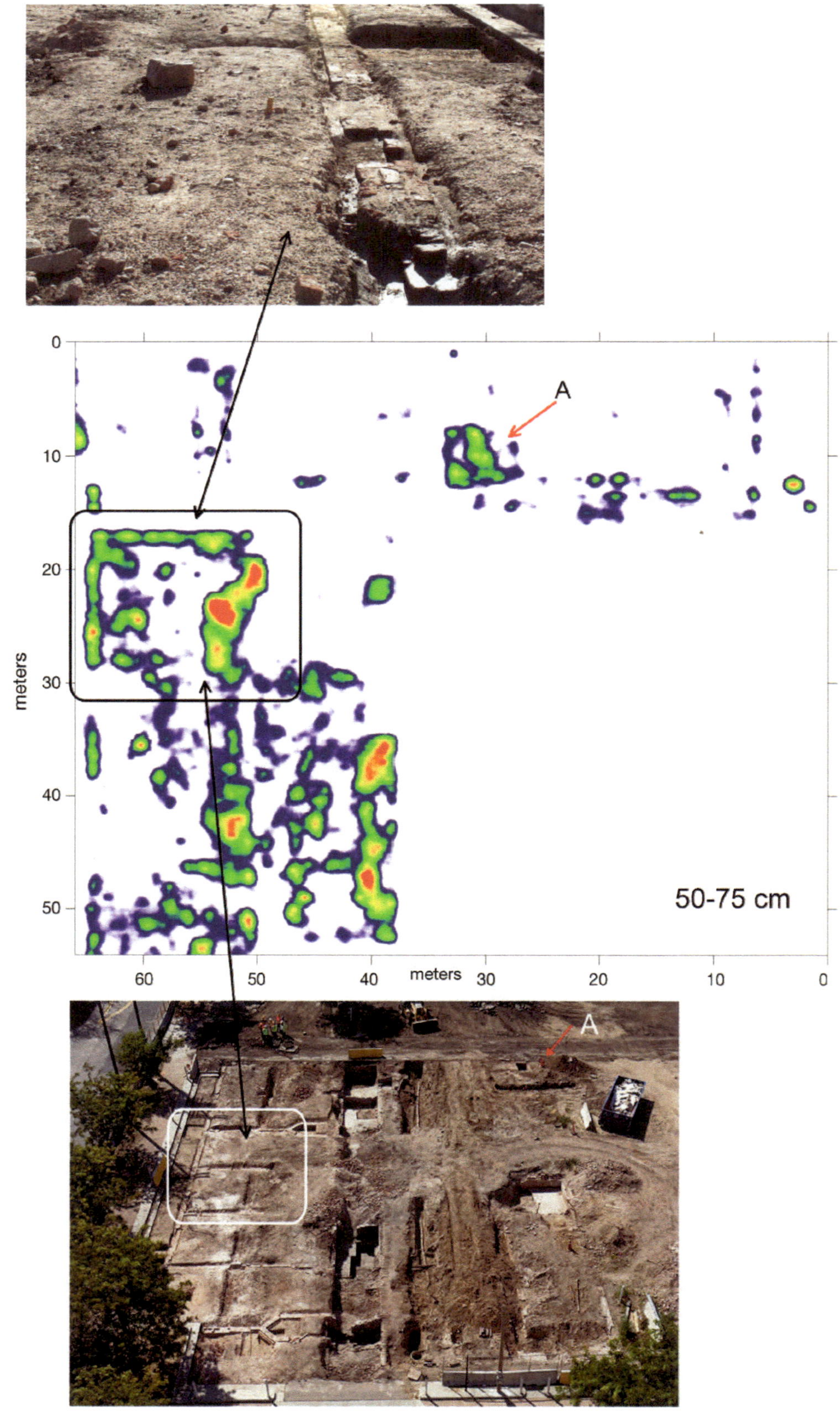

Figure 7-11: Amplitude map of foundations of urban row houses built in the late 18th century in Denver, Colorado. When excavated, the foundations' orientation and extent almost exactly matched the GPR images. Data collected using 400 MHz antennas. *Courtesy of Jennie Sturm.*

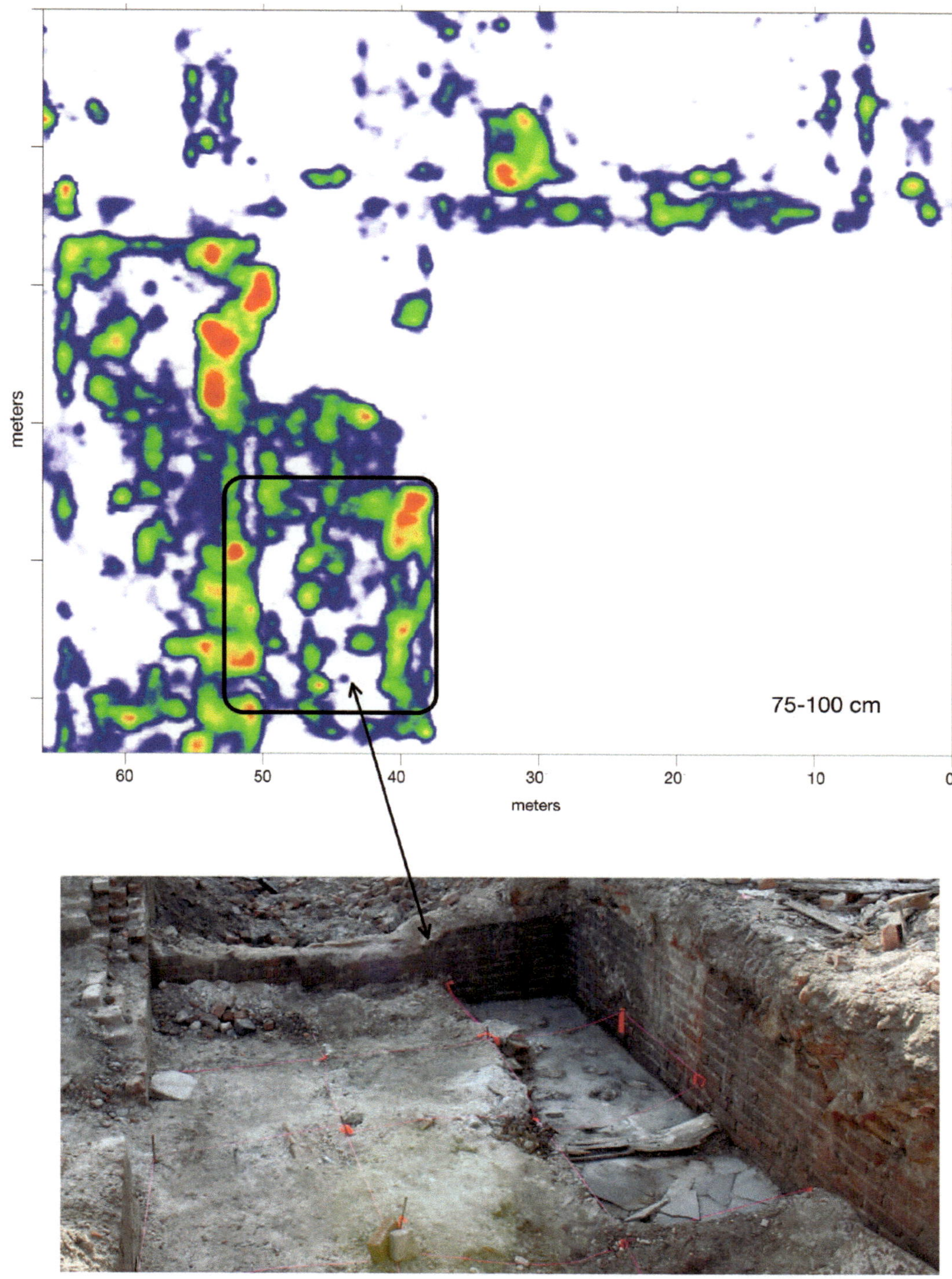

Figure 7-12: Amplitude map of basement foundations of urban row houses that had small storage and utility rooms in their rear. Data collected using 400 MHz antennas. *Courtesy of Jennie Sturm.*

shape. Knowing that perfectly square features rarely occur in nature, and that most gridding and mapping programs tend to produce more oval or circular features with their interpolation algorithms, I still had some confidence that the square feature was cultural. Then late one afternoon, when we were cleaning up the face of the excavation that was placed on top of the square feature, one of my observant students discovered a small piece of flat sandstone within the alluvial sediment we had been digging in. Knowing that this type of sandstone does not outcrop in the drainage of Cherry Creek, I could confidently assume that some person had placed

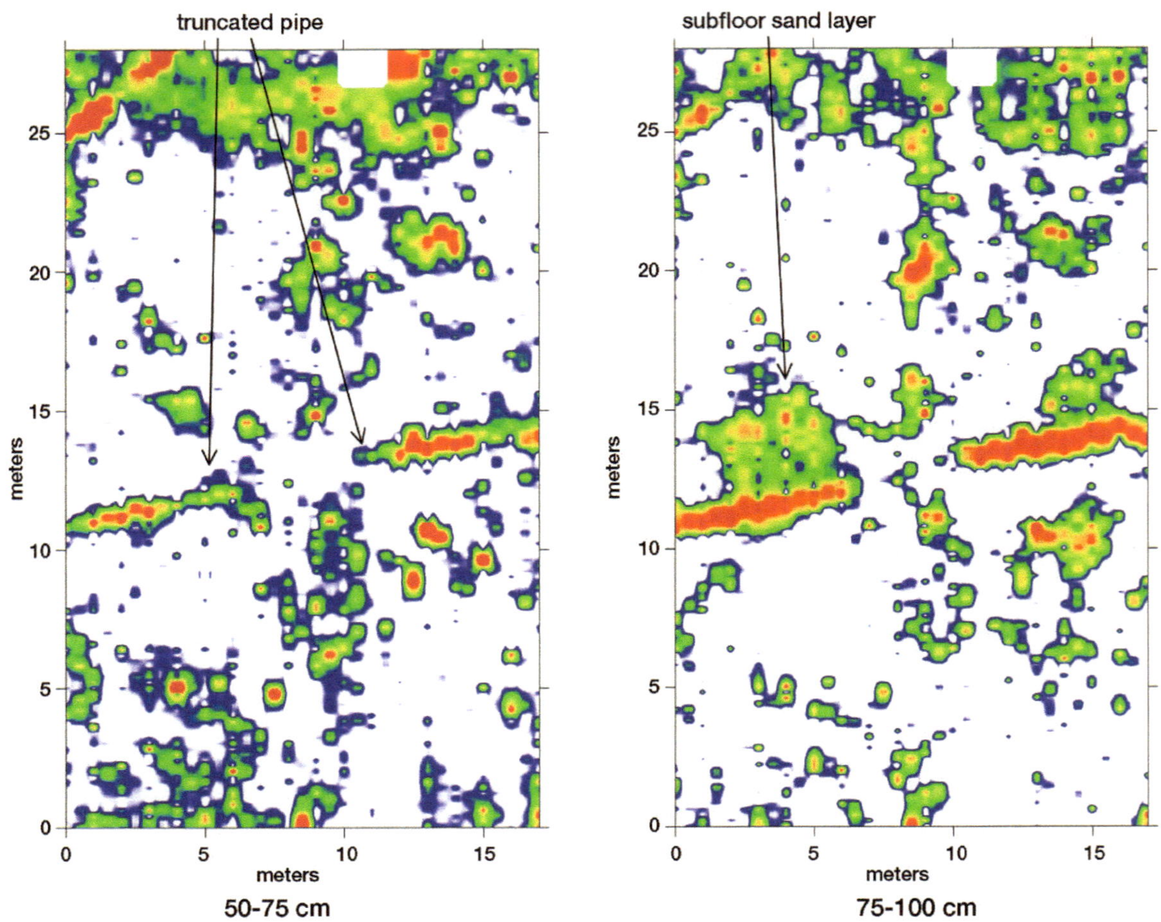

Figure 7-13: Amplitude maps in an orchard showing a square floor, produced from 400 MHz reflection data collected at Four Mile House, Denver, Colorado. The square feature is visible in the 75–100 cm depth slice. This reflection was produced from a 2 cm thick sandy layer, which was at one time the sublayer of a flagstone-floored building.

it there. We dug under that stone and found a 2 cm thick sand layer that had escaped our attention during excavations. Further excavations found more small pieces of this type of hard sandstone, locally called "flagstone," all resting directly on that thin sand layer. It was this thin sand, surrounded by alluvial sediment consisting of mostly silt and clay, that had reflected higher-amplitude waves and was producing the amplitude feature in the slice-map at that depth.

It is common practice in Colorado to use pieces of flagstone as flooring in sheds and other structures where simple compact dirt floors are not as functional. Usually sand is laid down first, and then pieces of the flat stones are placed in the sand to level and stabilize them. What we had found in the GPR map was the remains of the sand subfloor of a square building. The good pieces of flagstone had been recycled and placed elsewhere after the building was no longer in use. Only a few unusable pieces of stone remained on the sand layer, which was what my student had seen in the excavations and had provided the clue needed to identify the otherwise unnoticed layer. It is likely that this building, whose function remains unknown, had been abandoned sometime prior to 1912, when this area along Cherry Creek was inundated with a massive flood. That flood, which occurred on July 14, originated in a huge thunderstorm in the upper reaches of the creek, which washed large amounts of sediment into areas such as the orchard at Four Mile House. The sand layer, which is almost all that remains of this historic structure, was then covered with more than 75 cm of sediment and preserved for us to find with GPR.

Using GPR Images in Conjunction with Historic Maps and Photos

In Albany, New York, a survey was required prior to the construction of a new high-rise parking garage in an area that was considered historically important. Historic insurance company maps of the area (Figure 7-14) showed there to have been a malt production company, automobile repair buildings, and rooming and apartment buildings during different times in the city's history, but there was no surface indication of any remains of this architecture. All buildings had been demolished in the 1980s and the area leveled and paved for a parking lot.

Sanborn Company fire insurance maps, used to determine historic structures over time, are an invaluable resource for studying the history of many U.S. cities, as they document very accurately the location and function of buildings at the times the maps were made. These maps were created by the Sanborn Company in most urban areas of the U.S., beginning in 1867, to help determine liability for fire insurance companies in the case of large-scale fires. During the 19th century, huge conflagrations were a regular occurrence in many cities, consuming hundreds and even thousands of the wooden and masonry structures. In the settlement of insurance claims, where agents needed to determine how much was owed to the insured owners of these burned buildings, the Sanborn maps were critical. They were also consulted by insurance sales agents prior to issuing policies, to determine what materials the buildings were made of, and any possible sources of fires, such as lamps, ovens, and furnaces. These maps were periodically updated as renovations occurred, and thus serve as an invaluable archive for historical archaeological research done within large U.S. cities. The Sanborn Company is still in business today and has branched out into the production of many other types of digital mapping applications.

From the Sanborn maps (Figure 7-15), we learned that the area surveyed with GPR had been occupied by a company that processed barley into malt for the production of beer and whiskey in the late 19th century. This firm, the T. McCredie Malt Company, was a prosperous business in the 1800s in New York's capital, Albany. In its factory, barley was first steeped in water in two large tanks for at least two days and then dried on the factory floor. That material was then malted by baking it in the three kilns, visible on the Sanborn maps, before it was

Figure 7-14: Collecting GPR profiles in a parking lot in Albany, New York. A number of interesting historic buildings were discovered.

ready for the production of alcoholic drinks, which took place elsewhere. The owner of this establishment in 1890, Thomas McCredie, was born in Glasgow, Scotland, and moved to the United States in 1838, where he worked in a different malting factory in Albany owned by Robert Dunlop, whose daughter he soon married. McCredie and his father-in-law soon formed a partnership in the malt business, and after Dunlop died, McCredie took over and managed the malting company at this location until his death in 1892, when his son James became manager. The Sanborn maps show that the McCredie Malt Company had three kilns in the factory and two steep tanks in use in 1890. The mapmakers also noted that the factory was still using whale oil as fuel for their lamps. Other buildings of unknown function were also noted in the vicinity in 1890 north of the Malt Company property (Figure 7-15).

Sometime after McCredie's death in 1892, the company ceased business in this location, and by 1908 the malt production building was being used as a furniture storage warehouse (Figure 7-15). The kilns that had been in service in the malting process were still standing at that time, but apparently were not being used. By 1934, this whole area had been transformed into a commercial center for automobile repair businesses; the other buildings standing in 1890, whose function had not previously been documented (perhaps because they had not purchased fire insurance), are noted to be a rooming house and apartment buildings.

The GPR amplitude slice-maps could be directly correlated with the Sanborn maps and the extant features visible below the parking lot mapped with GPR (Figure 7-16). The most prominent feature in almost all the amplitude slices is the one kiln that was outside the malt house in 1890. In the deepest slice, 75–115 cm below the ground, there appear two pre-Sanborn-map houses in the area north of the McCredie malt factory. These may be the remains of simple dwellings that existed in this area of Albany prior to urban and industrial construction in the mid-19th century (Figure 7-16). They must have been demolished by the late 19th century, when the malt business took over much of this area. After Thomas McCredie's death in 1892, it seems that his son James did not fare as well in the malt business as his father, as by 1908 the factory was being used as furniture storage warehouse and the kilns were inactive (Figure 7-15). By 1934, the furniture warehouse had been renovated and was being used as an automotive repair garage, which was duplicated in many of the other buildings in this area. An area in the southwest portion of the study area was also being used as a stone quarry. That quarry was apparently expanded into the area previously occupied by the two indoor kilns from production days, as a much expanded quarry area is filled with architectural debris visible in the GPR amplitude maps (Figure 7-16). That filling episode likely occurred in the 1980s, when all the buildings in this area were demolished for a parking lot and the area was leveled.

In the GPR amplitude maps, the remains of these interesting urban transformations can be readily identified. The outdoor kiln produced a prominent high-amplitude feature in all the slices below 45 cm. There are also other interesting architectural features in the deepest slices, which appear to be buildings that are pre-1890, which is when the first Sanborn map was produced (Figure 7-16). These buildings appear in the slice-maps up to about 45 cm below the present surface but do not appear in any of the historic maps. They must surely have been present, or their remains still partially standing, during the operation of the malting factory, but there is no description of them in the fire insurance maps. Perhaps they were already in ruin by that time.

All that remains of the malting factory proper, visible with GPR, is the northern foundation. The steep tanks were likely removed sometime soon after 1890, as were the two indoor malting kilns. The outside kiln appears to have been more substantial than those inside the factory walls, and perhaps used at a higher temperature or perhaps for a longer time and therefore baked the ground under and around it, as it still remains a prominent feature in all the deeper GPR amplitude maps.

The apartment building north of the malt factory had a substantial foundation and is visible in all the GPR amplitude slices between 45 and about 100 cm in depth. The other rooming houses must have been much less substantial structures or were more totally removed during demolition, as there is no evidence of them in any of the GPR maps. It appears that during the 20th century, the stone quarry, noted on the Sanborn maps as the "stone pit," was enlarged to encompass the area that used to hold the two indoor kilns. This area was then filled in with building rubble in the 1980s during demolition to produce the flat parking lot. The fill in this old stone quarry shows up as very high-amplitude reflections in all of the depth slices. The uppermost slices show the distinctive scars produced by heavy equipment, which was used to demolish all the buildings in this area, similar to

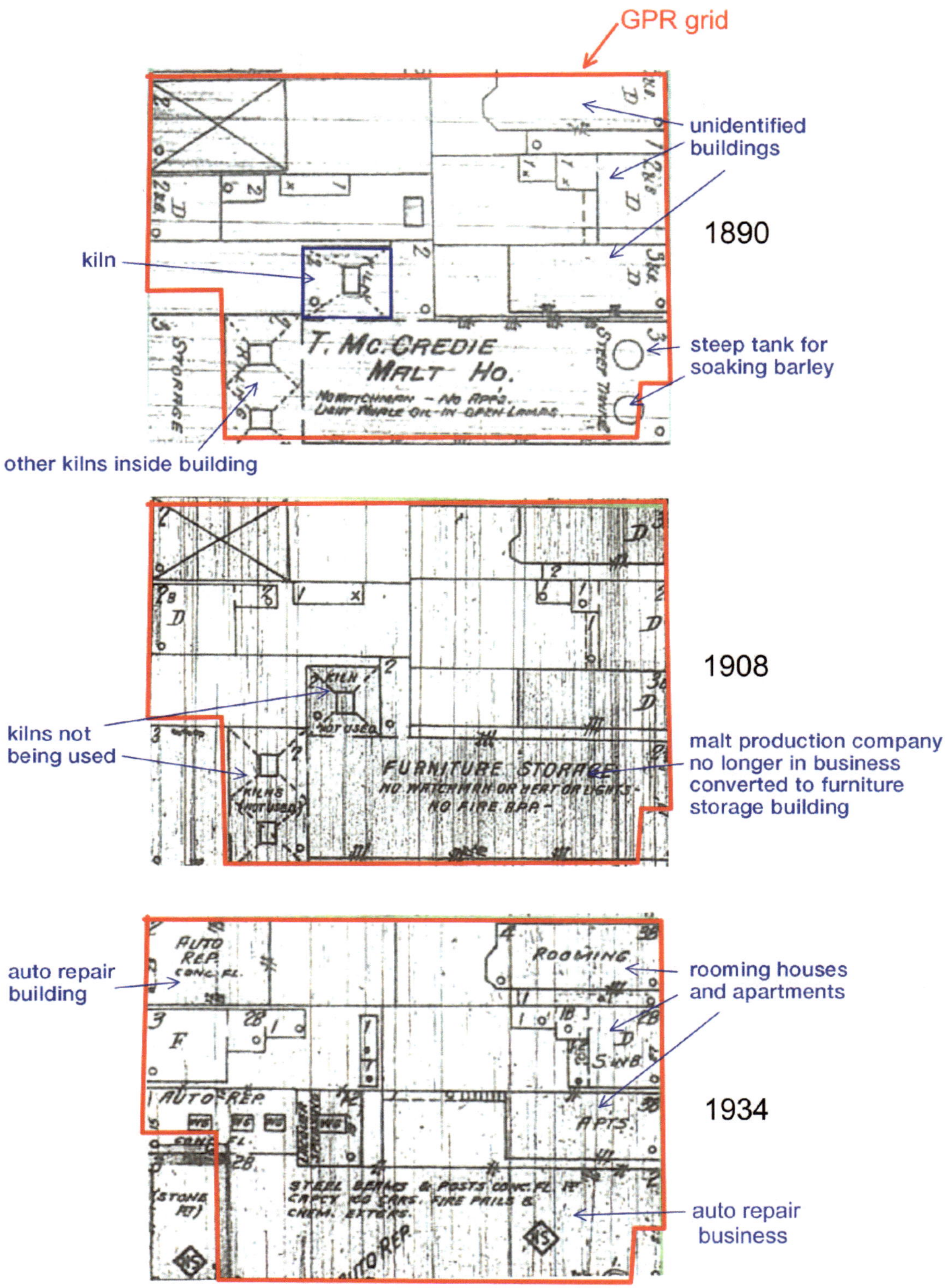

Figure 7-15: Sanborn fire insurance maps. These maps show what was located in the GPR study area over time. The uses of this block changed from malt production in about 1890 to furniture storage in 1908, and auto repair facilities and rooming houses and apartment buildings in 1934. All these structures were demolished in the 1980s, and the area is now a high-rise automobile parking facility used by employees of the New York State government bureaucracies nearby.

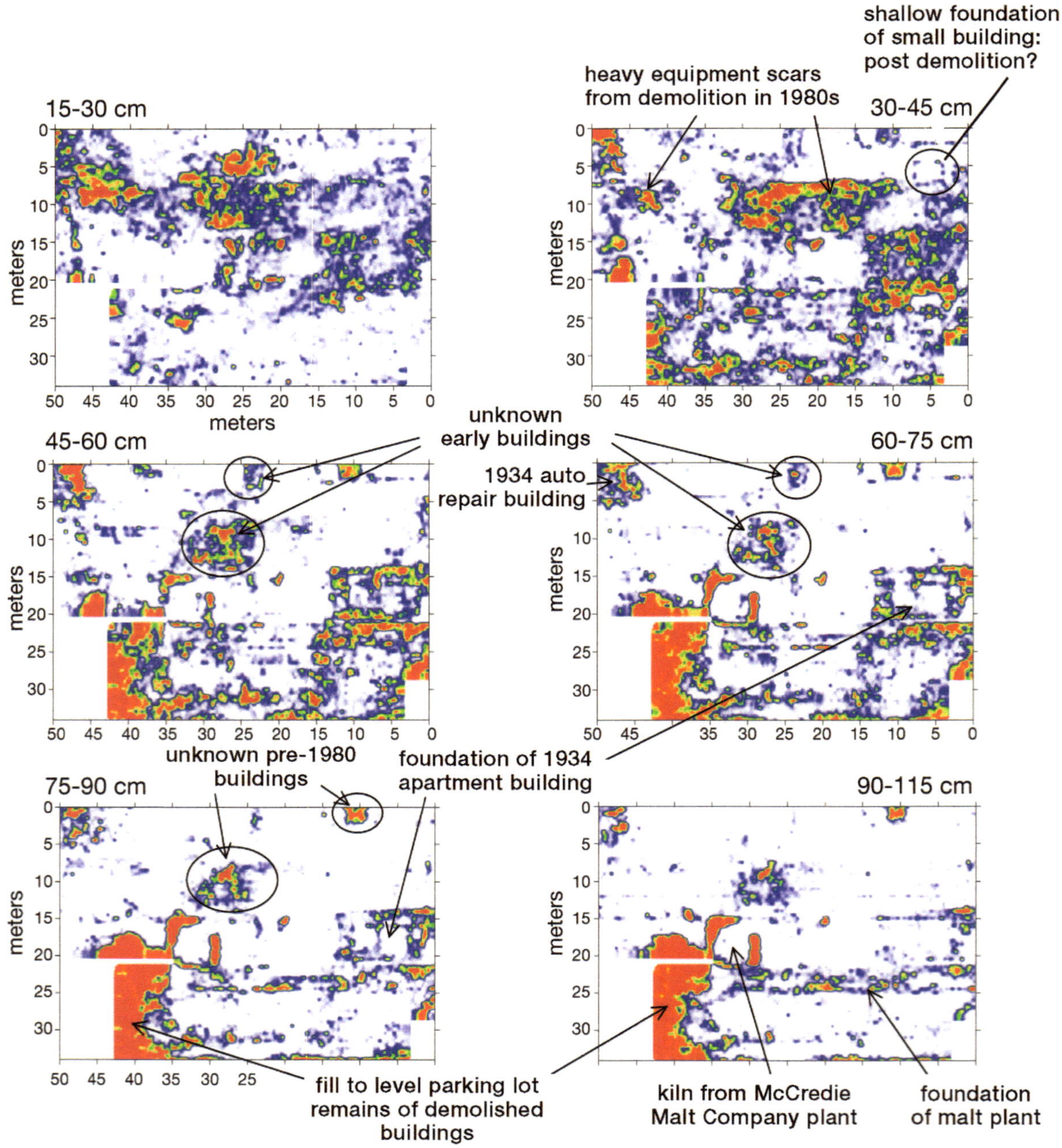

Figure 7-16: Amplitude maps compared to Sanborn fire insurance maps. The GPR amplitude maps of a small plot of land in downtown Albany, New York, show an interesting progression of land uses, which can be integrated with historic documents and the Sanborn fire insurance maps shown in Figure 7-15. Data collected using 400 MHz antennas. *Courtesy of Michael Grealy.*

those scars discussed in Chapter 5 at Goodnight Barn (Figures 5-15, 5-16). One lone building foundation in the northeast corner of the grid from 30 to 45 cm in depth is all that remains of a small unidentified structure that existed there after all the remaining buildings were torn down. Its function is unknown. Today the area is used solely as a parking lot for government employees and others who work in downtown Albany.

This study in Albany demonstrates the utility of integrating GPR amplitude maps with historic maps and other historic information. When this is done well, it is possible to document a chronology of land use and changes in economics, and how these transformations occurred. In this fashion, the GPR images become analogous to "time slices" in the truest sense of the phrase, as each depth slice showing the amplitudes generated from the remains of some buildings can potentially be given a time frame showing the transformation of the city over many decades. In this small area of Albany, what was once a rural farming community with two small structures (perhaps wooden cabins?) was transformed into a small industrial community. The historic documents tell an interesting story of a Scottish immigrant to the United States, who did well, married the boss's daughter, took over the company, and then left it all to his son, who quickly lost it. The transformation of this area to other, less productive uses, such as for storage, then occurred until the advent of the automobile made this area productive again with automobile repair shops. The growth of government and the need to park the cars of commuters who worked for the state quickly superseded the automobile repair business, and this area was then transformed into a car-park for state workers and other commuters. A need for even more automobile parking has changed the usefulness of the land once again, as the proposal of a high-rise parking structure prompted this study. This sequence tells an interesting story: of rural life modified and transformed to serve industrial capitalism, which declined in value, rose again to meet the needs of the car driving public, until finally becoming a metaphor for a 21st-century, large government bureaucracy in need of still more parking space for the cars of its employees.

HISTORIC GARDENS

Soon after the Japanese bombed Pearl Harbor, Hawaii, on December 7, 1941, the U.S. government began a program of incarcerating into "relocation camps" people of Japanese ancestry living in the United States. The fear was that some of these people might be working in collusion with imperial Japan and were potentially a threat to the security of Pacific Coast areas. Many of these relocation camps were situated in very isolated areas throughout the western states. One such camp, called Camp Amache, was located just west of the small farming community of Granada in Colorado, south of the Arkansas River (Figure 7-17). The camp was surrounded by a fence, with guard towers, and served as an internment camp for the duration of World War II.

Archaeological excavations have been conducted in the remains of Camp Amache for a number of years, and historic documents as well as recent interviews of Japanese-Americans who remember spending years in the camp have yielded some very interesting insights into the lives of the internees. For example, it was common for some of the people who lived there to attempt to duplicate traditional Japanese vegetable and ornamental gardens in the arid plains of Colorado. They sometimes constructed ponds with goldfish and small bridges over the ponds, as well as more utilitarian vegetable gardens, which were termed "victory gardens" at the time. Many of these gardens, constructed between and adjacent to the standard U.S. Army barracks where they resided, were partially sunken into the ground and fenced so that the thin soils would retain moisture and also so that the plants would not be buffeted by the strong winds that blow across the plains (Figure 7-17).

In a search for these gardens, GPR grids were collected in areas where the historic documents and oral histories indicated there were likely preserved features of this sort. The 900 MHz antennas were used for maximum resolution to a depth of about 1 m in this very dry, sandy ground. Reflection profiles showed concentrations of very high-amplitude point-source reflections, which appeared to be metal or other reflective objects such as concrete bricks used to make foundations for the barracks, and miscellaneous trash (Figure 7-18). Upon further analysis of the profiles, these concentrations can

Figure 7-17: Photo of Camp Amache, Granada, Colorado, from 1944 showing the fenced vegetable gardens adjacent to the barracks. *Picture courtesy of Bonnie Clark and the Amache Preservation Society, McClelland Collection.*

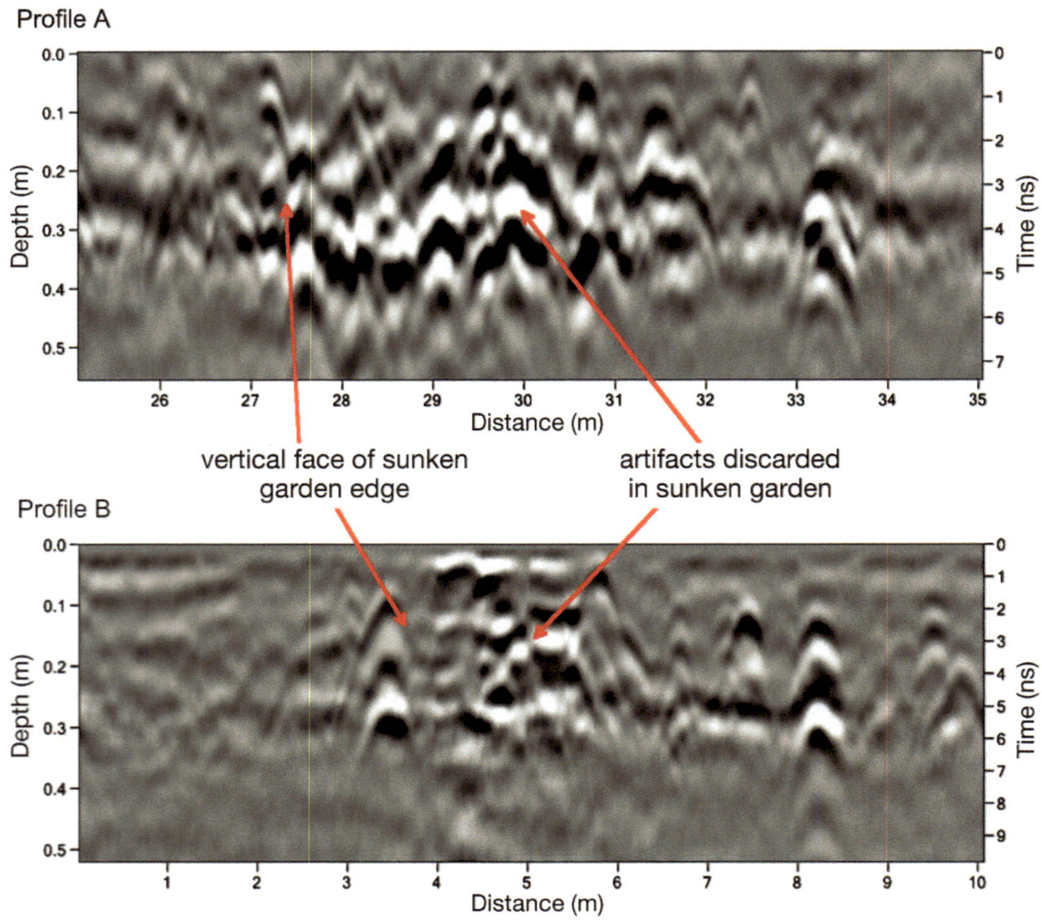

Figure 7-18: Reflection profiles at the Amache Relocation Camp showing sunken garden beds. These reflection profiles display high-amplitude hyperbolic reflections generated from metal objects and other debris disposed of in sunken garden beds. Data were collected using 900 MHz antennas. The location of these two profiles is shown on the amplitude maps in Figure 7-19.

be seen directly adjacent to nonreflective sandy ground at vertical interfaces, which suggests that the incised garden areas had been filled with this material, likely after the camp was abandoned in the autumn of 1945. Soon after that, much of the area was leveled. It seems likely that these incised features filled with materials were the historic gardens, which were then covered with architectural debris and other items during the camp's demolition.

The amplitude maps also show some square-edged features, visible as right angles between zones of abundant high-amplitude reflections and areas of no reflection at all (Figure 7-19). The nonreflective areas are the typical, undisturbed sandy ground in this area.

CONCLUSIONS

The variety of historic sites that can be studied with GPR is vast, and only a few examples from the United States are compiled here. Walls, floors, foundations, and gardens are features found all over the world in many cultures over time, and lessons from these U.S. examples are easily applicable elsewhere. The ability to compare the results of GPR mapping with historic documents and maps offers us a new avenue in the study of human activity and behavior. As GPR can map fairly large areas containing many buried features in three dimensions, it gives us access to stratified sites in powerful ways. Stacked amplitude maps of stratified historic sites can be used almost as "time slices," where time is actual years in the past.

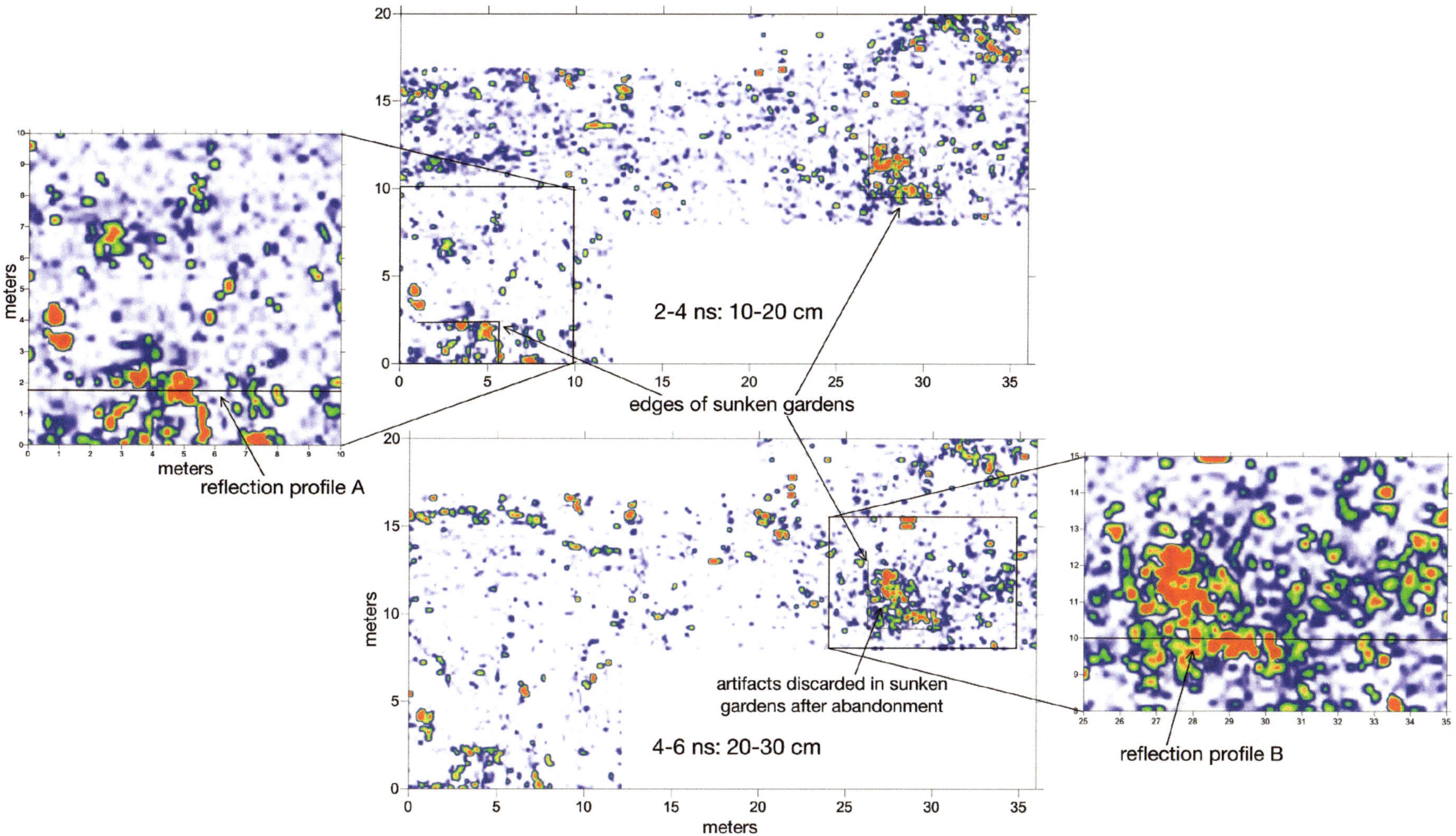

Figure 7-19: Amplitude maps (with detailed inset maps) of square deposits of trash thrown in sunken garden pits after the Amache Relocation Camp was closed in 1945. Data collected with 900 MHz antennas.

8
Graves and Cemeteries

People all over the world have long been interested in the use of geophysics to identify graves. Human interments can be found in many different environments and conditions: within formal cemeteries, in mass graves of disaster victims or from other actions, in out-of-the-way places where, for example, murder victims may have been hidden, and in the burials of prehistoric Native Americans or other groups that buried their dead in ways that may not be completely familiar to us today (Conyers 2006d). My students and I have conducted many hundreds of studies of interment sites and produced an array of images using GPR, some of which I discuss here. The variety of burial types visible with GPR is wide, and only some general themes are presented here, along with my interpretations of what the produced images can tell us about graves in a number of different contexts. All these examples of burials are from the United States and Australia; unfortunately, I am lacking any data from very old burials such as those from Bronze Age Denmark or Medieval England.

Recently I was part of a vigorous informal e-mail correspondence between geophysicists all over the world discussing different geophysical methods they had tried in the search for graves, mostly in Europe. I was struck by how few of the geophysical archaeologists who participated in these discussions had employed GPR, and even those who did were still primarily relying on the interpretive method of "anomaly hunting." With methods that do not allow for three-dimensional analysis (which is not the case with GPR), I suppose all one can hope for is that the anomalies detected are indicative of something different in the ground, which might be a grave. I do not follow that procedure here. The examples I present rely on interpreting GPR data using profiles and amplitude maps, and placing reflections into space once their origins are known. As readers will quickly see from the examples below, the array of grave types is vast, the reflections created from them numerous, and the GPR results often difficult to interpret without figuring out what it is in the ground that might be producing these reflection features. Mapping anomalies is not enough.

Formal Cemeteries

Collecting GPR data in cemeteries is often relatively easy, as the ground is frequently flat, and in formal cemeteries it is mowed and manicured. On the other hand, in some cemeteries, collection procedures are likely to be complicated by the presence of headstones, monuments, fences enclosing some plots, and a variety of other obstructions at or near the ground surface (Figure 8-1).

Reflections produced from burials are not usually simple, and it is a rare and joyous occasion when GPR data show simple point-source reflections from each casket or burial, with little variation in depth or burial type. When cemeteries of this sort are studied, interpretation is relatively simple and straightforward. A military cemetery at Fort Vancouver, Washington, holding the burials of American Civil War veterans who were given standard wooden coffin burials, all placed at the same depth, shows this type of uniformity (Figure 8-2). Each standard-issue coffin for these war veterans looks exactly the same in reflection profiles and can be readily differentiated from the more elaborate metal-lined coffin in the adjoining plot, which was not government-provided.

In the United States, there is a good deal of interest in using geophysics for grave mapping for a variety of reasons. Sometimes cemetery owners, both public and private, have lost records of burials and are unsure of where new graves can be dug safely. In other cases, we have located graves that were intentionally placed in out-of-the-way places, with little documentation, because the cemetery managers lacked funds for proper burials of the poor souls who had no family to bury them or wealth to pass on. We have also been asked to map graves in some older cemeteries so that headstones and other surface markers could be relocated following erosion, vandalism, and general neglect, as a way to help renovate the burial grounds. A few studies involved interesting historical research, such as trying to find the burial of a special mule, a mascot of some sort in a silver-mining town in Nevada in the late 19th century. We have located "secret" burials of American Indians who wanted to make sure their relatives' graves could be marked but not vandalized on traditional lands now in the hands of others. One Navajo Indian in particular was buried with his horse in a secret spot in the late 1800s so it would not be desecrated by European settlers during the time this tribe was being relocated to reservations.

Figure 8-1: Collecting 400 MHz reflection data at San Jacinto cemetery, Texas. Antennas had to avoid the monuments, which complicated profile collection. *Image courtesy of Jennie Sturm.*

Sometimes interesting mysteries from the past can be analyzed with GPR if there are extant historical documents that can reveal something about the individuals buried in the graves under study. One such story involved a man named Hartsville Jones, an important businessman, saloon owner, and liquor distributor in Denver, Colorado, in the 1860s. He had a complicated marital history, and GPR studies of his cemetery plot

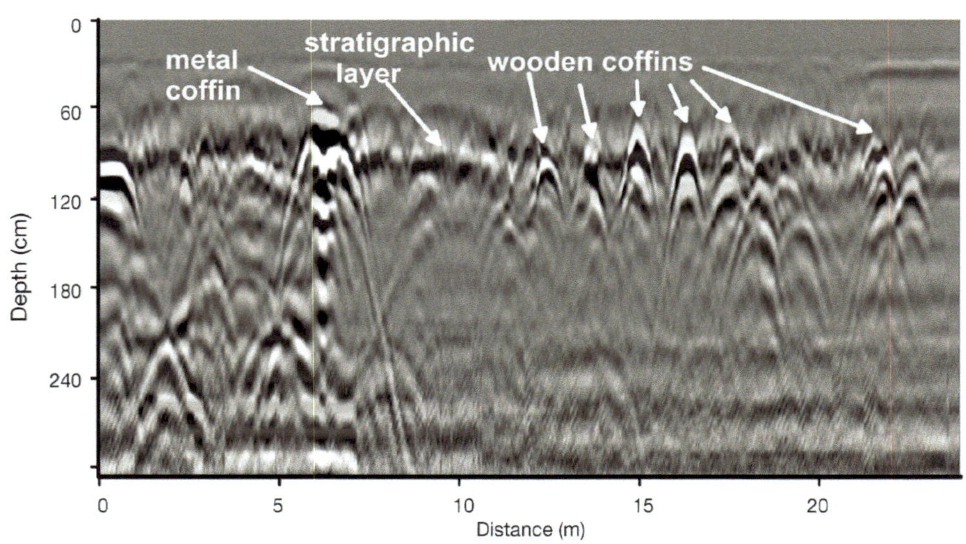

Figure 8-2: Reflection profile over coffins, with six burials. This profile was collected in the U.S. military portion of the cemetery at Fort Vancouver, Washington. The wooden coffins were buried at the same depth, and there is little variation in reflection types, in contrast to the non-military burial on the left, which is more elaborate and includes a coffin that contains metal, producing high-amplitude reflections. Data collected with 500 MHz antennas.

can add a few interesting details to that story. He bought a plot of land for his family's burials in 1868 in Riverside Cemetery and then constructed a large mausoleum (Figure 8-3), which appears not to have been used. There is no record of any interments in it, and the GPR profiles we collected on its floor and walls indicated no burials there. In the plot adjoining the mausoleum, the GPR amplitude maps showed the burial of Jones's first wife, Mary, who died in 1888, in the correct location marked by a "dual headstone" that had been carved for both Mary and Hartsville. After Mary's death, Mr. Jones was soon remarried to a woman named Sarah, who promised to have both their remains placed inside the mausoleum after their deaths. This was apparently never done. When Hartsville Jones died in 1900, a third Mrs. Jones appeared at the funeral and a dispute over his body ensued. Apparently he had been married to Sarah (his second wife) and the other woman (third wife?) simultaneously, to the shock of both when they appeared at the funeral. There is no cemetery record of the burial of Mr. Jones in this family plot, and our GPR shows this to be the case, as he was never interred next to the burial of Mary under the dual headstone. What happened to the remains of Hartsville Jones is still a mystery. In addition, there are many other unmarked burials in the Jones plot, which have been speculated to contain the burials of members of the third Jones family, from the "mystery wife" who showed up at the funeral and caused such a ruckus. Sarah Jones might also have been buried in this plot, but there is no record of her burial there in the cemetery records. In this historical mystery, the GPR was able to confirm some aspects of the historical record concerning the dispute between the two wives at Mr. Jones's funeral, and poses questions about the identities of all the other people who are buried in the Jones plot.

In general, the use of GPR for grave detection and mapping is often challenging, as ground conditions can make burials quite variable for a variety of reasons. Graves tend to be small "targets" for geophysics, and if there has been weathering and decomposition of remains and burial goods over time, they can be especially hard to identify. I have attempted to correlate GPR results with burial locations provided by "cadaver dogs," who have been called in to run around cemeteries barking and sitting on graves during data collection (Figure 8-4). This can be great fun but has proven less than successful, as the dog owners usually want GPR interpretations to be

Figure 8-3: Overlay of the GPR amplitude map on the Jones plot in Riverside Cemetery, Colorado. The Jones mausoleum is in the background, made of red sandstone. The mystery about where Mr. Jones's body was buried remains unsolved, as GPR shows no burial next to his wife Mary.

Figure 8-4: Porter, the cadaver dog, testing the results of GPR data. *Picture courtesy of Bonnie Guzman.*

generated as quickly as their dogs can point to certain locations. Real-time interpretation of graves during data collection, with results being correlated with pointing dogs, is a fool's errand and can only lead to frustration and a poor outcome. Reflections in profiles and amplitude maps need to be analyzed and visualized for a long time during interpretation. While I have not been involved with "witching" or other marginally scientific methods involving copper rods and forked sticks, I know others who have, and they have had equally marginal results as our dog tests. There might be something to these cross-discipline location methods, so I am keeping an open mind for future collaborations and also have plans for Porter the cadaver dog (Figure 8-4) in future projects.

Burial depth can vary a great deal in cemeteries. Depending on the time of year, gravediggers might expend more or less labor (especially in winter when the ground is frozen) in digging a burial shaft. Children or infants are often not buried as deeply as adults, and sometimes we have found "stacked" graves, where two or more individuals were interred at the same time and only one shaft was dug to save time, money, or space. Grave shafts in the United States typically are dug to 6 feet (about 2 m), but after caskets are placed in the shafts, the upper surface of the coffin that reflects most of the transmitted GPR energy is often calculated at 1.2–1.5 m below the ground surface, or even shallower, depending on how lazy the excavator was or how hard the ground.

In this chapter, I offer examples of a number of different burials in different ground conditions where burial practices are known or can be inferred. I lack any good data on very old burials in humid areas, but it is my understanding that in those graves, all human remains, and most wooden caskets or other items, have usually decomposed and weathered appreciably. Where I have been confronted with that type of problem, all that remains to indicate a burial is the shaft itself. In some cases, a reflective layer might be visible from the human remains that are chemically different from the surrounding ground and could produce a weak horizontal reflection. If the vertical shaft was cut through layered soils or sediments and that material was homogenized and then used as backfill, the incision might be visible to locate a grave, but little else. In cases like this, the cadaver dog method, or some other geophysical technique that can detect subtle soil differences, might be preferable to GPR. As GPR images are almost always displaying differences in water due to changes in the constituents in the ground, if there are no differences among the materials in the ground, GPR will most likely show no reflections that indicate burials. If that ground is also not layered with sediment beds or soil units, and human remains and associated burial artifacts have decomposed, GPR is probably an unsuitable method for detecting and mapping burials. In younger graves where coffins are still intact, the void spaces within them generate very high-amplitude reflections, and the graves are readily visible, if not the human remains within those coffins.

As with most GPR images, an understanding of how radar energy is transmitted, reflected, and recorded is the key to a successful interpretation of burials. As with most buried features, radar waves can be reflected from some surfaces back to the surface recording antenna, but from others it can be attenuated or reflected away from the surface at an angle and remain unrecorded. Void spaces produce good reflections, but a complication arises as the energy passing through them speeds up, distorting any reflections below due to velocity variations, which tend to "pull up" reflections (Figure 8-5). An analysis of the surrounding ground is also integral to interpretations, as reflections from truncated layers can point to the location of vertical burial shafts. A fairly simple recent burial is shown in Figure 8-5, where a vertical shaft contains stratigraphy that is layered differently from those units adjacent to it, which remain undisturbed. The wooden casket in this example has an arched lid, and only its very top produced a reflection, at the contact of the void space and the lid. The lid's sloping sides reflected energy away, and those waves were not recorded. The reflection from the base of the casket is shown directly below the reflection from the crest of the lid, and that reflection was pulled up as that energy passed through the overlying void space, was reflected, and moved back to the ground surface at the speed of light (Figure 8-5).

Other complications can occur when coffins are interred in shafts that are lined with some impermeable material and covered with a slab (Figure 8-6). In this case, the void space inside the narrow crypt allows air waves to be generated from the interior corners at the intersection of the sealed interment and the emplaced container that holds the human remains. The vertical lining of this crypt does not produce a reflection, as all radar waves transmitted from the surface antenna pass by those surfaces without being reflected or are reflected away from the ground surface. The container for the human remains is also almost invisible, and any waves that might have been reflected from it have been overwhelmed by the corner reflections and air waves produced inside the crypt (Figure 8-6).

Caskets that have not weathered and collapsed will retain void spaces, and those air pockets can often be identified in reflection profiles. Transmitting antennas generate pulses of energy that are propagated into the ground as a sine wave. The initial deflection of a wave, after it has been reflected from an interface where there is a velocity decrease, can be viewed in an individual trace as deflecting either to the right or the left of the mean (Figure 8-7). The first deflection, which defines what is called the *polarity* of the wave, can be in either direction, depending on the antenna electronics and which antenna is being used as a transmitter and which as a receiver. As a result, there is not a standard, "normal" polarity common in all GPR system antennas. But a "local normal" polarity can be defined by looking at the deflection of the first wave recorded (the ground wave)

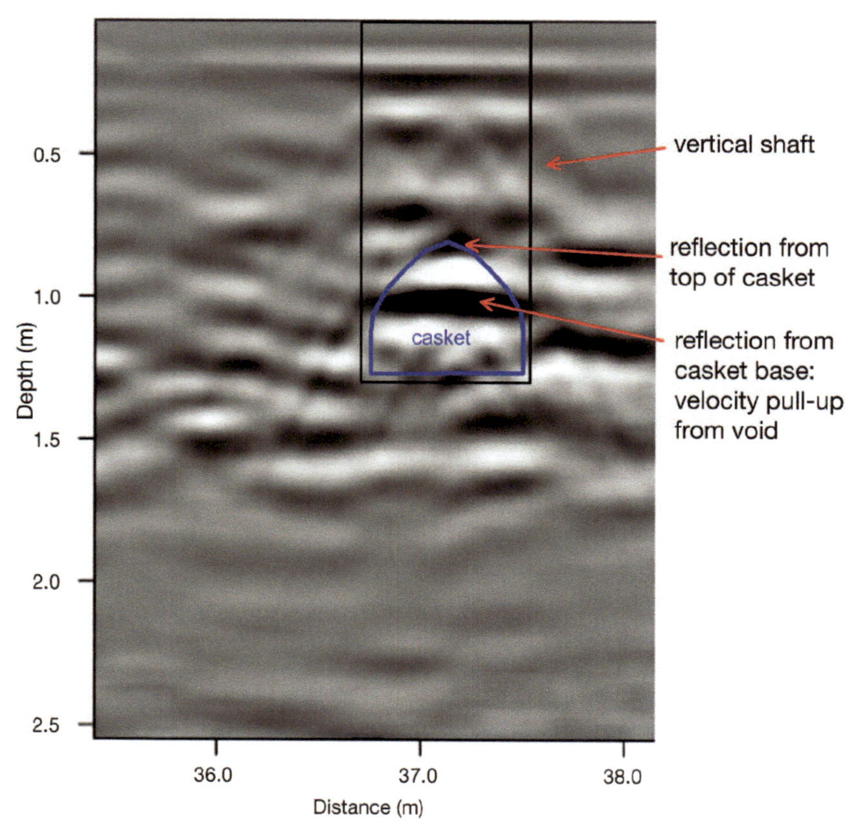

Figure 8-5: Reflection profile showing a reflection from the crest of a casket's arched lid. The arched lid deflected energy and a velocity pull-up from the casket bottom was produced. Data collected with 400 MHz antennas at Rose Hill Cemetery, Commerce City, Colorado.

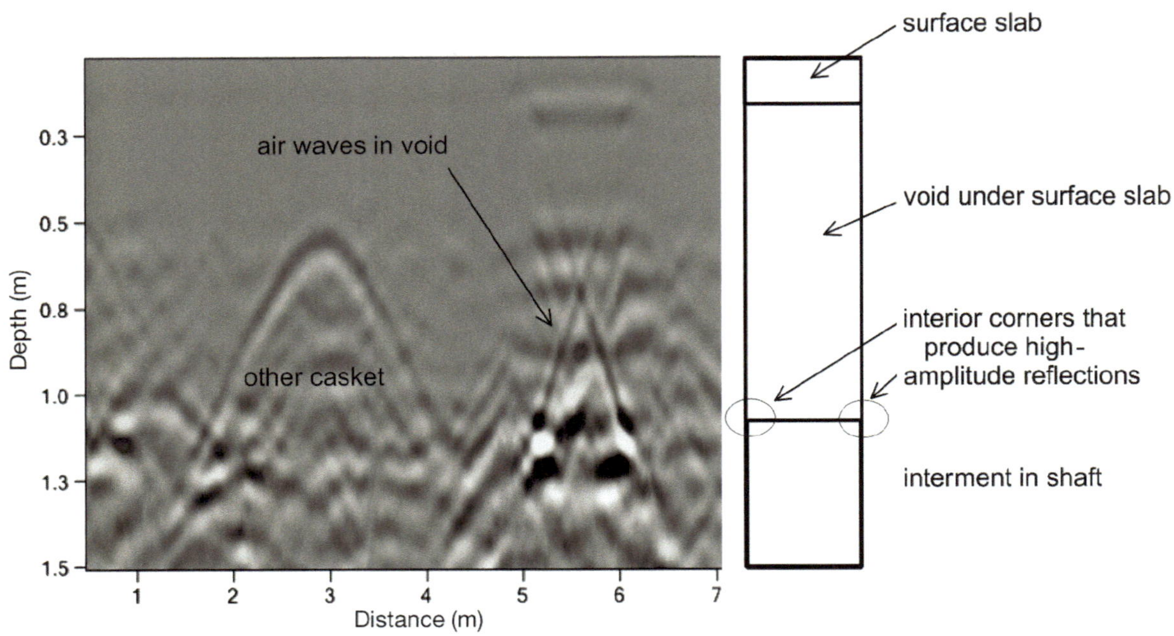

Figure 8-6: Reflection profile showing a complex series of reflections generated from the interior corners of a void space within a crypt. Data collected with 400 MHz antennas at the Pala Cemetery, Fallbrook, California.

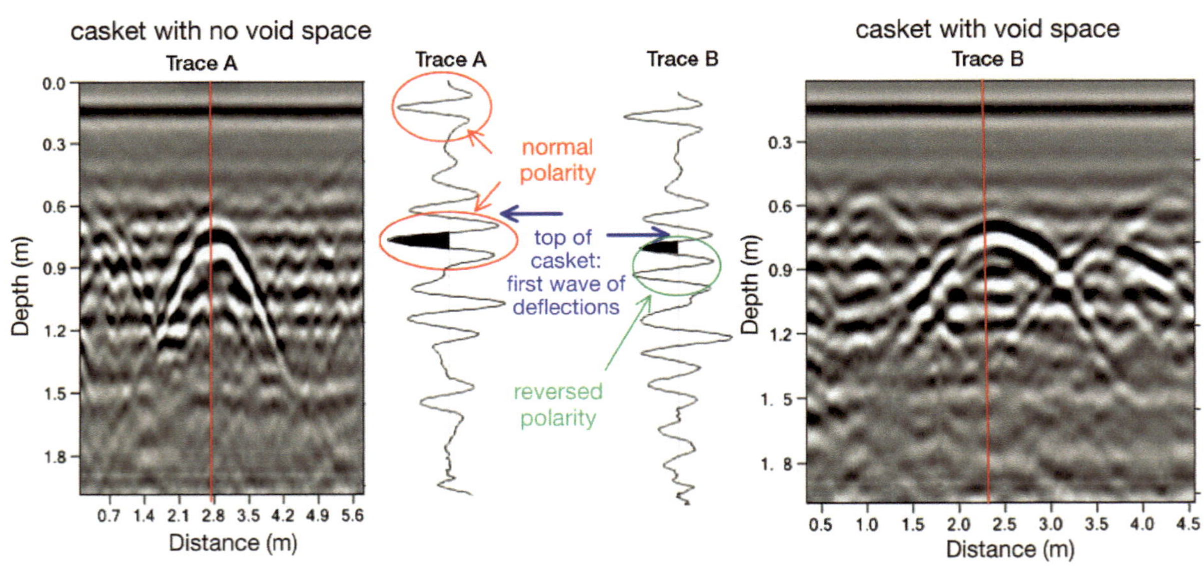

Figure 8-7: Examples of normal and reversed polarity of reflected waves. Depending on whether the wave was created at a void space (profile on right and in Trace B) or by some material where the wave slowed down, as with a collapsed casket (left profile and Trace A), there will be normal or reversed polarity of waves. Data collected with 400 MHz antennas.

in any trace in a reflection, and that can be defined as "normal." Every time a reflected wave is produced in the ground due to a velocity decrease across a boundary, this same polarity can be seen in the first deflection of the reflected wave in the trace. However, if an interface in the ground causes the propagating radar energy to speed up (as in a void space), the reflected wave will exhibit "reversed" polarity, and the first deflection of the recorded wave will move in the opposite direction. In cemetery studies, an analysis of the polarity of the reflected waves can be used to determine whether a casket contains air or has collapsed or filled with sediment. Void spaces in a casket cause the speed of the waves to increase to the speed of light, and the interface between the casket material and the void space will be recorded as a reverse-polarity wave. The reflected waves from a collapsed casket would have normal polarity.

An understanding of what causes polarity differences in reflected waves can be easily seen in an example of two different caskets in a cemetery, one of which has collapsed, and the other that retains a void (Figure 8-7). In Trace A in Figure 8-7, the local normal polarity can be seen in the ground wave, and the reflection from the top of the casket shows a similar polarity, indicating a collapsed or filled casket with no void space. As the reflection profile program is assigning colors to the deflection of waves that move either to the right (as white) and or to the left (as black) and shades of gray in between, the normal-polarity waves generated from the collapsed or filled casket are visible as white–black–white from top to bottom. In Trace B, the reverse polarity of the reflection can be seen in the trace and also in the reflection profile as black–white–black. Once normal and reversed polarity have been determined, a visual analysis of the waves in reflection profiles can define which caskets retain a void space and which do not. This is good information to have, as it can be used to determine the age of burials or the type of casket construction.

In one cemetery, we were able to differentiate the recently buried caskets that still contained void spaces from those that had collapsed, their void spaces probably filling with sediment (Figure 8-8). The tops of the collapsed caskets can sometimes be seen as "sagging" in this profile, and the reflected waves are normal polarity (white–black–white). In contrast, the recently buried casket to the right in Figure 8-8 shows a reversed-polarity reflection (black–white–black), and its lid is intact and has not "sagged."

Metal, which reflects all energy that encounters it, produces multiple reflections in the ground, and caskets that are lined with metal or are composed entirely of metal produce distinctive reflections (Figure 8-9). Lead-lined caskets at one time were common and produce similar reflections.

Sometimes a casket can produce very distinctive reflections from its top as well as its bottom, if it is composed of a material that allows good radar energy transmission (Figure 8-10). The interface between the

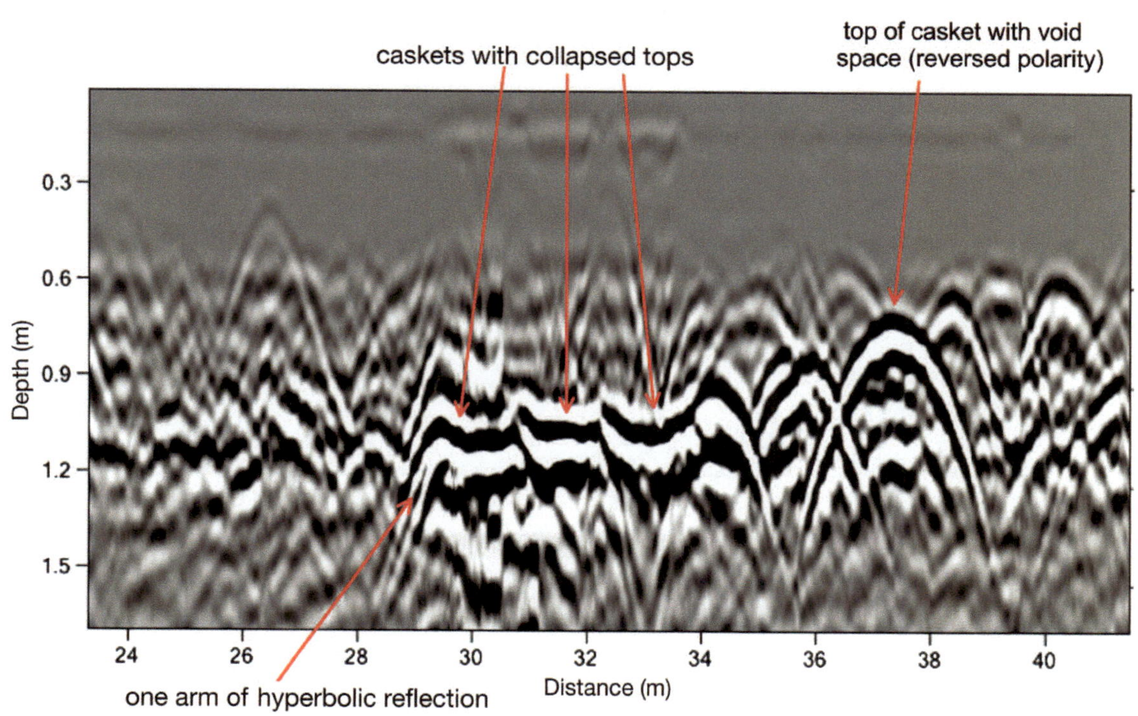

Figure 8-8: Reflection profile of multiple recent burials. A family plot produces reflections from coffins placed very close together. Data collected with 400 MHz antennas at the Pala Cemetery, Fallbrook, California.

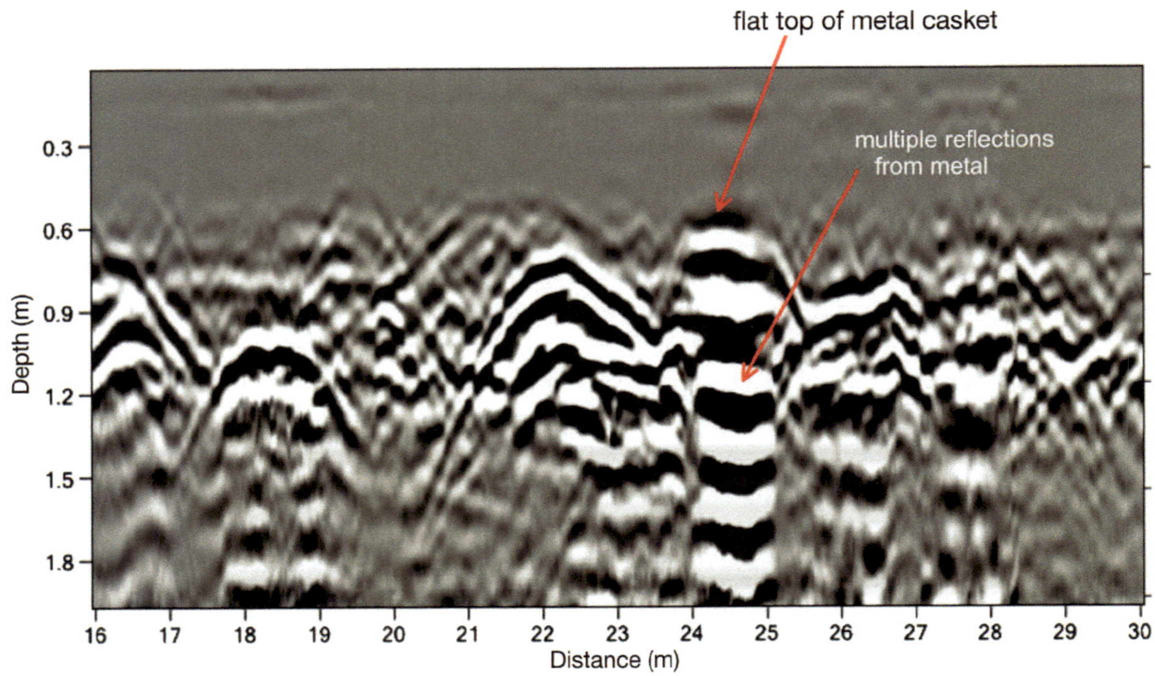

Figure 8-9: Multiple reflections from a metal casket. Data collected at the Pala Cemetery, Fallbrook, California.

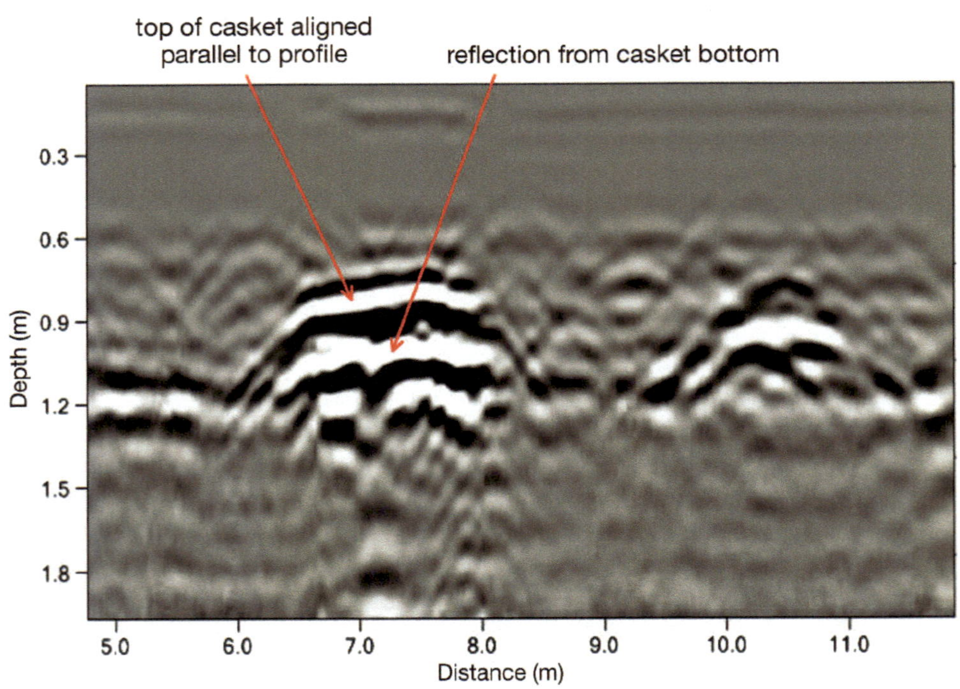

Figure 8-10: Reflections from the top and bottom of a casket. Data collected at the Pala Cemetery, Fallbrook, California.

casket lid and the void space produces one reflection, and another is produced at the interface between the void space and the base of the casket. Theoretically, human remains within that void space would also be visible, but I have never been able to document a reflection from bodies within a casket.

It is not unusual to find concrete crypts that are placed in the ground to hold coffins, or to find surface concrete slabs that cover some crypts. These outer burial container crypts are especially common in some Jewish cemeteries in the United States. The tops of the crypts, if buried, are quite visible as distinct high-amplitude planar reflections (Figure 8-11). Surface concrete pads often distort any reflections below them that might have been produced from crypts or caskets below (Figure 8-10). It is also not uncommon, in cemeteries where burials are very densely interred, to record many reflections from burials that are not in the plane of the antenna transect. Reflections from these "out-of-plane" features are usually lower in amplitude and often are recorded as a dizzying array of reflections (Figure 8-11). It can be difficult to determine whether the lower-amplitude reflection hyperbolas were produced from burials directly below the antenna transect and are just older and composed of more decayed remains, or are from some distance adjacent to the profile. If many profiles are collected parallel to one another in a grid and are spaced 50 cm or less apart, when all reflections are analyzed in amplitude maps, the resampling and gridding of the reflections tend to display the reflections derived from burials directly below, as they have the highest amplitude. The weaker reflections from out-of-plane can be effectively filtered out during this data processing and display.

GPR faces a number of challenges in older cemeteries where caskets have collapsed and human remains are mostly decomposed. If all remains are gone, the only feature still present that could be visible with GPR may be the subtle settling of the surface soils created as the caskets slowly collapsed (Figure 8-12). Sometimes totally collapsed caskets are still visible as aerially restricted planar reflections produced from only a thin layer of the human or other material remains. This type of burial feature is very common in historic cemeteries in the United States, where simple wooden coffins were used and collapsed and deteriorated soon after burial (Figure 8-12).

If collapsed coffins, which produce only one planar reflection as shown in Figure 8-13, are recorded by the surface antenna moving over it at an oblique angle, the reflection can dip and not appear as a "normal" hyperbolic or planar reflection. These types of reflections can be difficult to interpret unless there are many other profiles of the same feature in a tightly spaced grid, which allow many reflections to be recorded in a number of profiles. Amplitude mapping can then produce a three-dimensional view of the buried feature.

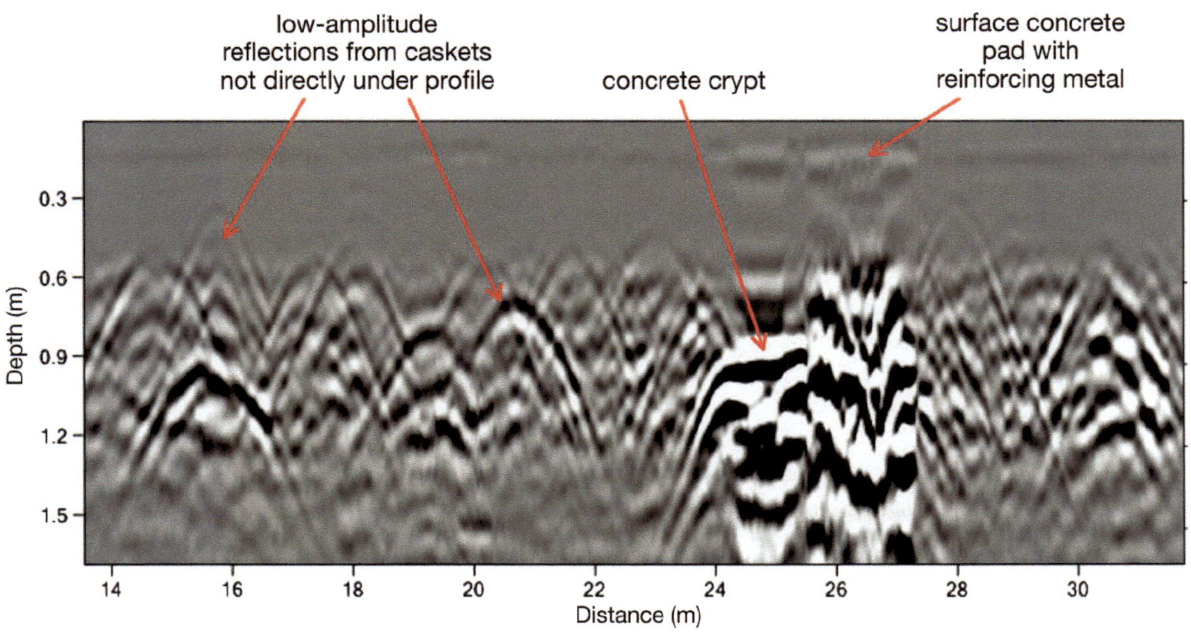

Figure 8-11: Two burials in concrete-lined crypts. One burial has a surface cover, and the other a cover of concrete that is buried about meter in the ground. Data collected with 400 MHz antennas at the Pala Cemetery, Fallbrook, California.

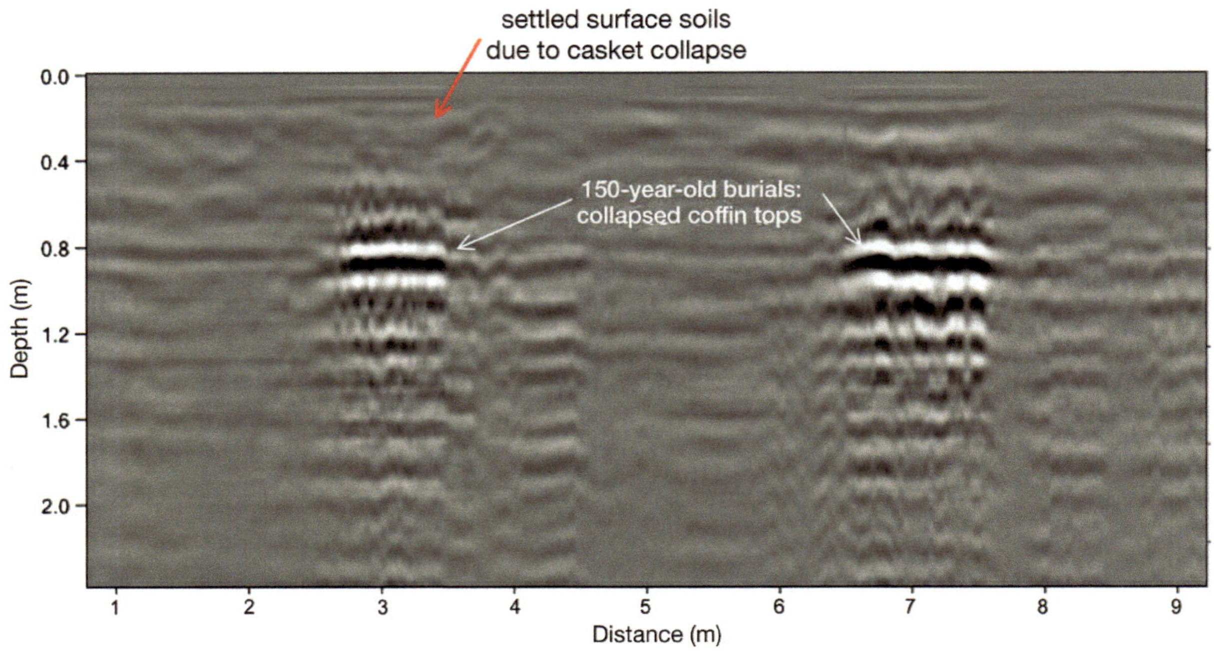

Figure 8-12: Horizontal reflections from collapsed coffins. Data collected with 500 MHz antennas, Paria Historic Cemetery, southern Utah.

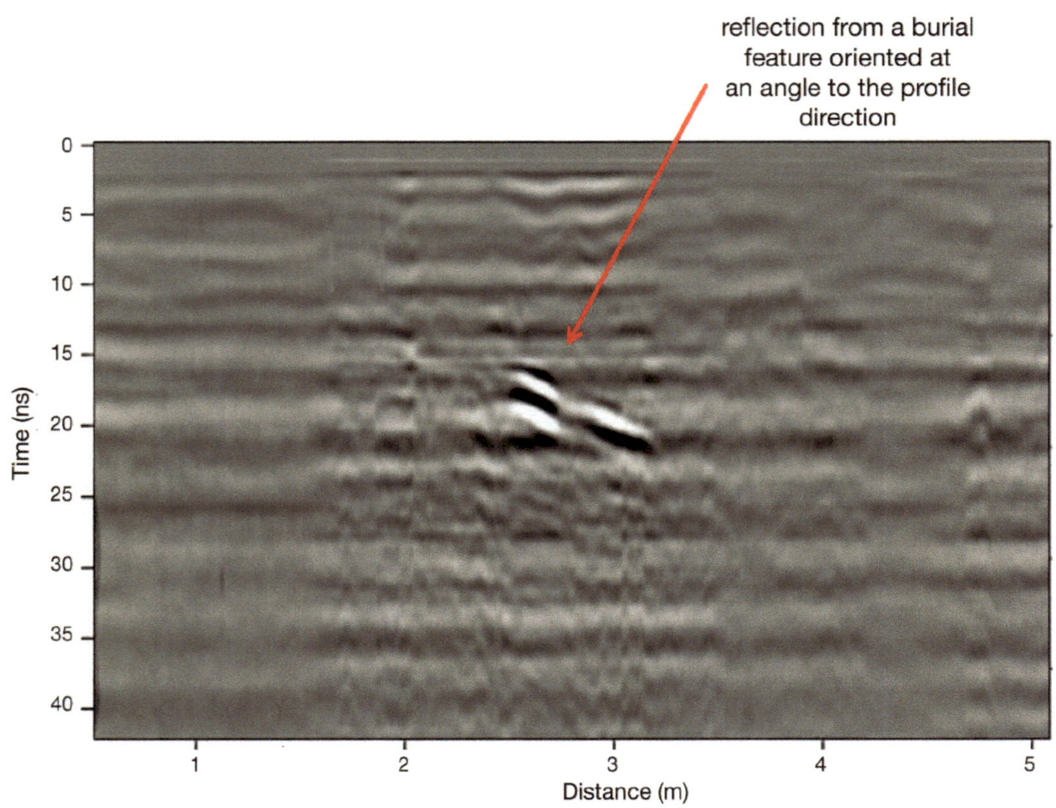

Figure 8-13: Dipping reflection from a burial out of the plane of the antenna transect, which produces this geometrically complicated reflection. Data collected with 500 MHz antennas, Paria Historic Cemetery, southern Utah.

Caskets that are almost completely deteriorated may still potentially produce low-amplitude reflections, some of which are planar and some hyperbolic in shape, depending on what remains in the ground to reflect energy (Figure 8-14).

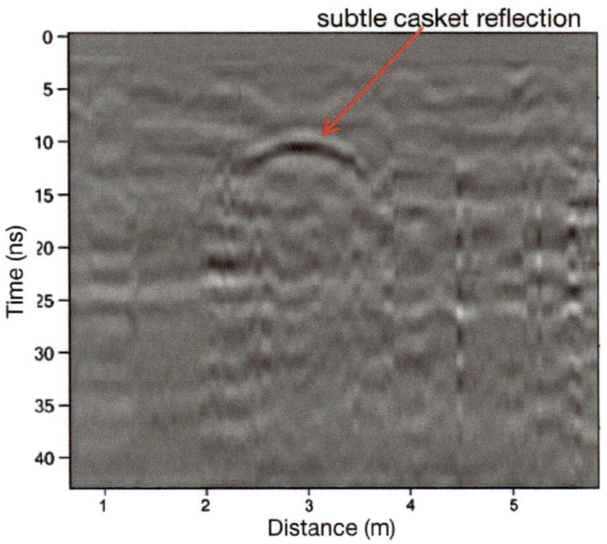

Figure 8-14: Very low-amplitude hyperbolic reflection from a mostly decomposed burial. Data collected with 500 MHz antennas, Paria Historic Cemetery, southern Utah.

When nothing physically remains of a burial, the only way to identify graves is by stratigraphic analysis and interpreting reflection profiles to look for differences in the ground between burial shafts and the surrounding ground (Figure 8-15). Slumped surface soils, caused by coffin collapse, and other difficult-to-interpret reflections within the vertical shaft (which could be decomposed remains or portions of a casket) will likely point to the location of a burial. I never feel good about putting an "X marks the spot" on features such as the ones shown in Figures 8-14 and 8-15, as there is no definitive reflection from the actual buried feature of interest. But if this is all that one has to identify a possible burial, then it is perhaps best to show where it is on a map and indicate your uncertainty in some other way.

All the examples shown above are of reflection data collected within an area known to be a cemetery, whether active or historical. Sometimes nontraditional burials studied with GPR can be interesting, as was found in northernmost Manhattan Island, New York. In a busy urban space in the Harlem neighborhood of New York City, some local researchers found records from the late 17th century that discussed both slave and free African

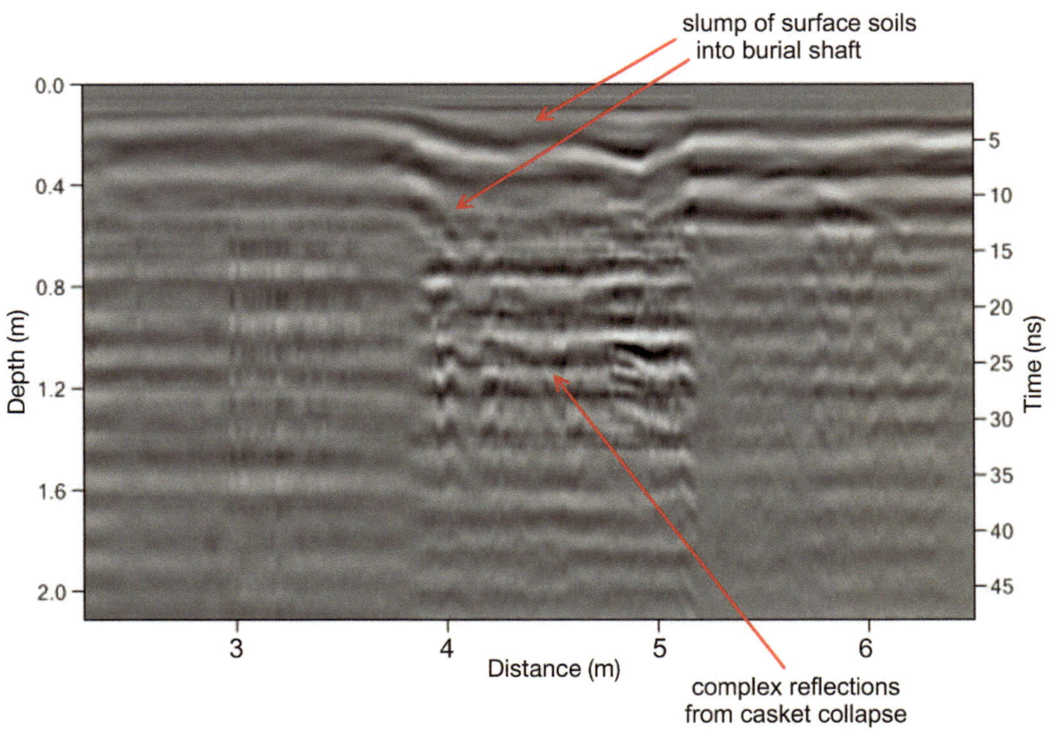

Figure 8-15: Surface soil units that have slumped into a burial shaft. This feature was caused by casket collapse below. Data collected with 500 MHz antennas, Paria Historic Cemetery, southern Utah.

burials in what was then the wilderness of northern Manhattan Island. This area later became incorporated into a cemetery associated with the Dutch Reformed Church. It appears from those records that all burial in this cemetery ceased sometime in the mid-19th century, but the actual location of the burial ground remains in question. Other records indicate that some graves might have been disinterred and moved elsewhere when the associated church was moved. No one knows where those human remains might have been placed, or if there are still human remains in their original resting spots. Some researchers say the most likely spot of the original burial ground is under an active bus depot, but the operators of the transit authority deny this (for no reason other than they don't want to be bothered). A likely area that might hold some of the original or reinterred graves is in a small grassy park next to the bus depot, which was chosen for a GPR survey (Figure 8-16). The test area appeared prospective, as this ground appears to have never been built upon and is on a river terrace above the Harlem River where flooding would have been minimal in the past.

The amplitude maps show two distinct parallel trenches that had been incised through the well-layered fluvial sediments of the river terrace (Figure 8-17). Those trenches are visible as areas of little reflection, as they are composed of homogenized, nonreflective sediment. Adjacent to the trenches are the more highly reflective intact fluvial sediments of the river terrace. Within the trenches are many distinct reflection features (Figure 8-18), which appear much like caskets studied elsewhere (e.g., Figure 8-8). The origin of these trenches, and the objects within them that generate reflections much like caskets, remain unknown. It is possible that trenches would have been dug to expedite the reburial of remains that were disinterred from the cemetery in the 19th century, and perhaps the caskets were then placed in those trenches. At present, the local historians are pushing to have this area, and the bus depot, studied more thoroughly in the hope of locating the missing remains of these people who were marginalized during life and whose final resting place remains unknown.

Organized cemeteries often tend to have burials in rows, with walking paths between them. Cemeteries of this type are quite easy to survey, as antenna transects can be located along the rows, often between headstones placed in rows. The only problem one might encounter at sites like this is the obstacle of fences around burials or family plots, which necessitates stopping and starting collection lines and having to leave some areas unsurveyed. The amplitudes of reflections from buried coffins often denote the types of materials used in coffin construction, with metal and void spaces producing the highest amplitudes, and lower amplitudes from either wood, collapsed coffins, or those that have been in the ground the longest and have deteriorated (Figure 8-19).

Older cemeteries, and those where burial was less formal, can produce a more random orientation of interments (Figure 8-20). At a pioneer cemetery dating from 1840 in

Figure 8-16: Collecting reflection profiles in a small park in Harlem, New York City. This open area is adjacent to the bus station in the background.

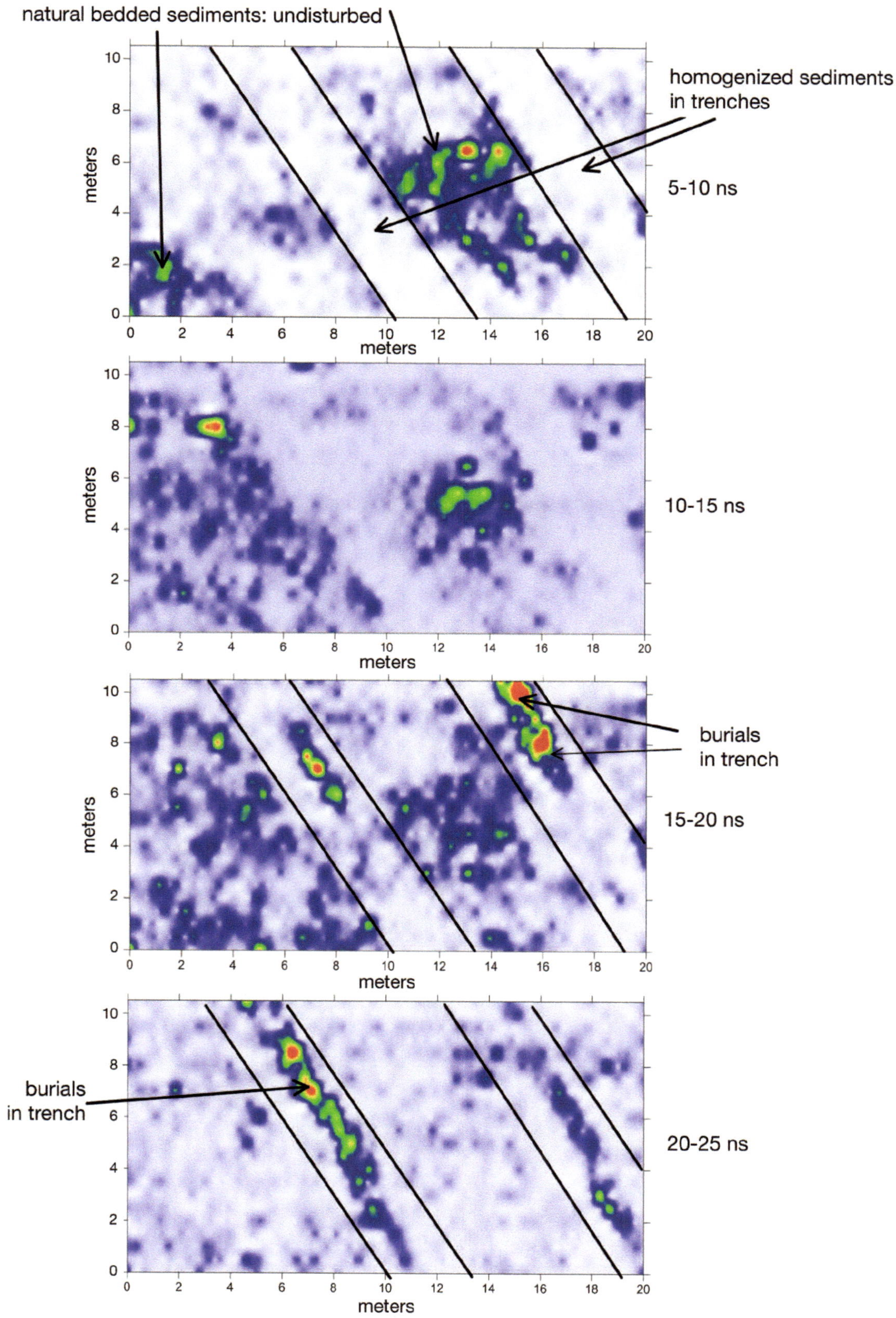

Figure 8-17: Amplitude maps showing two parallel trenches with burial-like reflections in them. Data collected with 400 MHz antennas, Harlem, New York City.

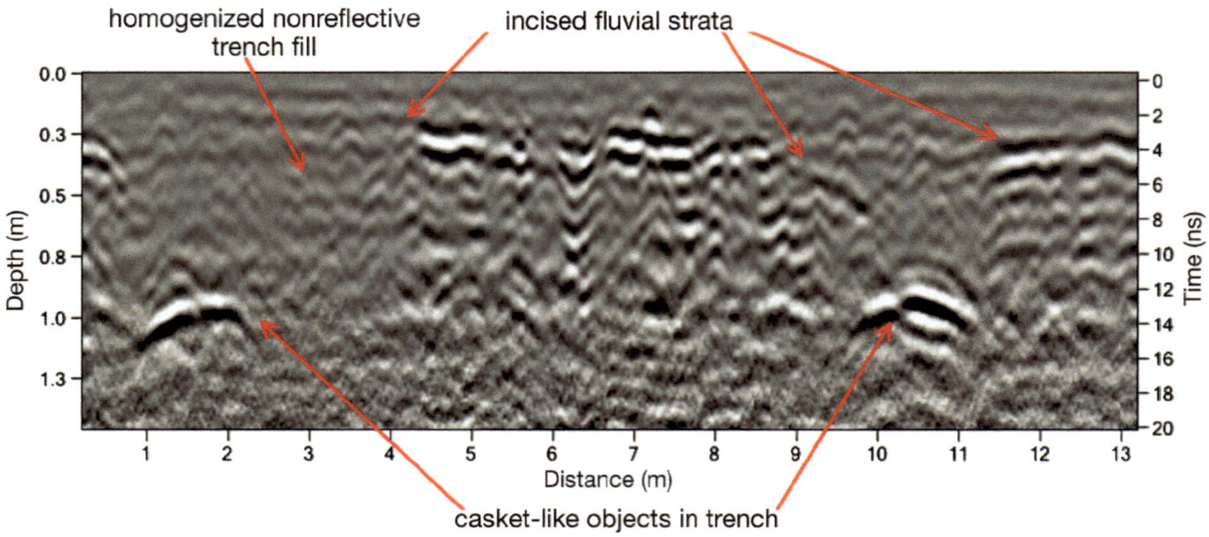

Figure 8-18: Reflection profile showing low-amplitude reflections in a homogenized trench fill overlying casket-like objects. Data collected with the MHz antennas, Harlem, New York City.

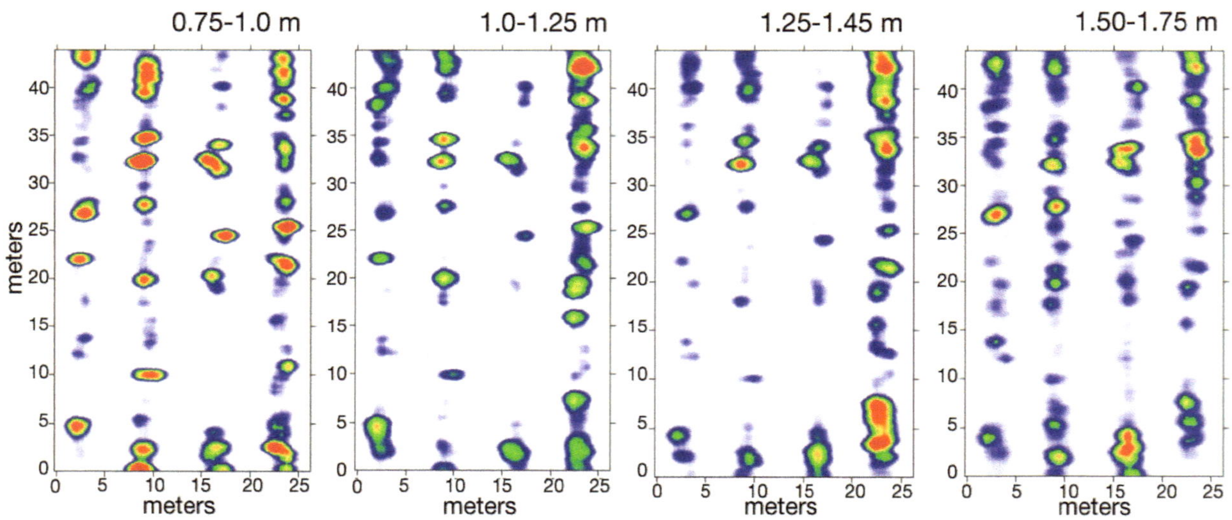

Figure 8-19: Well-ordered graves in a formal cemetery, visible as distinctive reflection amplitudes. Data collected with 400 MHz antennas at Rose Hill Cemetery, Commerce City, Colorado.

Lawrence, Kansas, there was a general orientation of graves in rows, but the GPR amplitude maps were complicated by the roots of a large tree just to the east of the cemetery. Its roots produced high-amplitude reflections in one portion of the grid that were readily mapped in horizontal amplitude slices. In this old cemetery, some of the graves were so deteriorated that they produced only very weak reflections and are shown as low amplitudes, colored light blue on the amplitude maps (Figure 8-20).

GRAVES BENEATH FLOORS

When I was working in highland Ecuador on a project that involved Inca sites, the local landowner of a very old hacienda asked me to survey the floor of his family's pri-

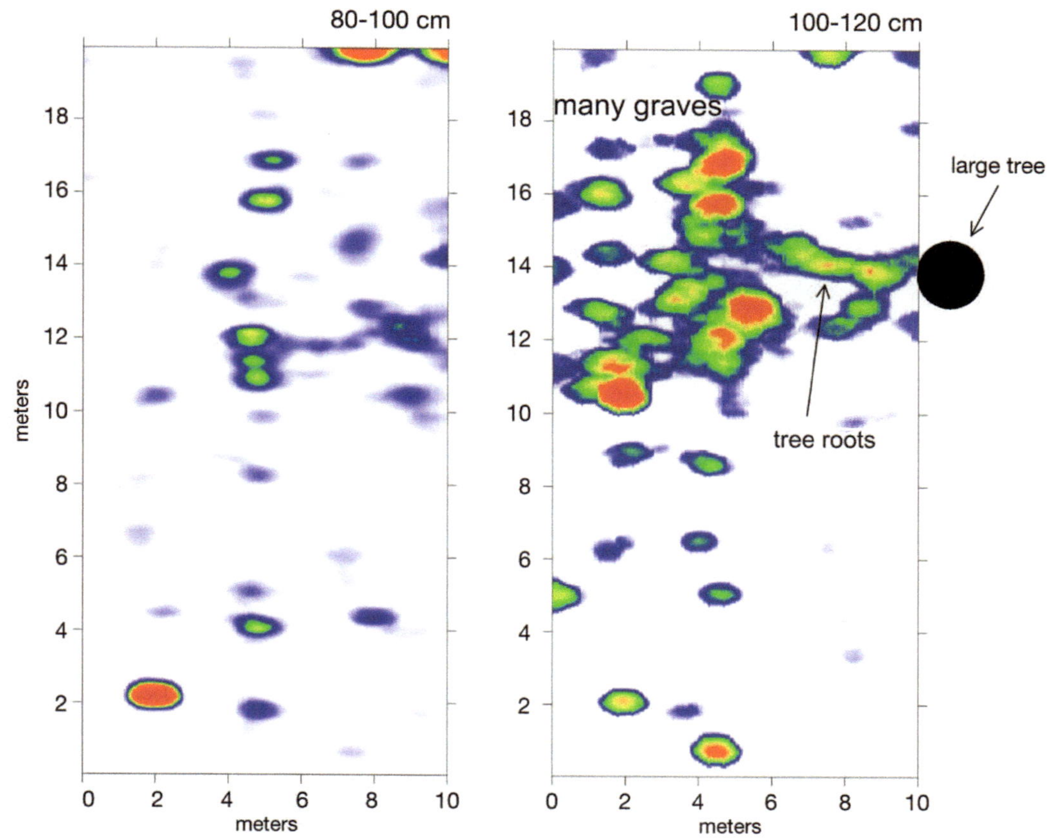

Figure 8-20: Amplitude maps of burials at a historic cemetery. This cemetery from the early 19th century in Lawrence, Kansas, shows tree root reflections within the graveyard. Data collected with 400 MHz antennas.

Figure 8-21: Collecting 400 MHz reflection profiles on the floor of an 18th-century chapel, Ecuador.

vate chapel (Figure 8-21). He had heard stories about his ancestors having been buried in this chapel but had no records of where they might be or if these stories were true. The church was reportedly constructed sometime in the 1700s, but its floor was composed of more modern cut stones.

The reflection profiles showed a few distinctive burials below the floor, but what was interesting was the amount of fill material that had been placed in one part of the chapel to level the floor, presumably during its construction (Figure 8-22). No records exist of how this chapel was built, but when I mentioned this floor fill to the hacienda owner, he recalled reading that the chapel was built on a spot that was considered sacred to the local indigenous people. This was a common practice when the Spanish were consolidating their power in the area and the Roman Catholic Church was attempting

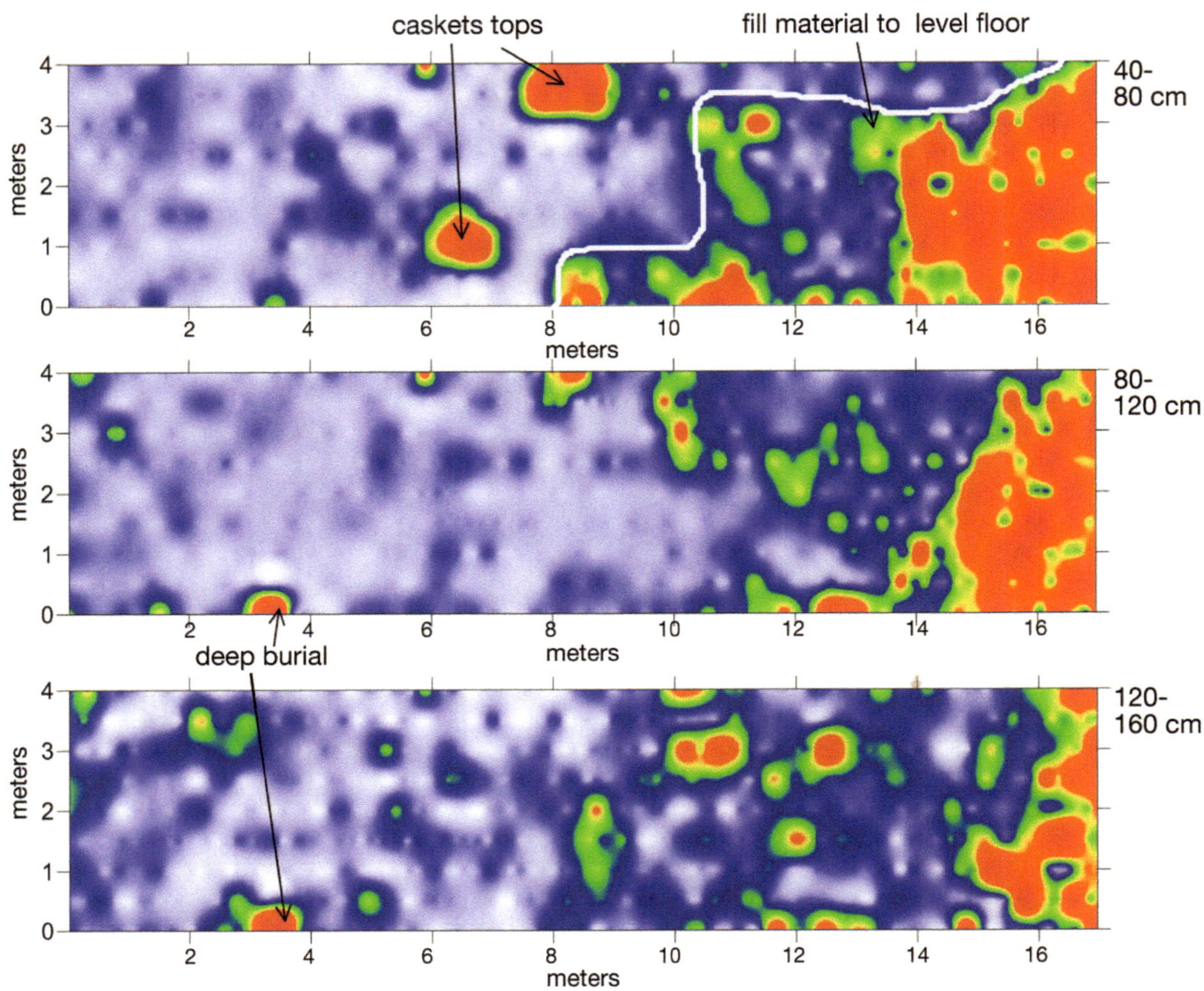

Figure 8-22: Amplitude maps from burials under the floor of a chapel (shown in Figure 8-21), Ecuador.

to supplant the religion of the people under their control. This information, and the leveling procedures used to build the chapel, suggest that if this story has any truth to it, the ancient sacred spot on which he church was built was a topographic rise (common in indigenous beliefs in South America), which needed to be expanded with fill in order to construct this church to the size necessary to meet the needs of the hacienda's Spanish occupants. The age of the burials beneath the chapel floor is not known, but could possibly be very old and could even predate the construction of the structure. Two of the graves are very shallow and appear to be elaborate, with void space still intact, producing the high-amplitude reflections with reversed-polarity reflections. These are probably not more than a few hundred years old. One deeper burial is much less reflective, and smaller in size, which suggests an earlier burial of indeterminate age.

BONES AND FORENSIC STUDIES

I have never been able to detect bones or actual human remains themselves with GPR. Others have reported to me that, in certain conditions of very minimal weathering or dry ground, and in permafrost areas, human bodies and even partially decomposed remains might be visible using GPR methods. In a cave in the Carmel Mountains of northern Israel, a number of Late Natufian burials, with well-calcified bones, had been excavated in the floor of the cave, and we knew of locations where there were still skeletons intact (Figure 8-23). Using 400 MHz antennas, I collected profiles over these known skeletons, which were just 30–40 cm below the surface, processed those data in every way I could think of, and still failed to produce any images of the skeletons that were discernible as different from the natural stratigraphy of the cave (Figure 8-24). In this dry cave,

Figure 8-23: Late Natufian burial in Israel used as a test for GPR identification of bones. The bones of this skeleton at Raqefat Cave, Israel, are not distinct enough from the surrounding matrix to produce radar reflections (see Figure 8-24). *Photo courtesy of Dani Nadel.*

the desiccated bones are chemically and physically much the same as the cave floor sediment; in terms of water saturation, there is no difference between them and therefore no radar reflection occurs.

I have been involved in a few forensic cases where it was hoped that GPR techniques could find recently murdered victims, whose bodies would conceivably produce very high-amplitude reflections. In none of those cases was I able to find the victim's body; presumably the criminals had not placed them in the locations I tested. Much effort was expended in digging up other materials that I could see in the ground in both profiles and amplitude maps, but no bodies were discovered. To test what a recent burial might look like, Sam Connell and I buried a water-soaked baby diaper, wrapped in plastic, in dry coral sand in Hawaii. We hypothesized that this target would have similar water saturation to a recently deceased human body, and contrast with the surrounding sand ought to produce a reflection. As expected, this feature produced the high-amplitude reflection (Figure 8-25) visible in profile.

I have read news stories of GPR being used to find murder victims buried in concrete, and I would expect that they looked much like the model we created in Figure 8-25. A notable example of GPR's success in such a

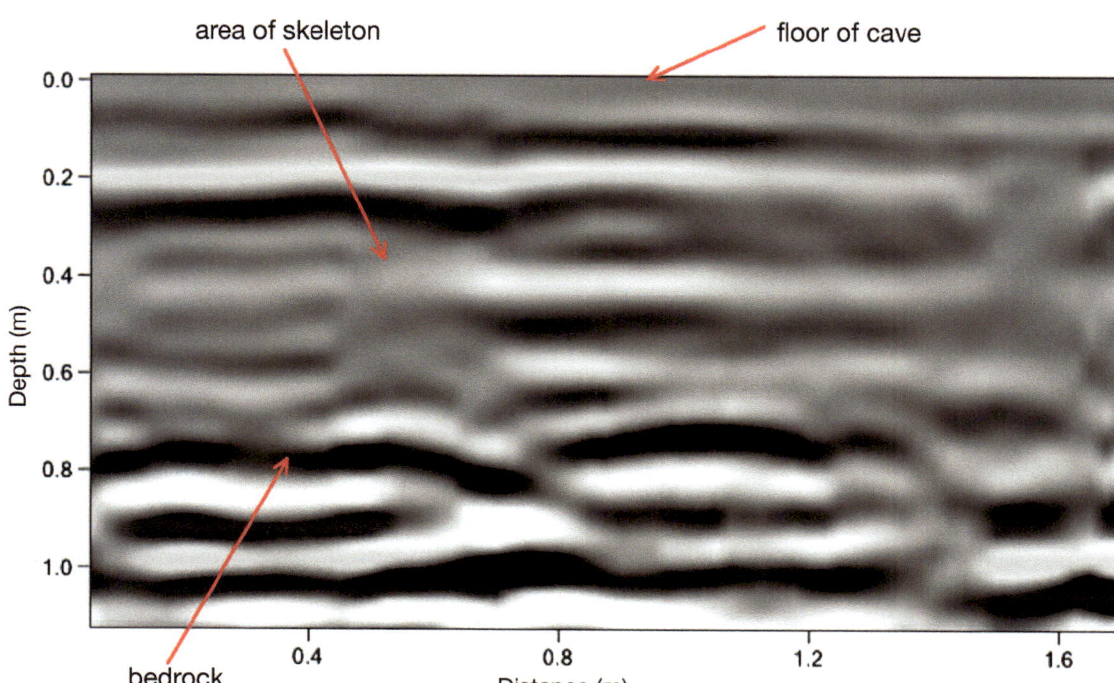

Figure 8-24: Reflection profile over bones in a Late Natufian–age skeleton, from a test of a skeleton at Raqefat Cave, Carmel Mountains, Israel (see Figure 8-23). No reflections were generated from it. Data collected with 400 MHz antennas.

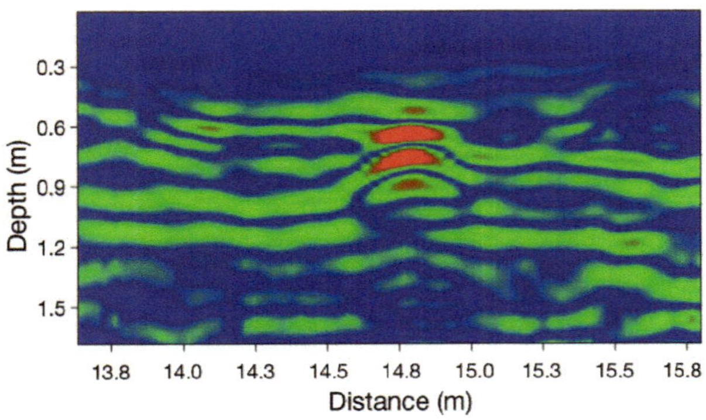

Figure 8-25: Reflection profile over a wet diaper used as a model for a human body. This saturated object was buried in coral sand as a model for a recently deceased human body. Data collected with 400 MHz antennas in Oahu, Hawaii.

search was a story from the late 1990s in Phoenix, Arizona, where a man's wife was reported missing about the same time that their neighbors noticed him pouring concrete in his backyard for a new patio. Her body was easily located using GPR below the newly poured concrete, and the husband was convicted of first-degree murder. Once I saw a television news story of a consultant to the police who was using GPR in an attempt to find the bodies of many missing people, all presumed to be the victims of the serial killer John Wayne Gacy in Chicago. I remember watching this story on a national broadcast and inwardly groaning when the police used the GPR profiles to identify buried "anomalies" and then dug them up, live on camera, only to find recent garbage, but no bodies. Here is how the *Chicago Tribune* (front page, Nov. 24, 1998) discussed that GPR work (with my comments inserted in brackets):

> Monday's radar test, done late in the morning by a man hunched over a yellow contraption he pushed slowly across the ground like a lawn mower, revealed two other "suspicious spots" in the fenced grass lot in the 6100 block of West Miami Avenue, police spokesman Pat Camden said. In a nearby funeral home, investigators ate lunch Monday and huddled over a printout of the new scan as they analyzed the results. They decided to dig in the grassy lot, proceeding to the parking lot only if they turned up something in the yard. In the lengthening afternoon shadows, police turned up only the one marble, the flattened sauce pan and some chunks of concrete. . . . Police would not release the name of the New Jersey company that conducted both radar scans. [LC: that was very prudent!] But experts in ground-penetrating radar systems said Monday that the confusion over what the radar scans showed—or didn't show —isn't at all surprising. The radar uses radio waves to detect buried objects in the soil, just as X-rays image the internal structure of the body. There are many variables to a good reading, however, from wet soil to the amount of metal in the ground. "Generally you cannot tell one object from another in specific terms. [LC: Very true.] If I look at a (radar) record and see a blip on the record, I couldn't tell you if it's a bone or plastic pipe or a rock in that one piece of information," said Peter Annan, president of Sensors and Software, which makes ground-penetrating radar units. While the radar measurement is very precise, say experts, it's the interpretation of the data that is inexact. [LC: That is for sure in this case.] A metal pipe shows up on the radar screen like a frown-shaped series of markings, said Steve H. Danbom, a geophysicist with Conoco, an oil company in Houston. [LC: Oh my, now GPR point-source hyperbolas are described as "frowns"!] But a human skeleton is far less distinct; in fact, bones typically don't register on the radar screen, but the disturbance in the soil from digging the grave does show up as "a gray blobby nothing," Danbom said. [LC: Not a great description of a GPR profile; this pains me to hear about "blobby nothing."] Several experts said Monday they wouldn't want to work under the pressure-filled conditions created by the Gacy dig. They said that an accurate interpretation of scan results requires a good mapping of the site as well as at least 24 hours to analyze the data. [LC: These GPR consultants likely fell into the "interpretation trap," giving out results too quickly. Timely, but also well-considered interpretation is important.]

This was not good public relations for the GPR method. I have tended to stay away from forensic applications of this sort, as they can be emotionally wrenching for all concerned, especially when GPR methods produce this kind of outcome. There are other GPR experts who specialize in high-profile forensic applications, and I wish them all the best in their endeavors.

I have never been asked to work on a mass grave, although others have reportedly used GPR to locate victims from conflicts in El Salvador, Bosnia, and Iraq. I

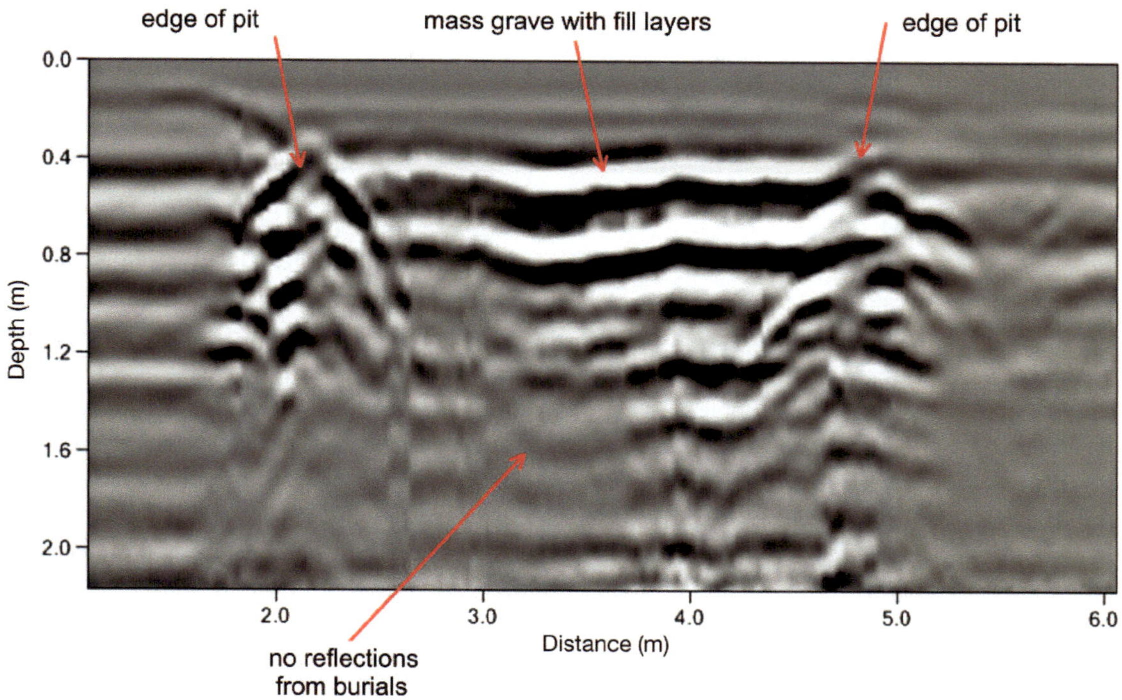

Figure 8-26: Reflection profile of a mass grave. Data collected in the Comstock Cemetery, Virginia City, Nevada, using 400 MHz antennas. *Courtesy of Jennie Sturm.*

have been told by scientists who conducted some of these surveys that much of what can be seen in reflection profiles are large pits, sometimes with point-source hyperbolic reflections within them. The point-source reflections were generated by the partially decomposed bodies, which still retained enough water to contrast with the surrounding ground, or by artifacts of various sorts that were thrown in the pits along with the bodies. The best example I know of is a multi-person grave from Virginia City, Nevada, which is likely the burial location of many people who died at one time. Virginia City was a silver-mining boom town in the late 19th century and at one time had as many residents as San Francisco. Living conditions were primitive and deaths common from mine accidents as well as communicable diseases, which periodically ravaged the populace. The cemetery studied is located just outside of town on a prominent hill and contains the burials of many tens of thousands of people, many of whom died early and tragic deaths.

In a survey of many plots in the cemetery, one grave was found to contain many miners who perished during one accident. The reflection profile across this burial shows it to be about 3 m across, with distinct, highly reflective fill layers (Figure 8-26). No caskets or other remains are visible in the profile, perhaps because they are below the energy penetration depth or are decomposed and do not produce reflections. This example is much like the ones that have been described to me by others working on mass graves elsewhere in the world.

INFORMAL CEMETERIES AND CULTURALLY DIVERSE BURIAL PRACTICES

In 2011, the city of Key West, Florida, was in the planning stages of a project to reroute Atlantic Boulevard, which runs next to the Gulf of Mexico on the south side of the island. The plan was to create a larger beach area, with other recreational opportunities for tourists. When they were in the planning stages of this redevelopment, a local historian found interesting records documenting how this area had been used as a burial ground in the 18th and 19th centuries. There were also many accounts of human bones washing out of the ground after storms in this general area from the time when this part of the island was called South Beach, and largely uninhabited.

In 2000, I had been asked to survey a small area nearby to locate the so-called African Cemetery, which was well known in local lore as the burial location of a number of African slaves who died during their transport back to Africa during the creation of the country of Liberia. That survey was successful in locating a number of those burials adjacent to the beach, but the GPR data suggested that many had likely been eroded away, as much of that burial ground showed evidence of erosion scarps produced by wave activity. The African Cemetery discovered in that project with GPR was fenced off, and a monument to those poor souls was placed next to the modern sidewalk.

In preparation for the Atlantic Boulevard relocation in 2011, I was asked back to the same general area, as the historian had found journals written by people who had further evidence of other burials nearby. One journal noted that human bones had been encountered when laying a water line to newly constructed Army barracks during World War II. Other earlier stories mentioned that this area had been a burial ground for pirates who were hunted down by American warships in the 19th century. They must have employed some lethal means in this eradication, as there were alleged to be many pirates buried here. There were other scattered accounts of burials, mostly of poor people who apparently lacked the means to be buried in the formal Key West cemetery in town.

My GPR study of the area was quite complex and necessitated, first, mapping the shallow bedrock, which was discussed in Chapter 4. When the intra-bedrock areas had been mapped, detailed grids of GPR data were collected in what would have been conditions conducive for burial, as the very hard bedrock layer was deeper there. The thickest sediment package between bedrock highs was located in the Key West Small Dog Park, an area popular with very rambunctious small dogs, who made data collection interesting, as many seemed to be fascinated with my equipment and personal scent and would not leave me alone. That grid of data, when sliced and the amplitudes plotted, readily located the buried pipes, which the journal from the barracks construction had mentioned; and adjacent to one pipe that crossed the thick sediment package, numerous graves could be identified (Figure 8-27). None of the burials appeared

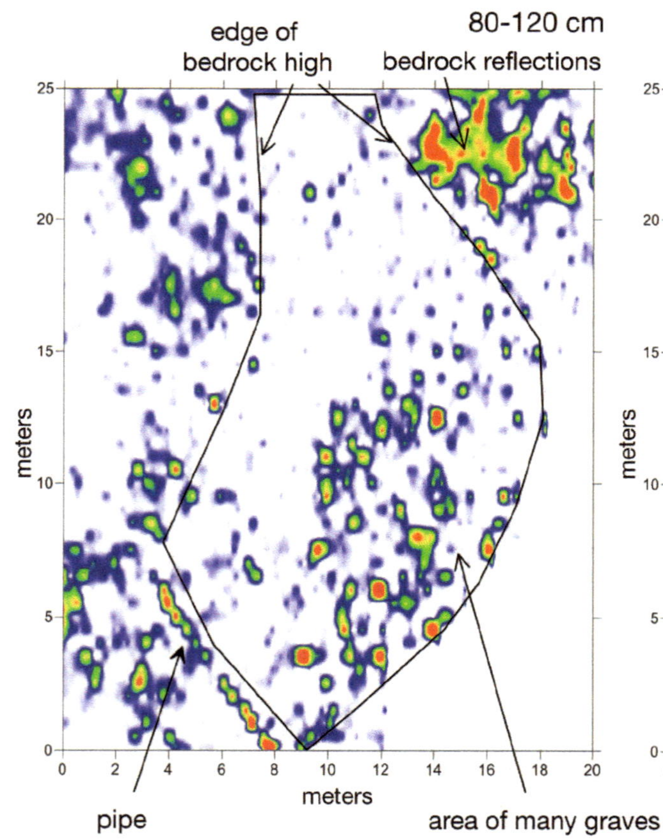

Figure 8-27: Amplitude maps showing a concentration of many graves. These maps were produced in the Small Dog Park, Key West, Florida, showing the concentration of many graves, transected by a pipe. Data collected with 400 MHz antennas.

to be oriented in the same direction, and there was no inherent order to the burial ground, as would be expected from the historical accounts of the informal burials that took place there.

We attempted to keep the discovery of this "pirate cemetery," located with GPR, secret from the general populace until a plan could be formulated to deal with the graves in the context of the planned construction. Word soon leaked out about what lay beneath the Small Dog Park, and the owners of those dogs were quite irate that they might no longer have access to this canine recreation area. The mayor of Key West consulted me on this problem, and she suggested that the city just "dig up all the pirates" and get rid of them, a suggestion I reacted to negatively. All I had to do was suggest the possibility of a public outcry from the "pirate descendant communities" who are "modern stakeholders" and she backed off that plan immediately. While it is doubtful that any residents of modern Key West can trace their ancestry to any of the pirates who might have been buried in this area, the residents of this community are known to be composed of eccentrics, artists, and general dropouts from society. My concern was that they might band together to protest the disinterment of any possible pirates, making all of our lives difficult. The city prudently decided to reroute Atlantic Boulevard away from the graves and the dog park, removing only the handball courts, which have a less numerous and vocal following than the dog park has. An open space will be constructed over the African and pirate graves, with some kind of commemorative marker.

In northern Queensland, Australia, throughout the 20th century and continuing into the present day, the indigenous Aboriginal community has been embattled in a contentious relationship with both the state government and local mining companies. This area on the Cape York Peninsula contains huge deposits of bauxite ore, the material from which aluminum is produced, which is presently mined by the Rio Tinto Mining Company. Prior to modern bauxite mining, the Queensland government, in order to make room for mining operations, evicted the inhabitants of the small town of Mapoon, burning many of their houses. Most fled and only returned much later to reclaim their property.

In 2011, a GPR survey was carried out at the request of the local council of Mapoon in the hope of mapping and establishing the boundaries of a historical cemetery in the town prior to this conflict (Figure 8-28). This was an initial step in formalizing land claims to this property by the local populace. The Mapoon cemetery dates at least to the late 19th century, when Methodist missionaries established a church and school nearby. Many people of European descent were buried there, along with many of the Aboriginal people associated with the mission. There are also historical accounts of human burials in this general area long before the mission was established, which may be why the formal mission cemetery was located in that spot. On the ground surface, there is little physical evidence to indicate the area's function as a cemetery. There are a few European-style grave markers, some of which can still be read (Figure 8-29), showing the location of burials of people associated with the Mapoon Mission. A number of other, less formal markers can be seen on the surface (Figure 8-30), which the local inhabitants informed us were used for more traditional Aboriginal burials. Those markers consist mostly of coral rocks that might have been put in a pile over, or outlining, the burials. These rocks have been scattered by horses that graze in this area, and the cemetery has fallen into general neglect over the years. The ground is composed almost wholly of carbonate sand, which is an excellent medium for GPR energy transmission and reflection. About 2–3 m below this sand, which is primarily aeolian in origin, is the bauxite layer, which is presently being mined nearby.

Figure 8-28: Collecting reflection profiles on very sandy ground in an Australian cemetery. This is the now-abandoned Mapoon cemetery, Queensland, Australia.

There are many accounts of traditional burial practices in this area, most of which consist of the deceased being wrapped in blankets or other fabric coverings and being placed directly in the ground. Many of the elders

Figure 8-29: Historic European graves, Mapoon cemetery, Queensland, Australia.

Figure 8-30: Traditional Aboriginal grave markers of coral stones, Mapoon cemetery, Queensland, Australia.

in the Mapoon Aboriginal community recounted stories to us of their family members and friends being buried in exactly this fashion, even after they were associated with the European influence of the mission. Burials of that sort would likely decompose readily in the extremely humid and rainy climate of northern Queensland.

The GPR profiles and amplitude maps readily illustrate the differences in grave types at Mapoon, with the European-influenced graves appearing as high-amplitude reflections due to the void spaces that remain in the caskets (Figures 8-31 and 8-32). The older graves, or less elaborate burials, which would likely have been of the Aboriginal deceased, are quite subtle and generally produced lower-amplitude reflections.

When all the graves are plotted in amplitude maps, the more elaborate burials, which are likely those of either European missionaries or others who were buried in a European style, are visible in distinct plots. The more simple Aboriginal-style burials also are found in certain areas of the cemetery, suggesting a segregation of burial practices throughout the time the mission was active. There are also many very subtle, presumably older, burials scattered throughout the general cemetery area, whose ages are not known.

CONCLUSIONS

The vast variety of burial types produces a similar multitude in how burials are displayed using GPR images. The features shown here can't always be easily displayed as "anomalies" in GPR amplitude maps. Often subtle, older graves barely show up as amplitudes in profiles and would be overwhelmed by the higher amplitudes from more recent graves in some formal cemeteries in typical amplitude maps. Only manual interpretation of individual profiles can identify the older graves. Graves that

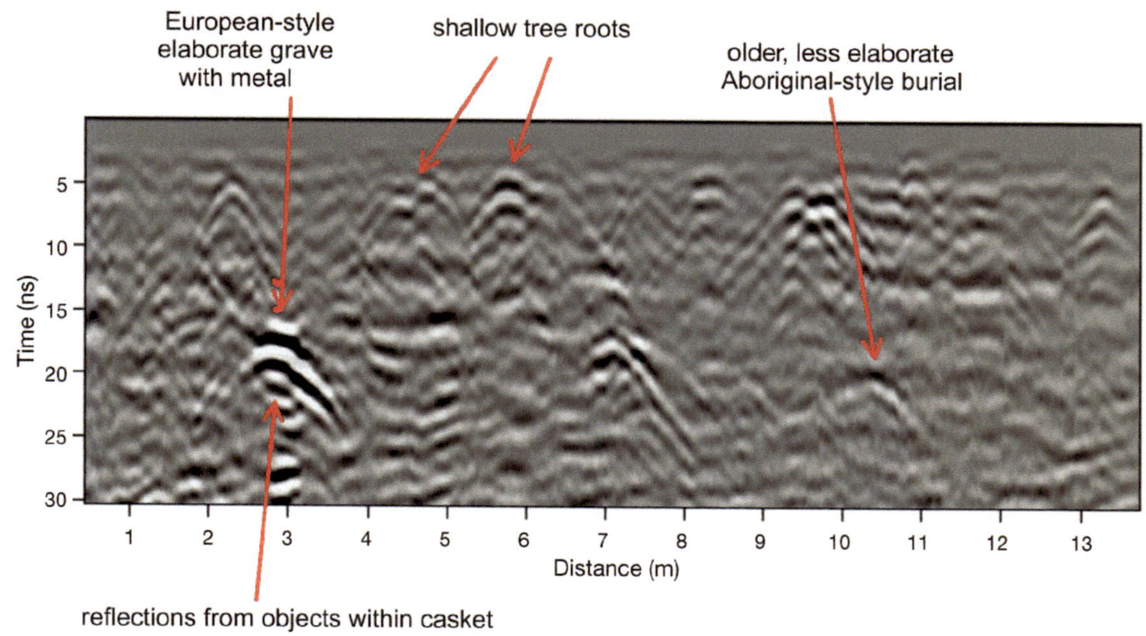

Figure 8-31: Reflection profile in the Mapoon cemetery, showing high-amplitude European-style burials and an older, less elaborate Aboriginal-style burial. Data collected with 400 MHz antennas, Queensland, Australia.

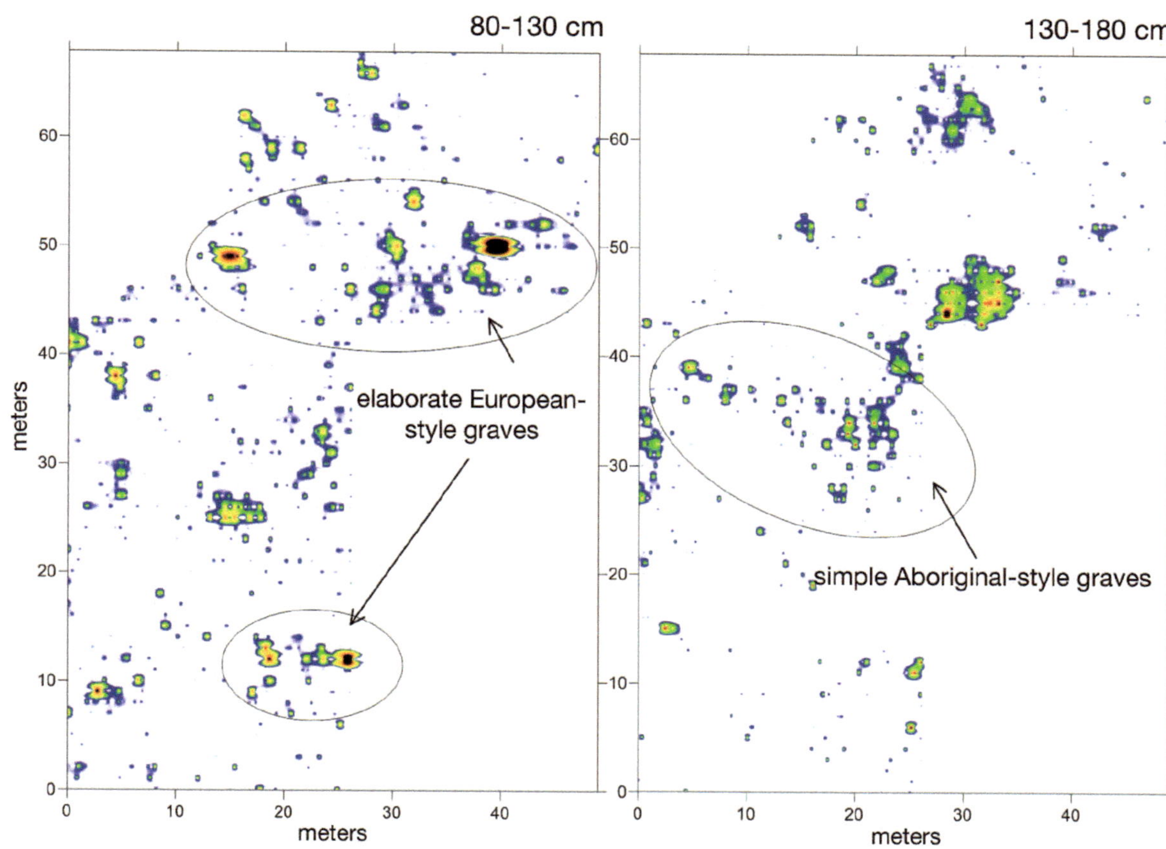

Figure 8-32: Amplitude maps at the Mapoon cemetery, showing concentrations of European-style graves and more simple Aboriginal graves. Data collected with 400 MHz antennas, Queensland, Australia.

contain caskets or other void spaces will produce a complex array of reflections and must be interpreted carefully, as there will be velocity pull-ups that will distort distance and thickness of materials in the ground.

Polarity changes in reflected waves can determine if void spaces still exist. The spatial placement of graves and clustering of different age burials across a large area can often be used to show changes in burial types, or in cultural practices with regard to depositing human remains. I am still waiting for some good examples of how GPR can be used to study those actual human remains, as what is usually visible in profiles and amplitude maps are the vertical shafts and associated materials, not the remains themselves. This may not be possible with GPR, as decomposed human remains usually do not differ enough from the surrounding material to produce radar reflections if there has been decomposition and weathering.

9

Prehistoric Sites

The range of features visible using GPR at prehistoric sites is much too vast for this chapter to be at all comprehensive. I have therefore attempted to include studies of some of the features that I have had success with, which will give perhaps some general examples of how GPR interpretations have been useful in prehistoric sites. Unfortunately, most of the examples shown here are limited to features produced by sedentary people—people who constructed relatively substantial dwellings and other associated features. Much as I tried, I could not find good examples of GPR data collected at sites once occupied by hunter-gatherers, people whose migratory lifeways left little in the way of substantial architecture that is visible in profiles or amplitude maps. The types of hunter-gatherer features for which I sought, but failed to find, GPR examples include lithic scatters (perhaps the artifacts are too small to reflect radar energy), tent rings (there are no examples in my data sets, but I imagine they would look like rings of stones), hearths (small bowl-shaped reflections), and butchering sites (animal carcasses would be as difficult to see as humans remains).

Floors and Pit Structures

The most easily recognizable of all buried architectural features, no matter the age of the site or the location in the world, are floors. These produce planar reflections generated from horizontal layers of clay, compacted earth, stone, bricks, or any other building material (Figure 2-5). Often associated with floors are subfloor features (Figure 4-1), supports for roof beams, hearths, associated benches (Figure 1-5), and a great variety of other items and features on or within the floors, all of which are readily visible in reflection profiles. Reflections from floor features of this sort can also be readily imaged and studied in amplitude maps (as in Figure 3-10).

Often floor reflections are not displayed in profiles as perfectly flat, due to subtle variations in the velocity of the overlying materials, which create minor pull-up or pull-down variations in the two-way travel time of the recorded reflections. The floor in Figure 9-1 is a rare example where the floor is almost perfectly flat, as it is overlain by very dry sand that has a uniform velocity.

In some ground conditions, even these types of floor reflections are not as distinct, and reflections can appear blurry in profile. Figure 9-2 shows a house floor that is much like that in Figure 9-1 but is less recognizable due to buried bed boundaries that are gradational instead of sharp. A less distinct reflection may also be produced where the floor and the floor covering material are composed of the same type of material, and therefore there is little velocity contrast at the interface to reflect energy. In such cases, the boundary reflections are not distinct and the amplitudes of those reflections generated from the floor are much lower. Nevertheless, such floors can still be found in reflection profiles with some diligent searching.

Cooking Features and Hearths

One very distinctive feature common to many sites all over the world is the ubiquitous outdoor baking oven. Ovens are even found in hunter-gatherer sites, where people harvested otherwise indigestible roots or other

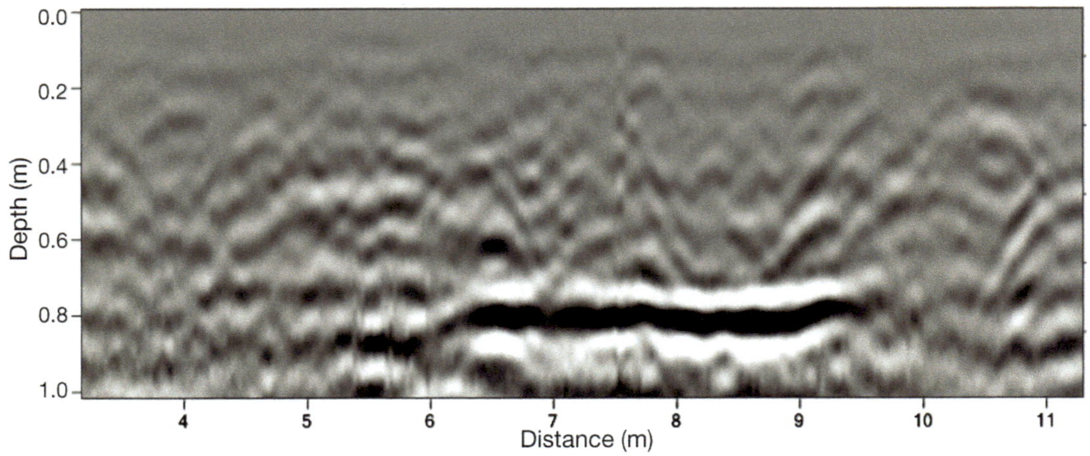

Figure 9-1: Almost perfectly flat floor reflection from a Classic-period Hohokam dwelling. Data collected at the Marana Mound site, southern Arizona, with 400 MHz antennas.

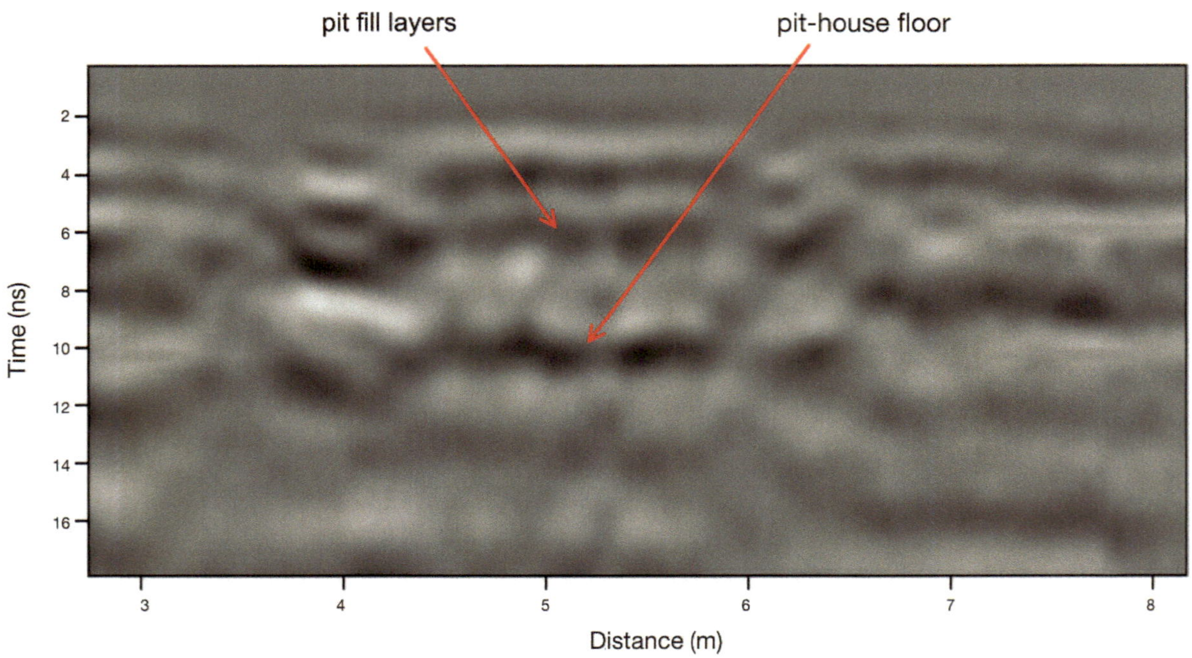

Figure 9-2: Reflection profile over a pit-house floor where the buried layers are composed of gradational changes in sediment. The floor also does not generate distinct high-amplitude reflected waves. Data collected near the Adamsville Hohokam site, southern Arizona, with 400 MHz antennas.

vegetable resources, and processed them into edible food by baking. Sometimes an oven was used regularly over many years, but in other cases they were constructed and used for just one baking episode. The heat of the fire tends to bake the oven bottoms, creating a surface that retains water and therefore produces a good radar reflection surface. One small oven was discovered at University Indian Ruin in Tucson, Arizona, which when excavated was found to have a number of fire-cracked stones sitting on the top of the burned layer. These stones are not readily visible in the reflection profiles (Figure 9-3), at least with 400 MHz antennas, as they are only about 5–10 cm in maximum dimension. Lower-frequency transmitted energy would probably be capable of greater delineation of the stones associated with features of this sort.

Baking ovens are also readily visible in amplitude maps (Figure 9-4). Any smaller stones or other materials associated with the ovens are not visible in these types of images, but the general outlines of the ovens can be readily mapped.

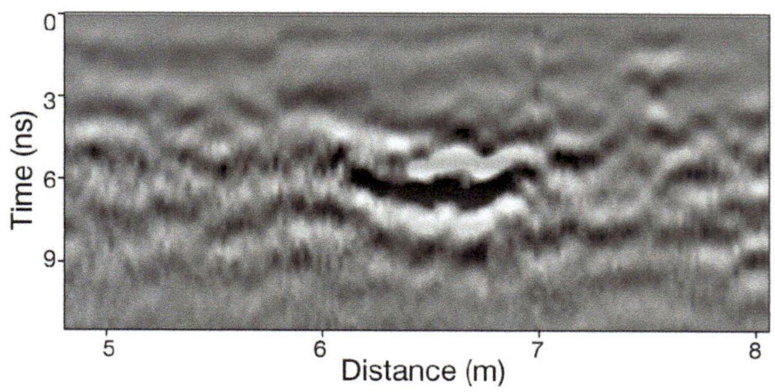

Figure 9-3: Reflection profile showing the base of a baking oven. The oven base produced a high-amplitude reflection. Data collected with 400 MHz antennas, University Indian Ruin, Tucson, Arizona.

WALLS

Walls, often associated with floors, are commonly visible in GPR images, especially if they were constructed from stone or other reflective materials. Near Caranqui, Ecuador, are the remains of the northernmost Inca palace, built by the last Inca emperor, Atahualpa, in 1495. In the backyard of a landowner who lives a few hundred meters from the palace, we found the remains of a standing wall which appeared to be an Inca construction, and we wondered if there were additional architectural features associated with it (Figure 9-5).

Most of the interesting reflections visible in the profiles collected near the Inca wall show what appear to be narrow linear bands, which were thought to be the foundations of walls. Some of these features are wider than others, and all are composed of point-source hyperbolic reflections generated from buried stones (Figure 9-6). These are located about 50 cm or so below the surface, probably buried and encased in the thick surface soils found in this well-maintained backyard. No distinctive

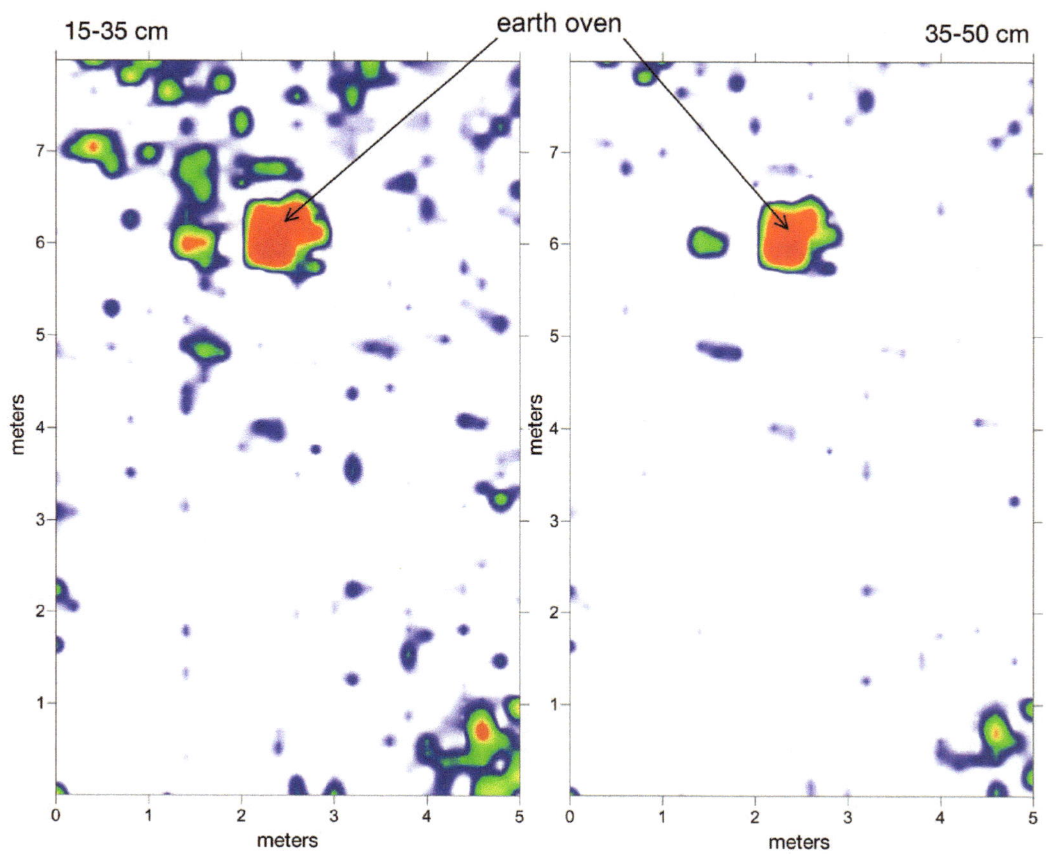

Figure 9-4: Amplitude maps of a baking oven. This is the same oven shown in profile in Figure 9-3 from Tucson, Arizona. Data collected with 400 MHz antennas.

Figure 9-5: A standing Inca wall in Ecuador. This wall was likely associated with a substantial dwelling near Atahualpa's palace in the vicinity of Caranqui, Ecuador. A grid of GPR data was collected directly in front of this wall.

floor features are visible in any of the profiles. None of these stones are more than a course or two thick, suggesting that if the walls were taller at one time, their stones were removed and recycled into other structures during the last 500 years. Some profiles show only randomly placed stones, which might be the remains of walls that toppled, their stones scattered.

From the amplitude maps, we see a much more definitive series of images of the walls of buildings that once were attached to the remaining wall standing today (Figure 9-7). The upper slice, from 0 to 40 cm in depth, indicates only a few linear stone alignments which are likely the very tops of the wall foundations that lie below. Most of the other nonreflective material in this shallow slice is the well-developed surface soil. The slice from 40 to 80 cm shows two different types of stone wall foundations, one type that is closest to the standing Inca wall and quite substantial, with bases up to a meter thick (Figure 9-7). These are likely the remains of the same structure as the extant standing wall. Attached to it were less substantial structures, whose outlines can be seen as the fainter, lower-amplitude reflections produced from much more narrow wall foundations (Figure 9-7). These stone foundations might have supported more simple adobe architecture.

While no excavations were conducted here, the GPR maps offer some interesting hypotheses related to differences in the usage of these buildings during the time the nearby Inca palace was inhabited. It is likely that this structure found with GPR was associated in some way with the nearby royal palace. The types of activities that took place in the more substantial end of the building were likely much different from those in the attached, less-substantial rooms. The fact that these two building types were found together and attached (or in very close proximity) suggests very different functions by perhaps different classes of people within the palace precinct. A comparison of the two, if excavated, would potentially offer interesting behavioral contrasts related to class differences, functions by two different levels of society, or differing types of activities within what would have been a royal household during Inca times.

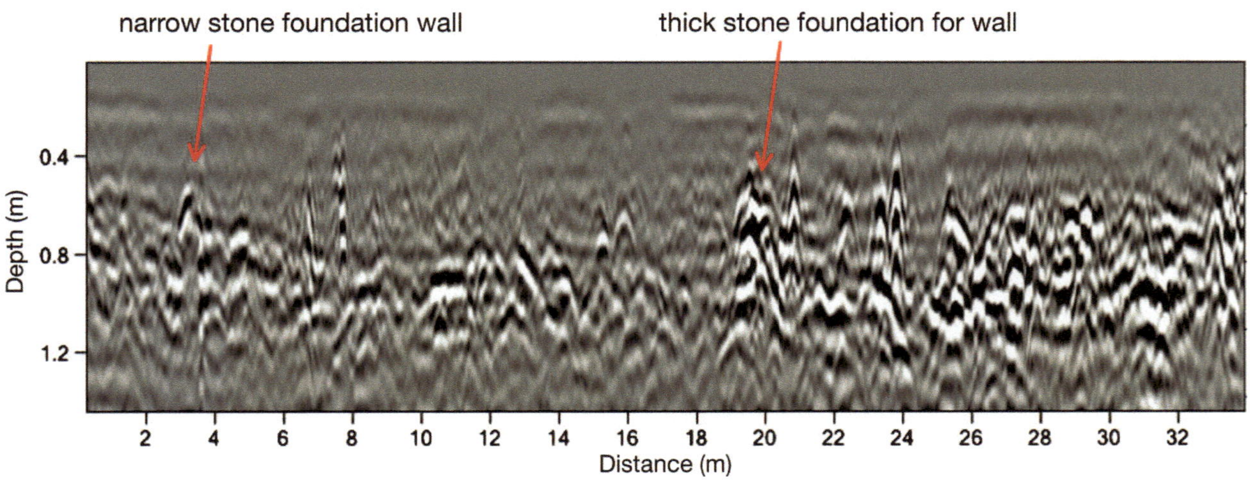

Figure 9-6: Reflection profile across Inca walls. The point-source hyperbolas generated from individual stones that likely made up the foundations for walls are visible. Data collected with 400 MHz antennas, Caranqui, Ecuador.

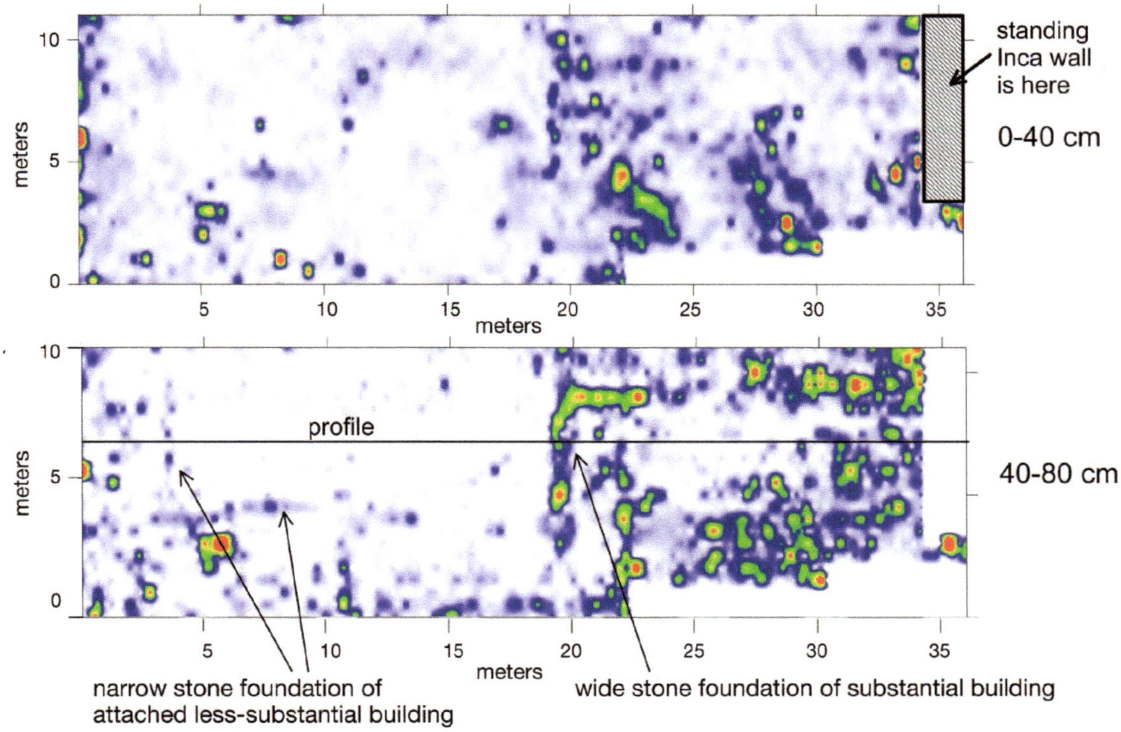

Figure 9-7: Amplitude maps produced from profiles collected adjacent to the standing Inca wall shown in profile in Figure 9-5 from Caranqui, Ecuador. Two different wall types are visible, suggesting that the inhabitants of this site near the Inca palace had two different functions . Data collected with 400 MHz antennas.

Walls were not often constructed with stones, especially in some areas of the world, but many were composed of compacted earth or adobe bricks. In southern Arizona, the Hohokam people constructed substantial multi-room buildings produced of "puddled" or "rammed-earth" adobe, some of which were many stories tall. Few of these remain standing today, as most began to erode soon after their roofs disintegrated, following abandonment. The earthen walls then "melted" into adjacent layers that were often interbedded with wind-blown sand. A slow accumulation of naturally deposited sediment and melted adobe adjacent to the walls ultimately produced large mounds. Few of these mounds retain any evidence of walls on their tops, but it only takes a minor amount of excavation, or even sweeping with a broom, to see color differences that distinguish the remaining earthen walls from the adjacent melt layers (Figure 9-8). One of these large mounds was used as a test for the GPR method to determine if earthen walls could be differentiated and mapped if the walls were more deeply buried (Conyers 2012).

Figure 9-8: Collecting reflection profiles on a mound where adobe walls are visible as lighter-colored earth, surrounded by darker adobe melt units. Adamsville Mound, Casa Grande, Arizona.

The reflection profiles over the mound show that the walls were almost completely nonreflective (Figure 9-9), a phenomenon observed at other nearby sites (Figure 1-4). The adobe walls of these buildings were constructed by quarrying earth nearby, mixing it with water and sometimes a binding agent such as grass, and placing it within forms, or just mounded on the ground in layers to be dried in the sun (Conyers 2011a). Over time, the walls were built higher and higher, sometimes reaching many tens of meters in height. As the earth was well mixed before the construction process began, it now contains no layers that retain differences between constituents that could effectively reflect radar energy. The walls are therefore visible in profiles as areas of no reflection, or very low-amplitude random reflections (Figure 9-9). Layers of the melted walls, which are interbedded with windblown sand and silt, are highly reflective, and each thick adobe melt layer adjacent to sand produces a high-amplitude reflection.

The amplitude maps produced from the 400 MHz reflection profiles take some time to interpret, as the features of interest are the nonreflective areas, and there can be a dizzying array of higher-amplitude reflections among the adjacent melt layers. The most substantial of these earthen structures, which are still partially standing today, were composed of earth that contained a good deal of calcium carbonate, which means they were well cemented and their walls much less affected by erosion over the centuries. Unfortunately, this carbonate material also retains water and is electrically conductive enough to impede radar waves. Energy rarely penetrates more than about a meter or so in these types of structures, irrespective of the frequency of the propagating energy.

I have concluded that it is best to use antennas in the 900 MHz frequency range for these types of earthen mounds, which can still transmit energy to about 50–75 cm in depth and produce much more precise reflection profiles. When many 900 MHz profiles are collected, with a profile spacing of 10 cm or so, high-resolution images of the buried walls and associated melt layers can be produced (Figure 9-10). At a similar mound site near Tucson, Arizona, the uppermost slice displays only random reflections from the surface soils and some well-mixed surface sediment on the top of the mound. The slice from the 20–40 cm depth shows distinct differences between the nonreflective linear walls and the adjacent adobe melt layers that generate high-amplitude reflections (Figure 9-10). Stones, which were often used as foundations for columns that held up the second and third stories of these buildings, are also visible as square high-amplitude reflection features. It takes a while to train one's eye to pay attention to the areas of no reflection on amplitude maps, as most GPR interpreters are accustomed to identifying the higher-amplitude reflections. In cases where the features of interest are the nonreflective areas, it just takes a little extra time and practice.

FEATURES ASSOCIATED WITH ABOVE-GROUND STRUCTURES

In southeastern Utah, along the rear of a large pueblo structure called the Bluff Greathouse, GPR data were collected (Figure 9-11) as a follow-up to some interesting excavations conducted by Cathy Cameron and her students (Cameron and Geib 2007). This is the same site where the great kiva was imaged, shown in Figure 3-8. The word "great" is used for multi-room houses (pueblos) and kivas built during a period of time when influential people from Chaco Canyon, in present-day New Mexico, had spread important ideas related to building "oversized" architectural features. This "overbuilding" appears to have been a way to impress others

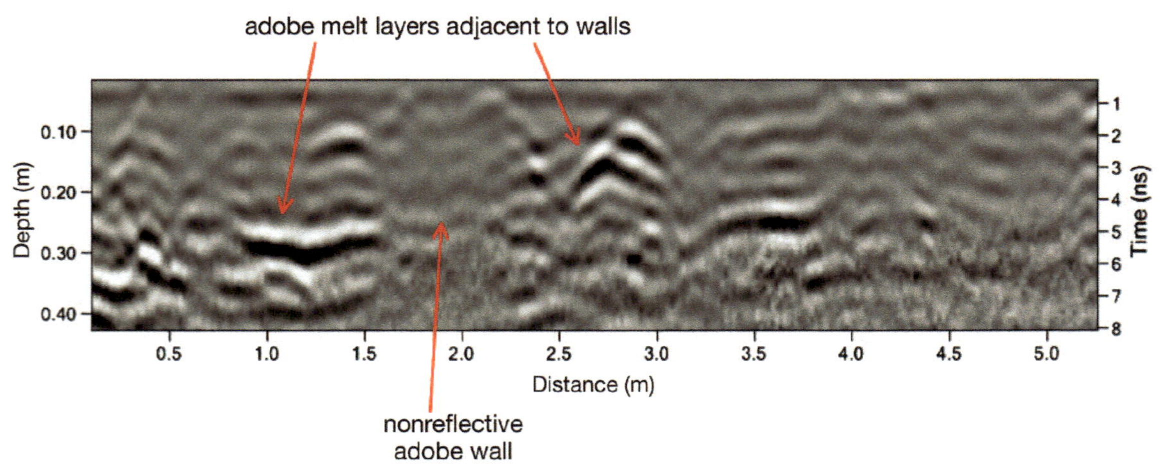

Figure 9-9: Reflection profile collected on top of a mound. Exposed walls in this profile, data for which were collected on the mound shown in Figure 9-8, could be directly compared with the amplitudes of reflections they generated. The adobe melt layers produced the high-amplitude reflections, while the walls are nonreflective. Data collected with 400 MHz antennas at Adamsville Mound, southern Arizona.

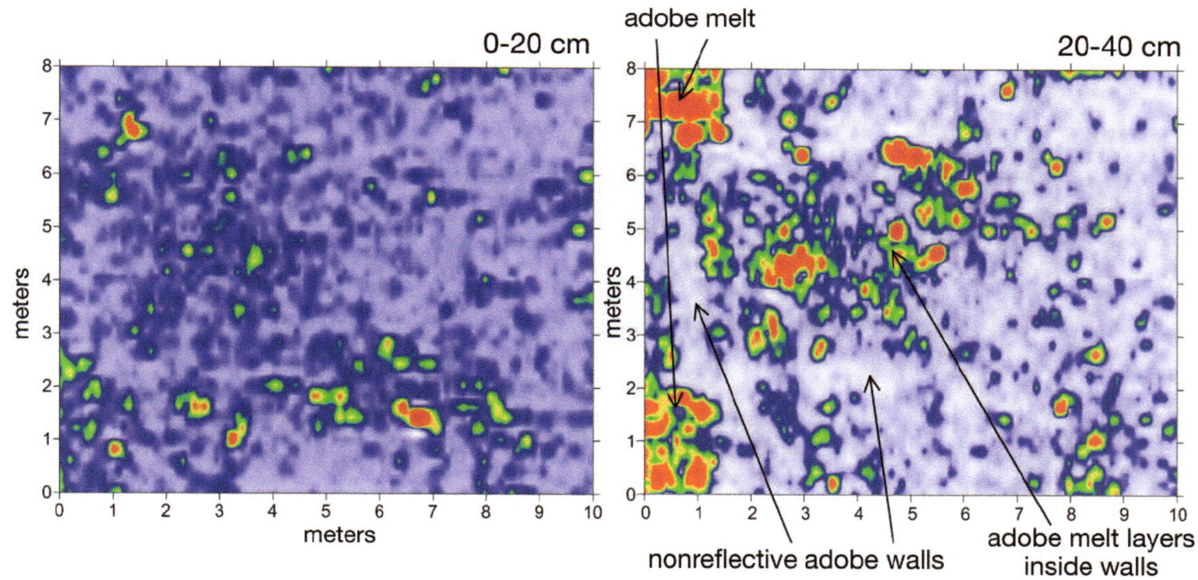

Figure 9-10: Amplitude maps on the top of a Hohokam mound. Data collected at University Indian Ruin, Tucson, Arizona, showing the areas of nonreflective adobe walls, bounded by the reflective melt layers within. Data collected using 900 MHz antennas with 10 cm transect spacing.

Figure 9-11: Collecting reflection profiles along the north side of a greathouse. This is the Bluff Greathouse (just visible on the right side of the picture) where 400 MHz antennas were used to collect data in order to map an adjacent earthen terrace feature. *Image courtesy of Sarah Lowry.*

with esoteric knowledge of many sorts. It was during the "Chaco" period, which lasted from about A.D. 800 to 1150, that most of the architectural features at Bluff were built.

In Cameron's trench (Figure 9-12), evidence was found of a compacted earth surface along the north side of this large greathouse building, which was interpreted as a formal terrace or perhaps a hard work surface. Elsewhere in the American Southwest, features of this sort are known to have functioned as places to perform everyday activities but, in other contexts, also served within ritual centers as a location for activities by visitors from nearby villages who came periodically to these central greathouse locations. The Bluff Greathouse site is known to have had connections to the important ritual and trade at Chaco Canyon for many centuries, where terraces of this sort were also constructed (Conyers 2010; Conyers and Osburn 2006). The function of the Bluff terrace was therefore of interest, as it could potentially help in understanding the transmission of ideas from that central place at Chaco to this more distant "outlier" where these functions appear to have taken hold. It is also known from excavations that the Bluff site was occupied long after the central place at Chaco was abandoned ca. A.D. 1150. Questions were then posed regarding the continuity of Chaco architectural styles after that influence had declined, and whether there were any architectural changes or perhaps functional usages of this terrace that were modified during and after the collapse of Chaco.

The amplitude maps of this terrace feature show a number of linear reflection features, some at right angles, which delineate walls of structures that had been built directly on the terrace surface (Figure 9-12). Many of these appear to have been constructed of material that was cut out of the terrace surface and then piled up and

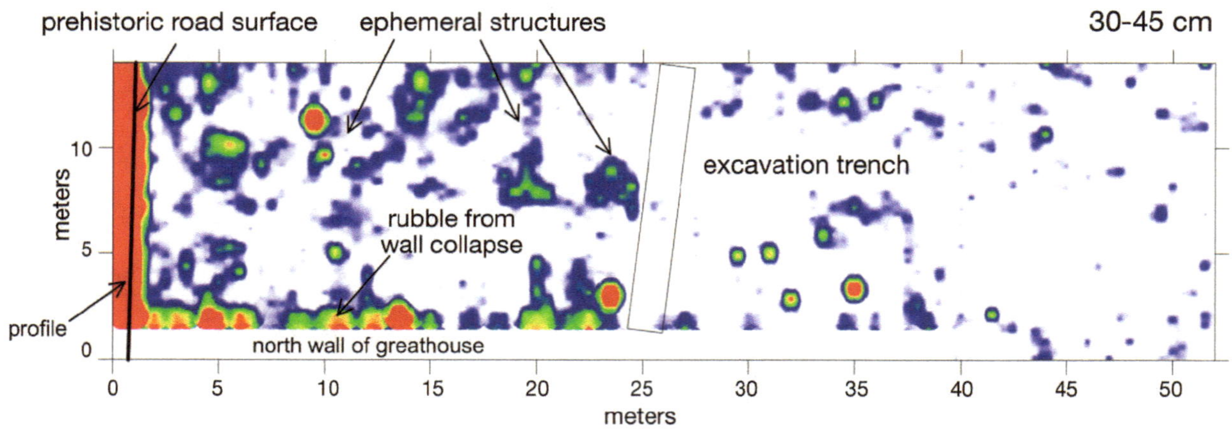

Figure 9-12: Amplitude map showing reflective features located on top of the terrace surface along the north face of a greathouse. A prehistoric road surface is visible along with a number of insubstantial structures north of the Bluff Greathouse. Data collected using 400 MHz antennas. *Courtesy of Sarah Lowry.*

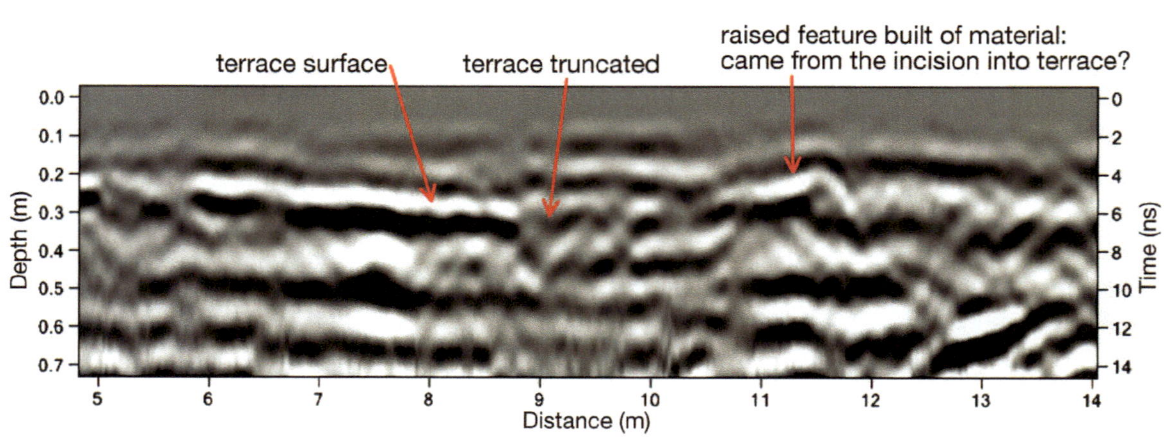

Figure 9-13: Reflection profile showing a terrace surface. This surface, north of the Bluff Greathouse, was cut into to build up a raised feature late in its use-life. Data collected using 400 MHz antennas. *Courtesy of Sarah Lowry.*

mixed with stones (Figure 9-13), perhaps to anchor walls of shade structures of some kind. Those features appear to have been built quickly on the already prepared terrace surface, with little concern for keeping the original terrace intact. Other subtle features, still visible with GPR, were built on the terrace and show up as low-amplitude point-source reflections, generated from stones placed directly on the terrace. All these features, visible in the amplitude map, are very small and were likely part of a short-lived building episode undertaken with no regard for the work that had gone into the earlier, more substantial terrace surface.

These very subtle architectural features would have gone almost unnoticed in a cursory interpretation of reflection profiles, and only became apparent after the amplitude maps were produced and analyzed. It appears these small features on the terrace are the remains of ephemeral structures (perhaps shade structures) built in this cooler, perhaps quieter area, away from the main entrance on the other side of the greathouse. On that other side, there are remains of another formal plaza, a number of small kivas, and the one great kiva, all of which suggest that that part of the greathouse was reserved for more formal types of activities. However, someone ordered this terrace surface to be built along the north edge of the greathouse, perhaps under the direction of influential people from Chaco early in the building's use-life. Later, perhaps after the abandonment of Chaco in about 1150, this terrace was transformed into an area of more utilitarian activities, as demonstrated by the features visible in the GPR images. It is likely that the GPR images are showing that transformation of influence, as the Bluff Greathouse, while still an important location for activities of the local people, was no longer under Chaco's sway.

Another interesting feature found associated with this terrace at the rear of the greathouse is the very high-amplitude linear surface along the western portion of the grid (Figure 9-12). This planar feature is at the same level as the top of the terrace, and appears to be composed of a different material, which produces a much higher-amplitude reflection (Figure 9-14). It is known that a prehistoric "road" led north from this village toward other contemporaneously occupied villages, and portions of this transit pathway can still be seen on the landscape. Elsewhere, roads of this sort have been extensively studied across the area that was influenced by Chaco, and they appear to have been used as ceremonial connections between important locations. To the southeast of Bluff, most roads of this sort led directly in a straight line to Chaco Canyon and indicate very close ties to that central place, documenting the movement of people and goods to and from that prehistoric city. The road at Bluff, visible in the GPR images, leads from the west side of the greathouse and heads north, where other important greathouses were located. It is questionable whether the ephemeral structures built on the terrace were being used at the same time the road functioned for more ceremonial activities. It is more likely that during this village's occupation, after the Chaco collapse, the road may have fallen into disuse and the terrace surface transformed into an everyday activity area. The GPR mapping here is telling us something about the reorganization of the ceremonial landscape after Chaco ceased to be a central place of importance.

Shell Mounds and Middens

Along the Crystal River, which is a broad, tidally influenced arm of the Gulf of Mexico in Florida, sits a large site containing many shell mounds dating from A.D. 300 until about 1500 (Thompson et al. 2011). Only some of the original mounds remain, as a number of them were removed in the past so that the shells could be used as road-surfacing material. The exposures along the partially eroded flanks of some mounds show that they are almost completely composed of bivalve shells which were gathered in the nearby estuary mudflats and consumed close by (Figure 9-15). While these mounds appear on first glance to be common shell middens, visible in many locations along this coastline, many of them have flat tops and are known to contain human burials, much pottery, charcoal layers, and other features suggestive of functions other than as trash heaps. Some mounds have been described as "temples," with little to support that contention except their somewhat larger and taller size. Some are oblong, some circular, and others have appendages of unknown function.

Many grids of GPR reflection profiles were collected on the mounds to see if they contained internal structures or stratigraphic layers that might help in determining

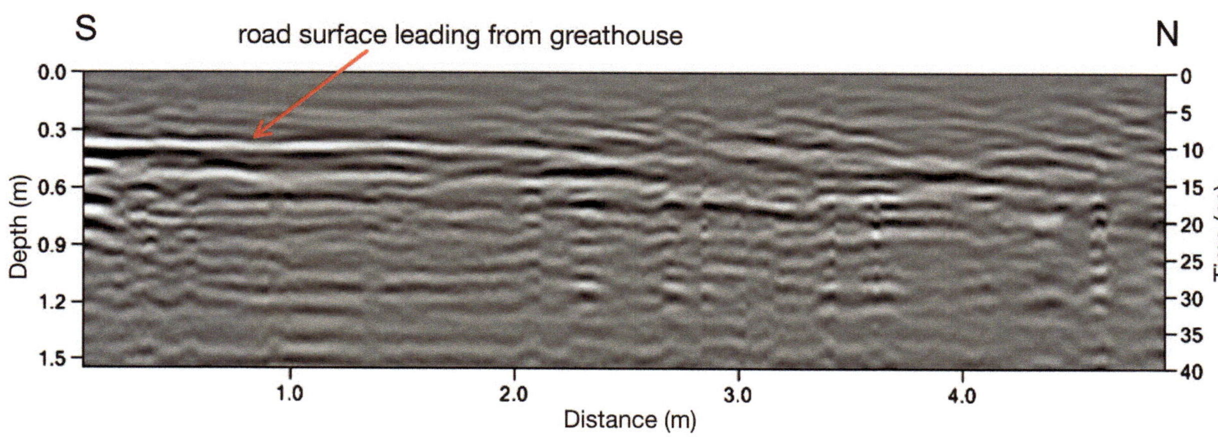

Figure 9-14: Reflection profile showing a prehistoric road surface leading away from a greathouse. Data collected with 400 MHz antennas north of the Bluff Greathouse. *Courtesy of Sarah Lowry.*

Figure 9-15: Exposure of shells in a mound in Florida. This is an exposure on the flank of one of the mounds at Crystal River, Florida. *Image courtesy of Victor Thompson.*

imported stones used for some type of construction directly on the imported sand layer, which was used to surface the mound. While it is not known what the function of these structures might have been, their future exposure and study will help in the understanding of the ancient activities that took place during one of the last occupational phases of this site. The GPR reflection profiles recorded here reveal reflections from a number of other layers that make up this mound and indicate these structures have a complex history of construction and renovation, which is just beginning to be understood. The GPR images definitively show they functioned as much more than places where shells were discarded.

On the bank of the Tennessee River, during bridge construction, GPR profiles were collected along a natural levee on the first terrace above the active river, and a number of prehistoric features were discovered. Earth ovens, semipermanent houses, and scattered lithic debris were found on a buried soil layer about 50 cm below the ground surface. In one area, a distinct planar reflection was found, with many small hyperbolic reflections and with nonreflective units above and below (Figure 9-18). The origin of these reflections was hypothesized to be a concentration of small materials of unknown origin directly on the known buried soil layer, exposed in nearby excavations. When the profiles were used to create amplitude maps of this reflection feature, it could be seen as a linear feature, parallel to the natural levee just to the west of the river channel (Figure 9-19).

This feature was excavated and found to be a shell midden, radiocarbon dated to between 1200 and 1800

their origin, use, and modification over time (Figure 9-16). Reflection profiles show four distinctive layers making up one of the mounds, with an abundance of point-source reflection hyperbolas directly on one high-amplitude planar reflection, which appears to be a compacted living surface (Figure 9-17). The individual shells used to produce these features are too small to generate their own reflections using 400 MHz antennas. This mound was cored, and it was found that the uppermost layer is composed almost wholly of shells. It is underlain by a carbonate sand unit about 50 cm thick, which was likely imported from a nearby beach or sand bar in the river. The boundary between these two units, which is interpreted as a living surface, appears in all reflection profiles as a planar surface comprising many larger stones that produce point-source reflection hyperbolas. Those distinctive reflections were likely generated from

Figure 9-16: Collecting 400 MHz reflection profiles over a shell mound at Crystal River, Florida. *Image courtesy of Victor Thompson.*

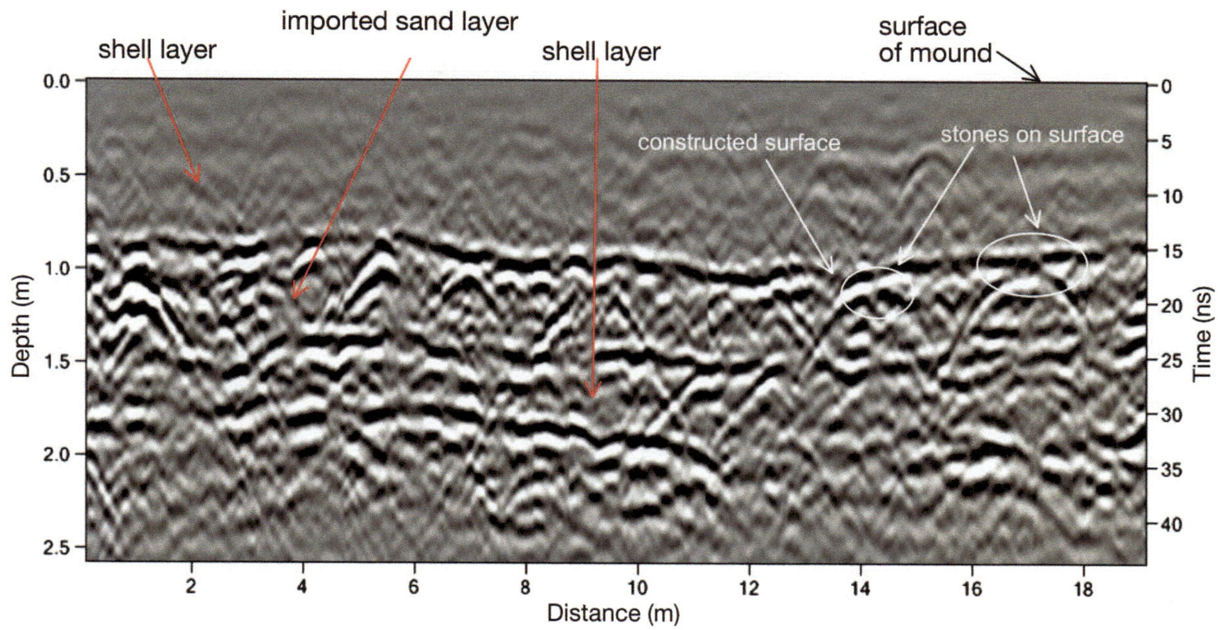

Figure 9-17: Reflection profile on the top of a shell mound. This profile was collected with 400 MHz antennas at Crystal River, Florida. *Data courtesy of Victor Thompson.*

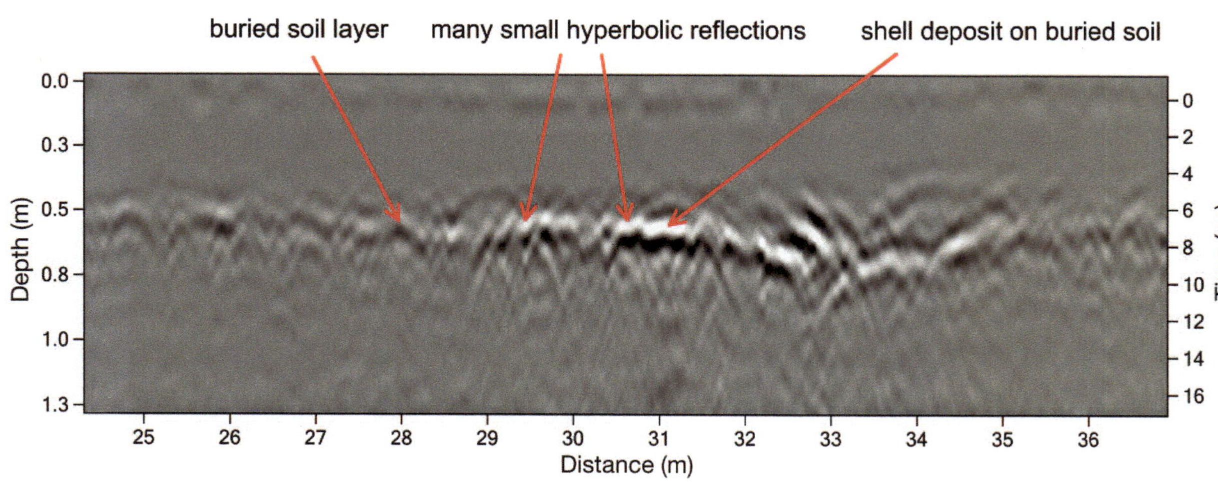

Figure 9-18: Reflection profile showing a buried soil horizon with shells and other materials on it in a river floodplain. Data collected on the bank of the Tennessee River, with 400 MHz antennas. *Courtesy of Shawn Patch.*

B.P., sitting on a distinct buried soil horizon. The midden is about 20–25 cm thick, somewhat mounded at its thickest point, and covered with alluvial sediment, which has been altered into a well-developed buried soil horizon (Figure 9-20). The midden appears to have been deposited in what was a low swale parallel to the ancient natural levee of the river, probably when the river was flowing at a higher level than it is today (Figure 9-21). Freshwater clams were consumed by people who lived nearby during part of the year, and the clamshells were discarded in this low area, to be covered by sediment from flood events. There is some evidence in the way these shells are stratified, with interbedded flood layers, to suggest that this deposit was produced from a number

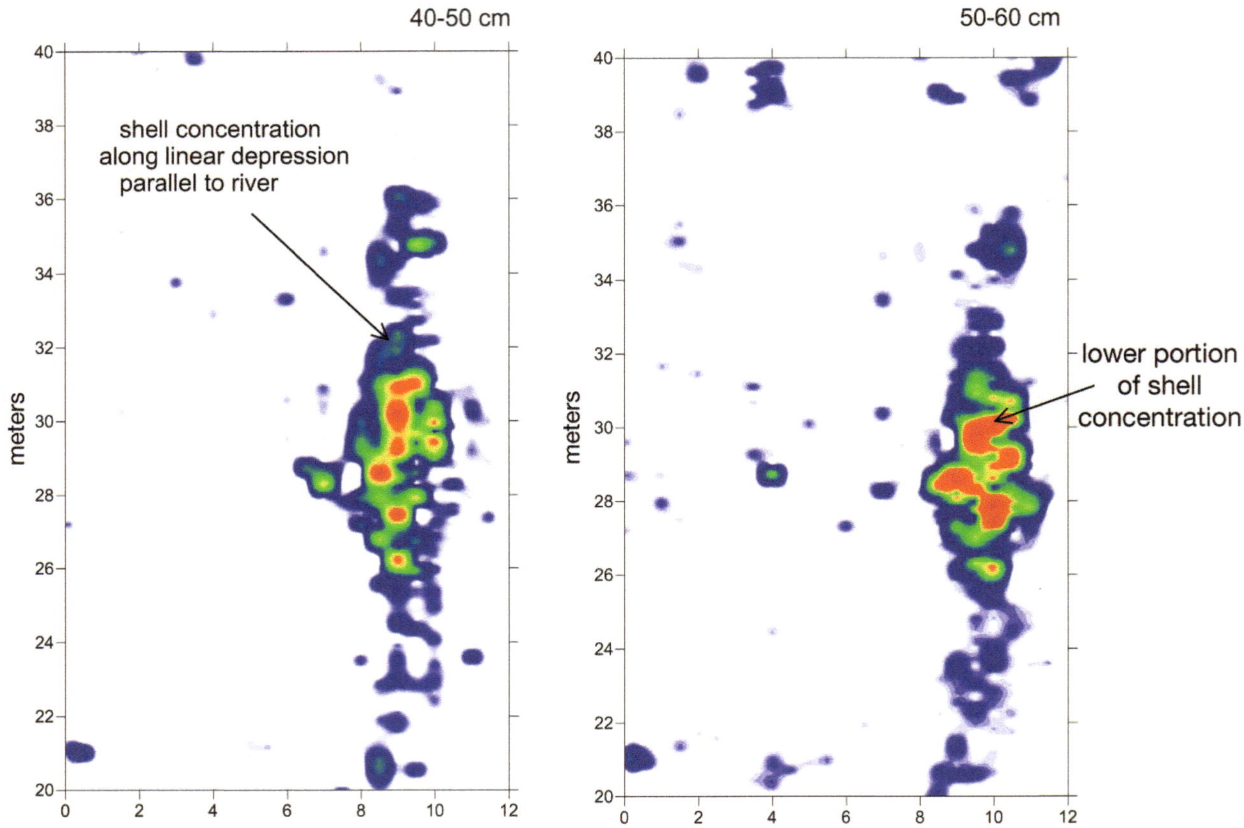

Figure 9-19: Amplitude maps of a shell midden deposited in a swale along a river terrace. This feature is parallel to the Tennessee River. Data collected with 400 MHz antennas. *Courtesy of Shawn Patch.*

Figure 9-20: Excavation of the shell layer deposited on a soil horizon. This is the feature visible in the profile in Figure 9-19. *Image courtesy of Shawn Patch.*

Figure 9-21: The linear nature of the shell midden that became visible in the amplitude maps (in Figure 9-19), showing the subtle mounding of this deposit.

of discard events, perhaps during seasonal use of the area by the same people over some long period of time.

IRRIGATION CANALS

One of the first groups to build extensive irrigation systems and practice large-scale farming were people appropriately, if unexcitingly, called the "Early Agricultural People." They inhabited the perennial watercourses of southern Arizona after presumably migrating from Mexico where they learned farming practices (Conyers 2010). As far back as 2000 B.C., they were redirecting water from the Santa Cruz and Rillito Rivers near Tucson, Arizona, into plant beds about 10 × 10 m in dimension. These were constructed with raised earthen berms around a central depression used to impound the water, which was necessary for plant growth in this hot, arid environment. The canals that brought water to these planting beds had to be periodically cleaned of silt and repaired, owing to erosion from thunderstorm-generated floods and everyday siltation. Many canals have been found superimposed on one another, as they were periodically repaired and reconstructed over time. Some appear to have had many tens of years of use-life before silting up and being abandoned, choked with sediment (Figure 9-22).

In order to study a known Early Agricultural canal system exposed in excavation trenches (Figure 9-23), about a meter of overburden sediment was removed and a flat surface produced on which to gather GPR profiles. This was necessary because the sediment in this area near the Santa Cruz River is known to contain electrically conductive clays that attenuate radar waves within about a meter of the ground surface.

The profiles crossing these canals are quite variable, reflecting many different contacts in these channel features. Where the canals are lined with clay and the overlying fill sediment is sand, a high-amplitude reflection was generated (Figure 9-24). In some places where the canal is filled with material that differs little from the surrounding sediment, only a very low-amplitude reflection was generated, rendering the canal barely visible. Depending on the orientation of the canal bank, energy was reflected from one side of the canal or the other, with only one side of the canal producing a high-amplitude reflection. Interesting bow-tie reflections were also commonly generated in the manner described in Figure 4-6. Exactly where the sediment was deposited within the canal, and how strong the reflections were, determined the particular geometry of these bow-tie reflections when viewed in reflection profiles.

The differing reflection types, orientations, and geometries make the interpretation of amplitude maps of these complex canals difficult (Figure 9-24). To map them correctly, a combination of images and interpretation methods must be employed, with amplitude maps producing only a general image of the basic trend. Each reflection profile then has to be interpreted individually and the location of the channel placed in space. Care must also be taken when interpreting the amplitude maps, as the arms of point-source hyperbolas generated

Figure 9-22: An Early Agricultural canal. This feature is located at the Rillito Fan site in Tucson, Arizona, and shows silt and clay layers that filled the canal either during its life or after abandonment.

Figure 9-23: Collecting reflection profiles at the Rillito Fan site in Tucson, Arizona, with 400 MHz antennas. About a meter of overburden sediment had to be removed, as the sediment at this location is electrically conductive and attenuates radar energy within about a meter of the ground surface.

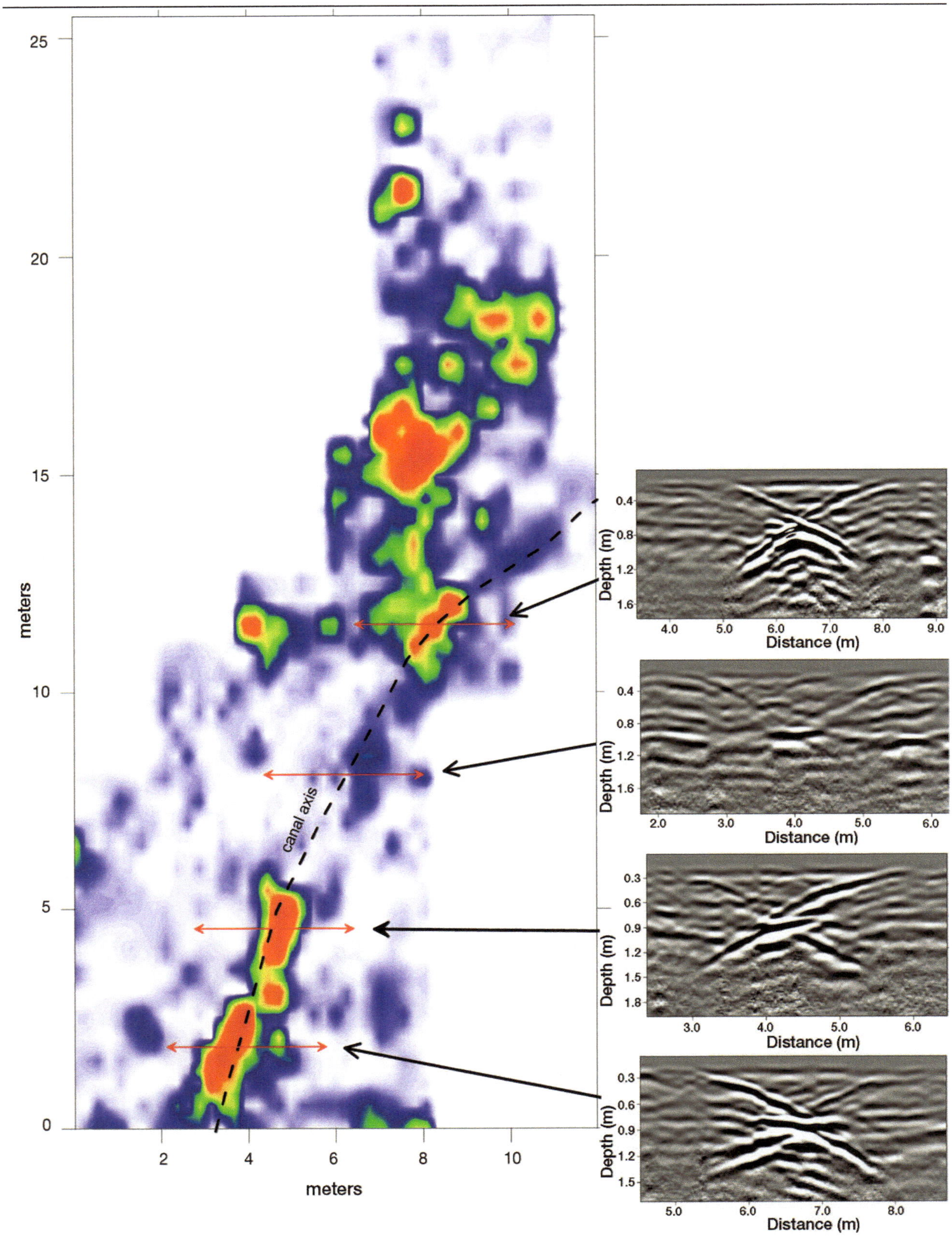

Figure 9-24: Amplitude map of an Early Agricultural canal with associated profiles. This is an image of a canal at the Rillito Fan site in Tucson, prepared with 400 MHz reflection profiles. Examples of the complexity of reflections along this short reach of the canal show how both profiles and amplitude maps must be used for an accurate placement of these linear features in space.

Figure 9-25: Exposure of the Early Agricultural canal shown in the amplitude map in Figure 9-24. The dark layers are moist clay layers that filled the canal; the fine sand and silt appear as lighter colors.

from the very bottom of canals will potentially produce amplitude features, as will portions of the bow-tie reflection features. Most important, reaches of the canal that had no bed contacts to produce high-amplitude reflections will not appear at all in the amplitude maps (Figure 9-24).

After the GPR data were interpreted, the study area was stripped using heavy machinery, and the top of the canal exposed (Figure 9-25). The general linear trend of the canal was visible, as was the complex layering of the different constituents that filled the canal.

Rock Shelters

In highland Ethiopia, along the East African Rift Zone, a rock shelter was discovered with a flat terrace leading from its drip line to a steep cliff. The total area within the drip line is small, and if this small area had been occupied in the past, that terrace surface must have been employed in some way. Nearby are many pecked-stone petroglyphs, and pottery sherds were found on the ground surface, all of which indicated the presence of people sometime in the past. The area is very rocky, with one large boulder visible within the overhang and a few large boulders projecting from the terrace surface in front of the shelter (Figure 9-26). A grid for the collection of GPR profiles was set up, with almost every profile starting and stopping at different locations within the coordinate system, making for complex note taking and data processing later on.

The reflection profiles show a number of interesting cultural features in this small grid, including a terrace surface, bounding walls, and accumulated sediments within the drip line (Figure 9-27). Within the drip line is a meter or so of sediment, which appears mostly unstratified. That sediment was held in place by a wall placed just outside the rock overhang, which probably acted as a barrier of some sort, or supported a wall used to enclose the interior portion of the shelter. On the terrace in front of the shelter is a flat surface, which looks to be composed of some reflective material, perhaps compacted earth or an imported paving material. Parallel to the cliff is a second wall, which enclosed this terrace and the hard packed surface built on it (Figure 9-27). Both walls were composed of large stones and are visible in most of the reflection profiles. Bedrock, which is in places exposed as boulders that sit above the present ground surface, is also visible in the profiles, projecting to near the surface in some areas of the grid. The gaggle of features visible in the reflection profiles show that this small site was constructed in a complicated area of uneven bedrock surfaces, which were filled, leveled, and then walled off.

The reflection profiles were much too complicated and "busy" to produce usable amplitude maps, so all interpretation had to be done by studying each profile individually and placing visible features manually on a base map. When this was done, the bounding wall showed that a good-sized use area had been constructed, probably for one family (Figure 9-28). The bounding wall may have helped retain animals or children, keeping them away from the nearby cliff at the terrace edge. One of our crew almost fell off this cliff during data collection, demonstrating how important a bounding wall would have been in the past.

Figure 9-26: Setting up a GPR grid in front of a rock shelter in highland Ethiopia. *Image courtesy of John Arthur.*

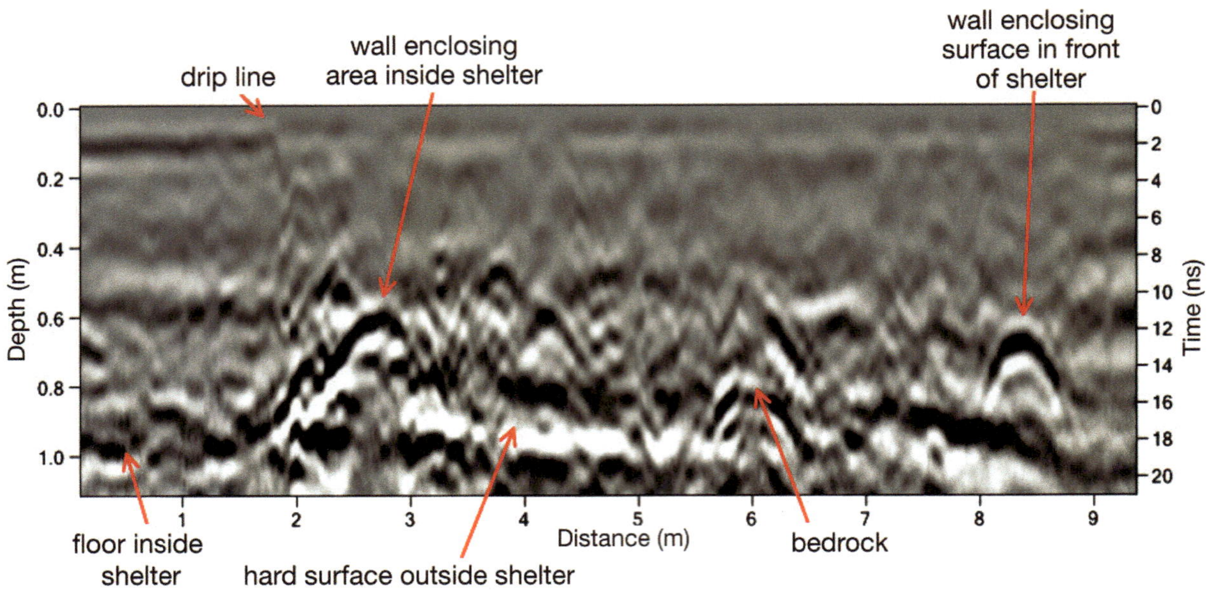

Figure 9-27: Reflection profile of features near a rock shelter. This profile shows the nonreflective sediment within the drip line and two enclosing walls of a rock shelter in Ethiopia. One wall enclosed the shelter proper, and the other was near the edge of the terrace cliff. Data collected with 400 MHz antennas.

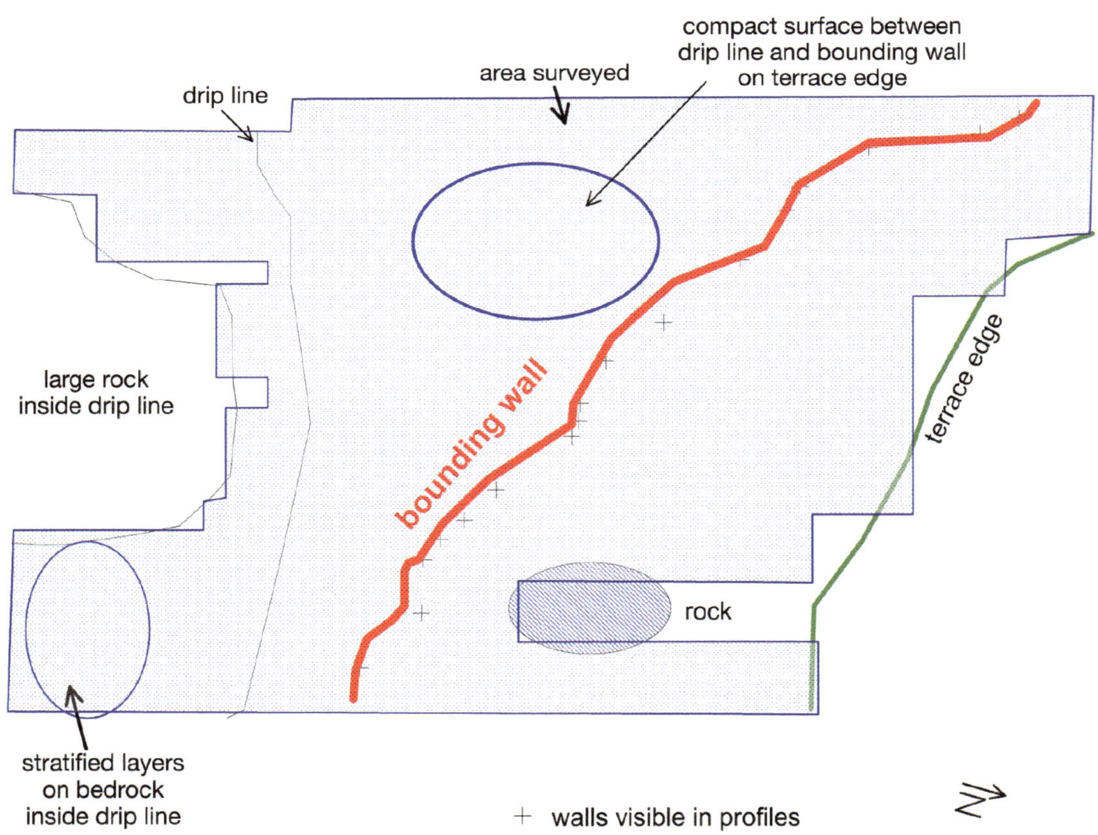

Figure 9-28: Annotated map of reflections of cultural importance near a small rock shelter in highland Ethiopia.

Conclusions

In most archaeological studies, there are good reasons for doing the research beyond just finding interesting things and describing them. The same can be said for GPR studies in archaeology, where there is usually an interesting research question or hypothesis that can be tested with the data acquired. GPR has not in the past been used this way, and most still consider it only an exploration tool. I always ask my students "So what?" when they produce interesting images and maps of buried sites using GPR, and often they look back at me with forlorn faces. For them, as for many GPR scientists, the production of images and a basic interpretation of what is below the surface are reason enough for doing GPR. I always push my students and colleagues further to ask what their discoveries have yielded about prehistoric people besides locating their buried features. Teaching as I do in an anthropology department, I prefer to focus on what GPR information can actually tell us about those people, not just what it can say about the ground. In Chapter 11, I present two studies where my students and I did just that with GPR images. Some of the examples presented there, and elsewhere in this book, are the product of my students' and colleagues' work where they asked the "so what" question of themselves.

In the example presented at the Bluff Greathouse (Figures 9-12, 9-13, 9-14), Sarah Lowry demonstrated some interesting use of the back terrace of the structure, which seems to confirm her hypothesis about the reuse of a Chaco feature in a post-Chaco time. In her study, the more formal Chaco terrace was shown to have been reconfigured into a location for ephemeral structures associated with the road that led north from the site. It appears that after Chaco collapsed, the "Chaco architectural features" such as the terrace were no longer considered as important, but still were used and modified for more mundane purposes. The Chaco road was probably still in use at that time, but maybe not in the same way as before, and the short-use shelters next to it support the idea of the recycling of previously important architectural components into more day-to-day uses.

At the Caranqui site in Ecuador, I was only able to use the GPR data to pose some interesting hypotheses regarding the classes or divisions that might have existed in the Inca palace site, which could be seen in the architectural differences in the structures (Figure 9-7). Targeted excavations are the next step to test both of the connected room features to determine if these differences were functional (different usage for each structure) or if the rooms were just inhabited by different groups of people living together near the palace. Both hypotheses are interesting within the framework of an Inca palace during its last decade or so before the Spanish conquest. At that time, there was a civil war between the northern and southern Inca polities, and the northern polity in Ecuador was still attempting to conquer and incorporate groups just to the north of this area. The GPR data from this small site could point the way to testing ideas about internal Inca social divisions and organization of the royal household during this dramatic time.

While I did not work at the Crystal River site in Florida, Victor Thompson's discovery of the three different construction episodes on one of the shell mounds is quite interesting within the framework of monumental construction during a time of postulated population growth (Figure 9-17). It appears that on one mound during one of these building episodes, the top of the mound contained structures with stone foundations or walls, which can be seen quite well on GPR profiles and are possibly indicative of a use of the mound during that time that was different from its use in previous periods. Perhaps these shell mounds ceased to be just everyday living surfaces or places where shells were discarded, and were transformed into ceremonial structures. The reconfiguration of the mounds, inferred with the GPR images, might point to a change in social structure or the incorporation of new ideas regarding classes of people and their motivations in building monumental structures. Only targeted excavations in the areas delineated by GPR can begin to test the ideas generated with the images presented here.

10

Caves, Tunnels, and Void Spaces

The contact between almost any material and a void space will create a high-amplitude reflection, which is readily visible in reflection profiles. It was this realization that led to one of the first applications of GPR by the U.S. military, where early systems were used to explore for tunnels constructed by the Viet Cong during the Vietnam War. Today, a number of researchers are using GPR in much the same way to search for smugglers' tunnels. I even got a call from someone in 2001 wondering if I could use this technology to find Osama bin Laden in a cave in Afghanistan. I told him that they would need to find someone braver than I to work on that project using GPR.

Most manufacturers of systems admit that much of their profit comes from the booming business in utility locating, and little money is made from archaeologists. The GPR method is almost the only geophysical technique that can find and map buried conduits for fiber-optic cables and other plastic pipes with no metal in them. With objects like this, it is the void spaces that produce the reflections needed for their identification.

A simple arched tunnel carved into stone, where the ceiling and sides are smooth and therefore produce few "scattered" reflections from that interface, is a simple case. The smooth interior wall produces one reflection that appears as a broad arch, which mimics the geometry of the top of this feature (Figure 10-1). Reflections generated at the interface with the void will display reversed polarity (Figure 8-7). When profiles cross tunnels of this sort perpendicularly, they will look much like large point-source reflection hyperbolas. If the transect along which the antenna is moved on the ground surface is parallel to the tunnel, only a high-amplitude planar reflection will be recorded from its crest. In this orientation, any energy that encounters the tunnel along its sloping sides will likely be reflected away from the surface antenna and not recorded (Figure 10-1). There are also other reflections recorded in this simple example, which appear to have been generated within the void, but it is difficult to determine their origin. It is not uncommon to record "stray" reflections from waves that enter a void at angles other than vertical and then reflect from discontinuities along the ceiling, walls, and floor of these features, and these can be difficult to interpret. Reflections of this sort, which have an unknown origin, can also be seen below the distinct arch reflection in Figure 10-1.

Complex Reflections from Void Spaces: Bridges and Catacombs

An arched bridge over a small drainage in central Crete, Greece, near the ancient city of Axos, provides a good example of GPR over larger arches, demonstrating the kind of complexity that can make interpretation interesting (Figure 10-2). A general interpretation readily shows that the most distinct reflections in the profile mimic the arches almost perfectly. The riverbed below the left arch produced a reflection that was pulled up due to the increased velocity of energy moving in air within the void below the arch, also as expected. On closer inspection, many other reflections were recorded that are not as readily interpretable. There were low-amplitude air waves recorded between the arch and the riverbed reflections, which were produced by energy traveling in complex ways within the void. Visible air

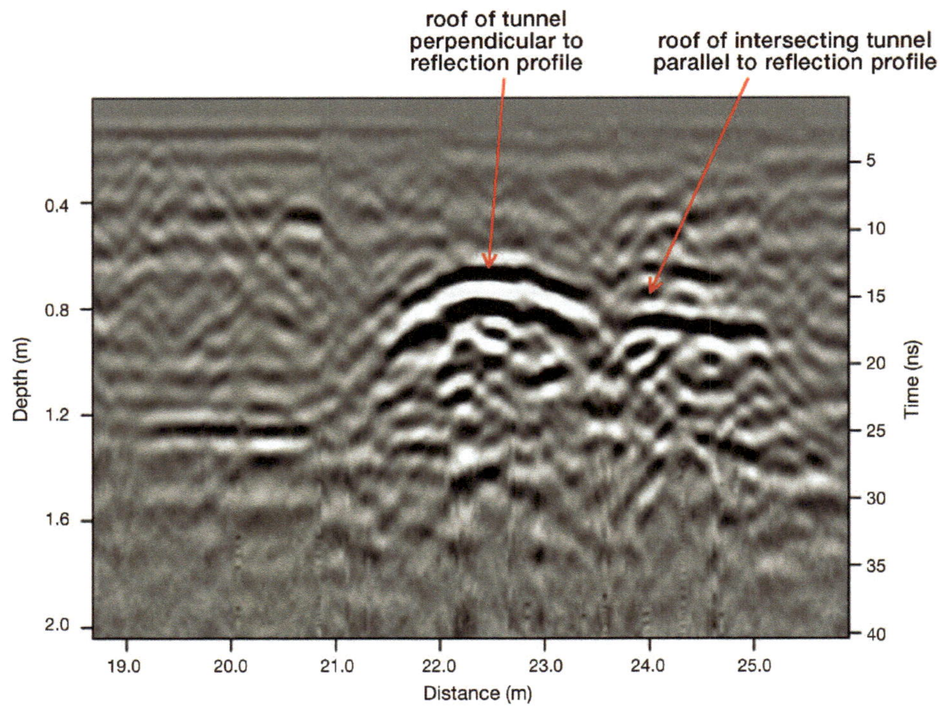

Figure 10-1: Reflection profile showing two perpendicular, intersecting rock-cut tunnels. One tunnel shows up as a distinct arched reflection, and the other, oriented parallel to the movement of the surface antennas, is visible as a high-amplitude planar reflection. Data collected near Jerusalem, Israel, with 400 MHz antennas.

waves within a reflection profile, bounded by reflections generated at more usual boundaries between stones or soil units, are good evidence that there is a void space below the ground surface. In the case of this bridge profile, the void space was known, but if a profile were collected elsewhere over a buried feature that was expected to be a void, air waves of this sort would be very good evidence of an open space below.

The arch on the left produced a number of discontinuous reflections at the boundaries of stones that make up the arch (Figure 10-2). This is more obvious in the left arch, but not every stone produced its own distinctive reflection, probably because a 450 MHz antenna was used in this project, which tends to "blur" some of the reflections recorded. Each stone would likely generate its own reflection if a higher-frequency antenna had been used, but there is always a trade-off involved in projects of this sort between energy penetration depth and resolution when deciding what antenna frequency to use.

The origins of all the reflections under the right arch (Figure 10-2) are unknown but may be from air waves and other reflections from the streambed, all interfering with one another. The left arch more clearly shows other complexities of interest. Between 30 and 40 ns, when the antennas were between the arches, reflections were recorded as energy moved at an oblique angle to the void space in front (or perhaps behind), which were recorded as if the feature that produced it were directly below (Figure 10-2). Those reflections are somewhat discontinuous and lower in amplitude than the reflections generated when the antennas were directly above the arches. This "false arch" reflection occurred because the radar waves that traveled along those oblique paths lost some energy due to attenuation, and traveled at slightly varying velocities along their path to and from the antennas. This seemingly simple example, which shows the arches distinctly, demonstrates how complicated the other, more subtle reflections can be when radar waves move through void spaces.

One of my first projects to be focused almost wholly on finding and mapping void spaces took place along the coast in Tunisia where there is an abundance of catacombs and underground Christian churches dating to pre-Roman times. A test was performed in one easily accessible catacomb near Salakta where we could readily measure the thickness of the rock above the void space and study its interior as a way to test GPR in features of this sort. The catacomb ceiling was less than a meter

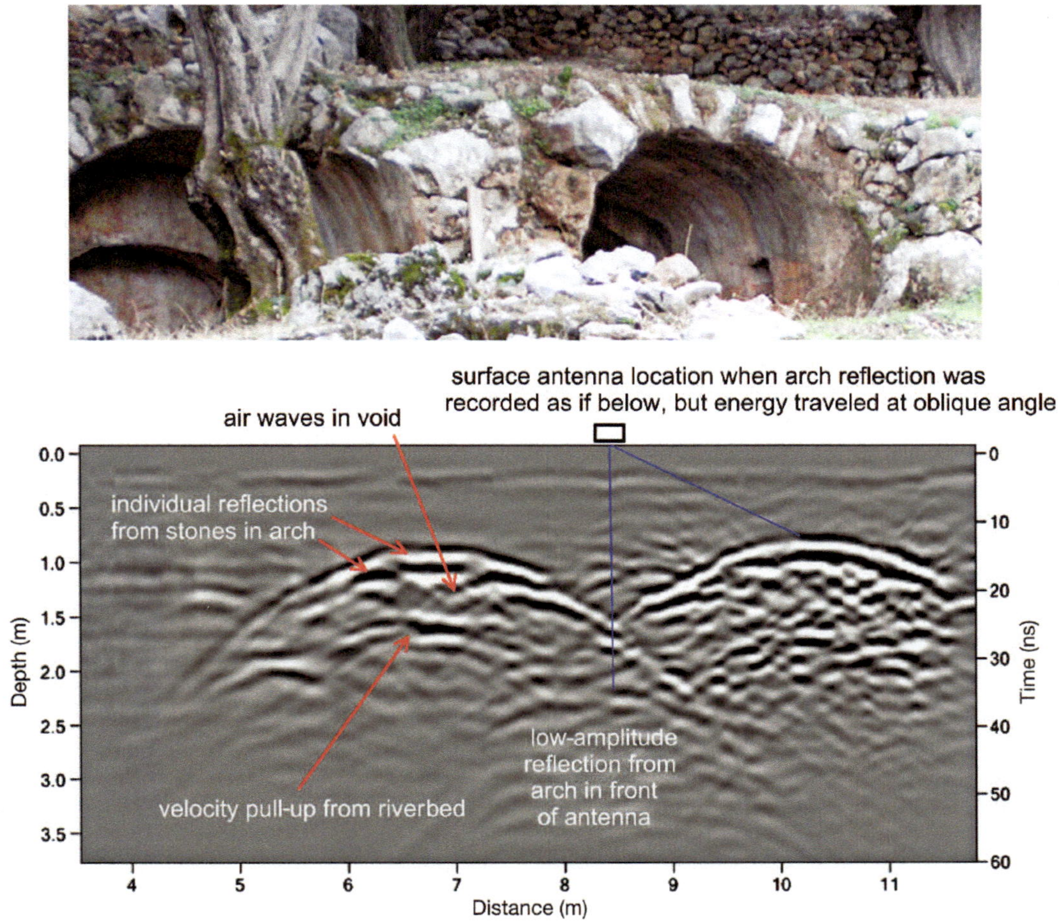

Figure 10-2: Reflection profile collected over a two-arch bridge at Ellinospita Mouri, near the ancient city of Axos, central Crete. The arches are quite distinct, but many other complicated reflections were also recorded. Data collected using 450 MHz antennas. *Courtesy of Apostolos Sarris.*

thick, and bedrock was composed of limestone (Figure 10-3). Inside this catacomb were many shelves that at one time held the remains of thousands of people, all now looted and destroyed.

I was initially quite confused when viewing the profiles, which had been collected using both 400 and 270 MHz antennas crossing this catacomb at a right angle (Figure 10-4). The arched ceiling of the void space was only faintly visible as a low-amplitude reflection, because much of the propagating energy that encountered it was reflected away from the surface antennas and not recorded. The void space reflection is readily visible as a reversed-polarity wave (black–white–black, as in Figure 8-7). The highest-amplitude reflections recorded were generated from the floor of the catacomb, which showed a distinct velocity pull-up. Another very high-amplitude reflection was also recorded in all the profiles, one that was initially difficult to explain (Figure 10-4). That high-amplitude reflection appeared to be almost straight, which seemed to suggest an air wave of some sort, but reflected from what? Only when I went back into the catacomb to locate the source of this distinct reflection did its origin become apparent. One particular shelf constructed along one wall of the catacomb was ideally oriented to act as a perfect radar reflection surface. It had focused all the energy that encountered it directly back to the surface receiving antenna when that antenna was located almost on top of the crest of the catacomb ceiling. That small focusing surface generated half of a point-source hyperbola from one small area on the wall of the void. This was my first encounter with the "stray" reflections that are common in void spaces of this sort and was a good reminder to think about radar reflections in three dimensions when interpreting data such

Figure 10-3: This test catacomb near Salakta, Tunisia, has less than a meter of limestone bedrock at the thinnest part of the roof and many shelves for bodies along its walls.

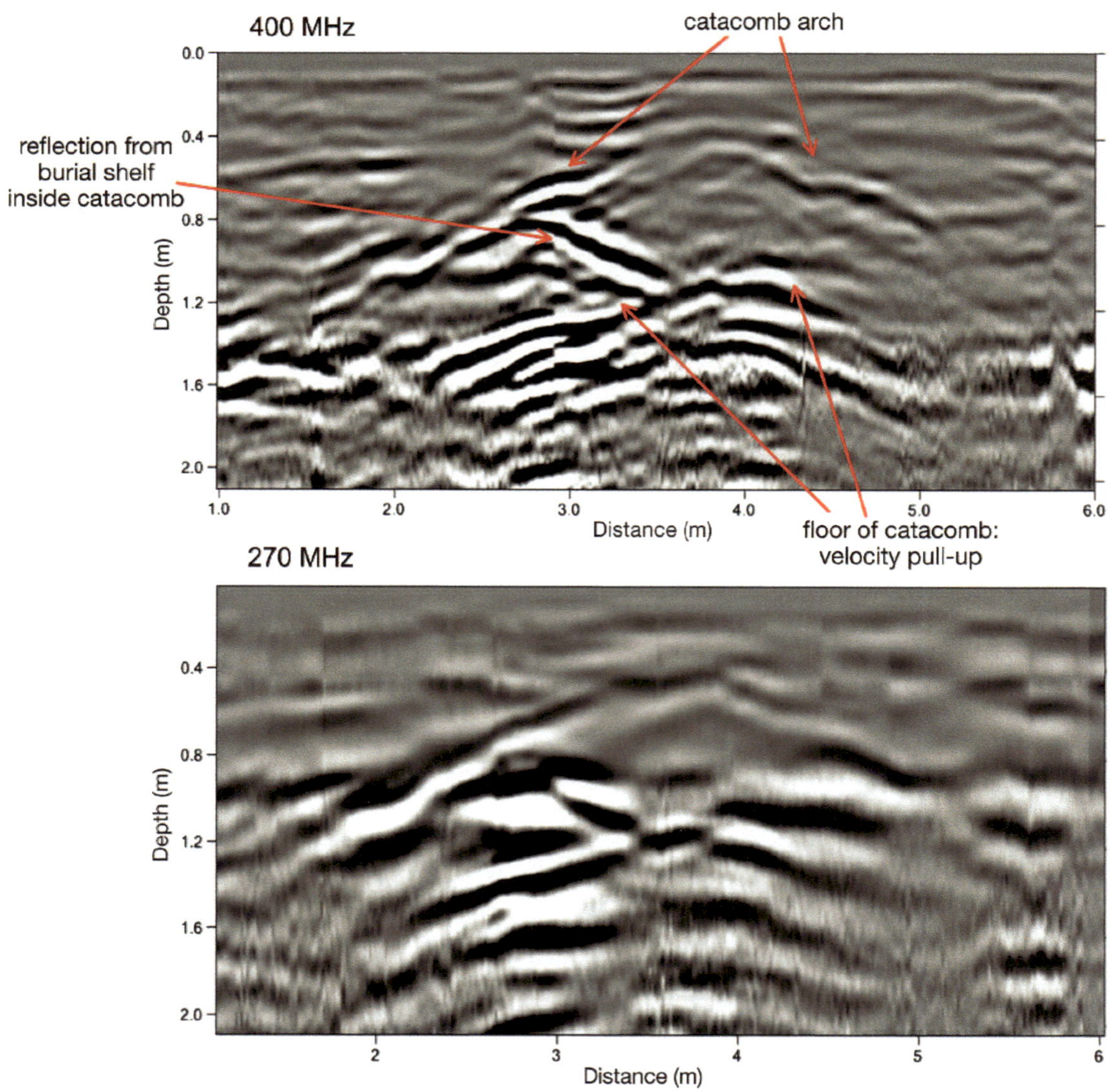

Figure 10-4: Reflection profiles over the catacomb at Salakta, Tunisia, collected with both 400 and 270 MHz antennas. The catacomb is the same as the one shown in Figure 10-3.

as these. It was a good learning experience, as it highlighted the types of complex reflections that can be produced from within a void space.

As energy enters a void, it speeds back up to the speed of light, and there is no attenuation of the propagating waves until they hit the next surface that will reflect energy. If that surface is the floor of the void, a velocity pull-up will be produced in the resulting reflection from the energy that moved vertically from the surface antenna downward. However, if some of the energy from the surface antenna moved into the void space at an angle, due to the wide transmission beam, energy could intersect other surfaces and be reflected, possibly from projecting stones in the walls of the void, or from the shelves in this case, which acted as a small focusing surface.

In more deeply buried void spaces, lower-frequency antennas would have to be used to get the penetration necessary to intersect the feature. In this catacomb test, the 270 MHz reflection profiles showed almost the same features as the 400 MHz, but with far less resolution (Figure 10-4).

Synthetic Modeling as an Interpretation Tool for Voids

Having a vague understanding of what catacombs and other void spaces might look like in the limestone area of Tunisia, we collected profiles in an area where a local resident suspected the presence of an underground Christian church. He had noticed one area in an olive grove where water never pooled on the surface during large winter rainstorms. This information was passed along to a local archaeologist, who notified us of his suspicions. Reflection profiles were collected, first using 400 MHz antennas, but it was quickly determined that energy of that frequency was not being transmitted to the depth necessary to intersect any possible void spaces, necessitating the use of 270 MHz antennas (Figure 10-5).

As an aid in understanding what types of reflections might be generated from an arched void space with a flat floor, a two-dimensional model was produced using the software program called GPRSIM (Goodman 1994). This ingenious program allows the user to generate a synthetic model of what might be expected using known properties of the ground and the geometry of expected underground features. I produced a simple model of an underground nave of a church. This model (Figure 10-6) assumed homogeneous limestone with an RDP of 10, and a hard, flat floor of the church with an RDP of 30, which might be expected of a wet clay layer deposited from centuries of rainwater percolation into the cavity. The synthetic reflections demonstrated that the arch would generate a reflection when energy intersected the void space, which was expected. The floor also generated an upward bowing reflection from the velocity pull-up in the void space in the model. What was unexpected, and not immediately visible in the model, were the air waves generated from the upper corners of the church, which are displayed as straight reflections cutting across the void (Figure 10-6). When this model was compared with the reflection profiles collected in the olive orchard, all these modeled reflections were visible. The 270 MHz profiles were sufficient to produce an image of the underground church, but without the resolution that higher-frequency antennas would potentially produce.

Complex Small-Scale Voids

When mapping a Roman site called Hippos-Sussita in the Golan Heights of Israel (Conyers 2006b), just east of the Sea of Galilee, we came across some interesting ground stones that appeared to be related to liquid processing (Figure 10-7). Large ground-stone artifacts of this type are well known in the Mediterranean area as collection and straining devices used during the processing of grapes into wine and olives into oil. Usually these stones sit below a crushing device, which could be a mechanical press where the liquids flow out of the bottom into a channel, or a series of channels that funnel them into a collection vessel. The presence of these stone

Figure 10-5: Collecting reflection profiles in an olive grove in Tunisia. The 270 MHz reflection profiles were used to test what was under the surface in an area where water did not pool on the ground during rainstorms, suggesting a void space below.

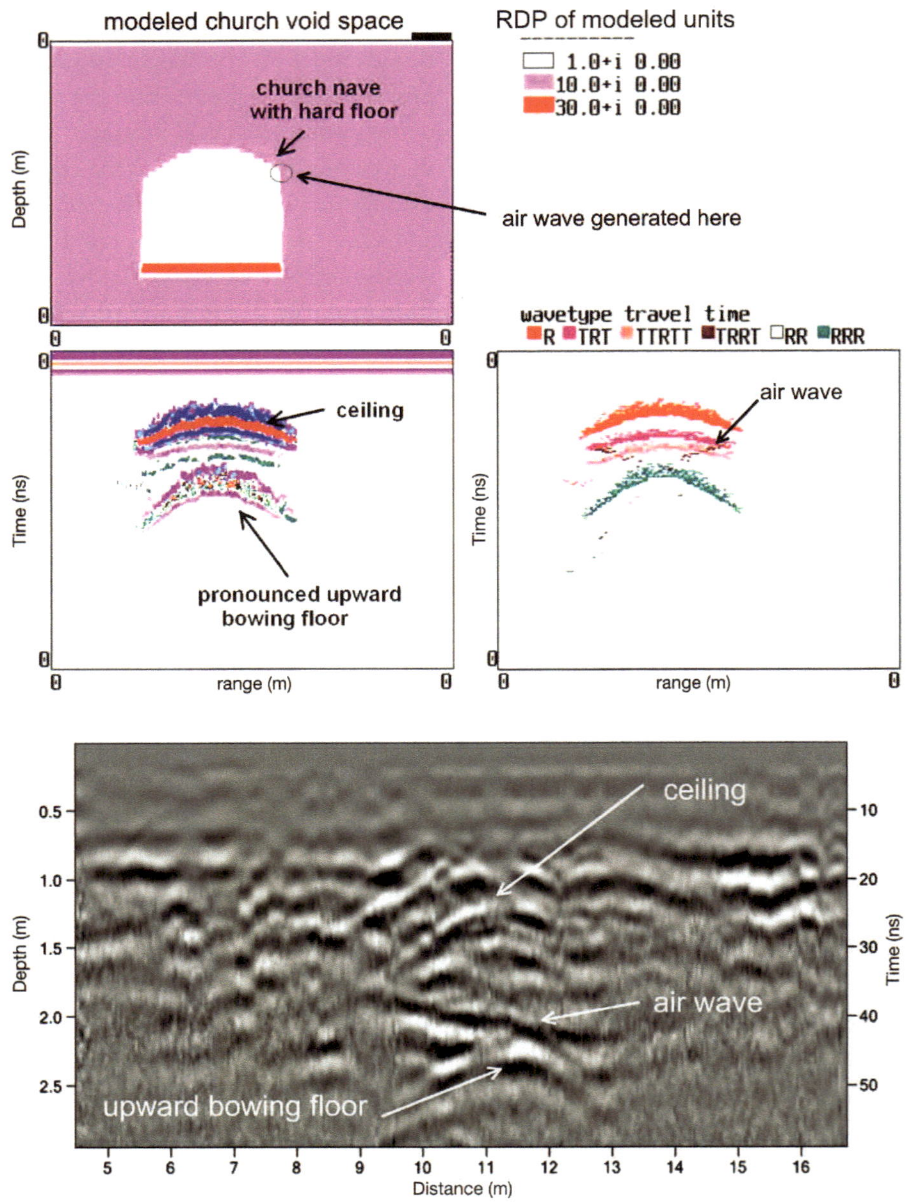

Figure 10-6: Comparison of a synthetic model of an underground church to an actual reflection profile. This is an underground church in Tunisia. All the modeled reflections are visible in this 270 MHz reflection profile.

Figure 10-7: Stone artifacts suggesting that grapes had been processed into wine nearby. This is the Hippos-Sussita site in Israel.

artifacts suggested that there had been wine-making activities close by, and a nearby flat area was suggested as the location where the press had sat. This platform consists of a small, flat, paved area, bounded by low stone walls, with a small opening in the middle of the floor, which appeared to be a likely candidate for the place where liquid was drained from the overlying press (Figure 10-8).

A 5 × 4.5 m grid was set up on the floor of the flat area, and 400 MHz reflection profiles were collected (Figure 10-8). The collection of this grid was an afterthought, at the end of a day's work mapping Roman walls; daylight was running out, so I only had time to collect reflection profiles that were spaced 50 cm apart. Also, I only had access to 400 MHz antennas for the collection of these data. I now know that much better results would have been obtained with higher-frequency antennas and denser profile spacing. However, I was still able to derive some interesting results from an interpretation of these data.

In the reflection profiles, a small-diameter pipe produced a distinctive reflection about 75 cm below the upper stone surface (Figure 10-9). In other profiles, a more complicated series of reflections was recorded from about this same depth, which appeared to have been generated from the interior corners of a square "collection box" for the liquid produced above. The interior angles in this box generated one arm of a point-source hyperbolic reflection feature (Figure 10-9: Profile 2). When all the reflection profiles were resampled and plotted in amplitude maps, the outlines of the collection box, directly under the screw hole apparent on the paved surface (Figure 10-10), were visible. Also, a linear reflection feature, which was generated from the pipe leading from that box at an angle, trends toward the southwest corner of this feature. We moved some of the architectural rubble away from the area where that pipe was trending and found a small stone-pipe outlet, which would have led to some kind of a collection vat for the liquids produced by the press placed on the flat surface above (Figure 10-8).

The amplitude maps of this feature produced high-amplitude reflections that were not immediately interpretable as a collection box for liquids (Figure 10-10). Instead, three distinct reflections were generated from the interior corners of that box, directly under the screw hole for the overlying press, visible in the 25–50 cm slice.

Figure 10-8: Flat surface hypothesized to be a grape-crushing platform, from the Sussita-Hippos site, Israel, where GPR profiles were collected. The Sea of Galilee is in the background.

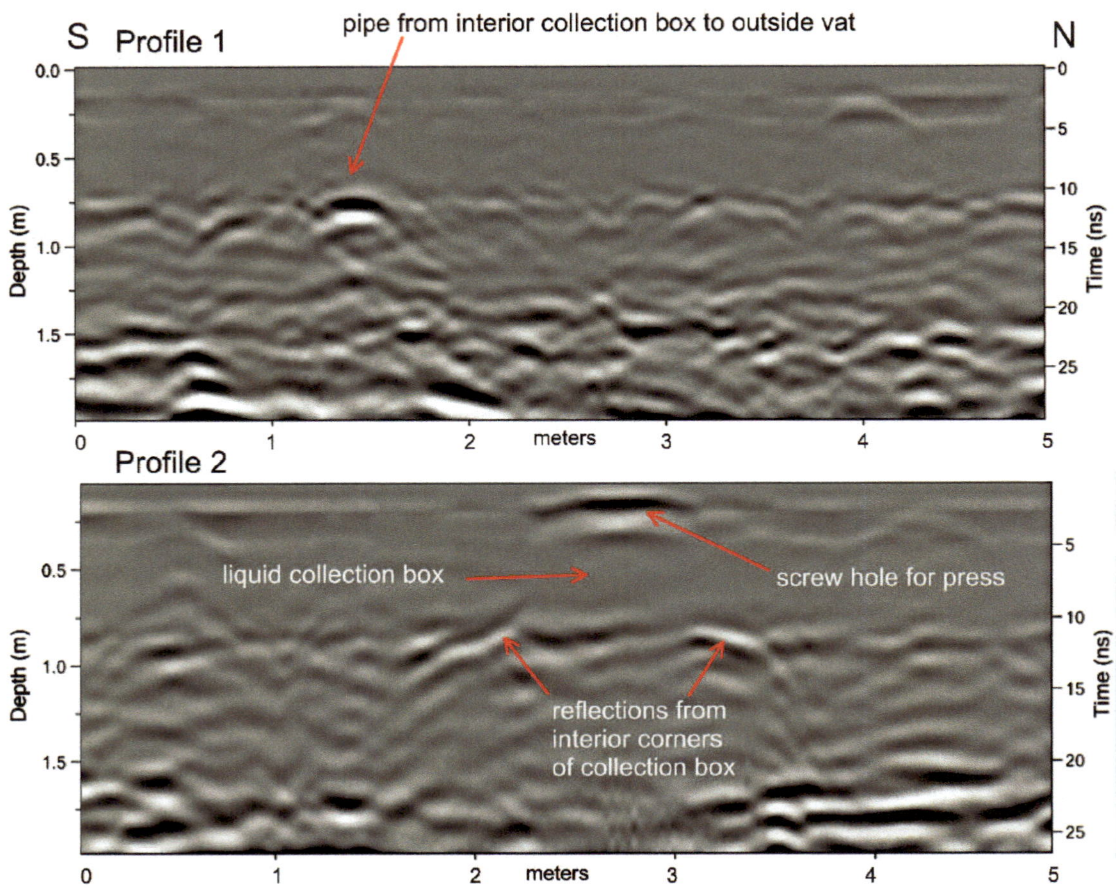

Figure 10-9: Reflection profiles crossing a wine-collection feature. This is the feature from Figure 10-8 at the Hippos-Sussita site in Israel. The collection box below the screw hole is visible as an area of no reflection in a void space, but the interior corners of the square-shaped void produced one axis of a point-source reflection. Data collected with 400 MHz antennas, with profile locations shown in Figure 10-10.

In the slice from the 50–75 cm depth, the pipe leading from that collection box projects toward the southwest corner of the grid, where its outlet was discovered (Figure 10-8). The origin of other high-amplitude reflections in these slice-maps is not known but are likely reflections generated from interfaces between courses of stone used in the construction of this wine-press platform and collection feature.

LAVA TUBES AND CAVES

Lava tubes, a type of cave common in volcanic areas, are produced in areas with abundant basalt flows. They form when the slower-cooling interior liquid lava in a flow drains downslope long after the surface of the flow has hardened. The flows then solidify and contain a number of sinuous linear "tubes" within them. In Hawaii, these tubes are commonly produced in lava flows called *pahoehoe*, a term used for very fluid basalt that forms under a congealing surface crust. These long tunnels, used as burial chambers by the ancient Hawaiian people, are considered sacred sites today. Often during modern construction operations in areas underlain by lava tubes, heavy equipment falls through the ceilings of these void spaces, potentially destroying and desecrating ancient burials. As a test of GPR to produce images of these features, a number of profiles were collected over known tubes on the Island of Hawaii (the Big Island) in an area where the lava flows are 5,000 years old or younger, derived from Mauna Loa volcano. The study area is located on the arid leeward side of the island, where there has been much less weathering of the basalt, and therefore little electrically conductive clay has been produced within the rock (Conyers and Connell 2007). All of the lava tubes tested were visible along ceiling collapse features, and their geometry could

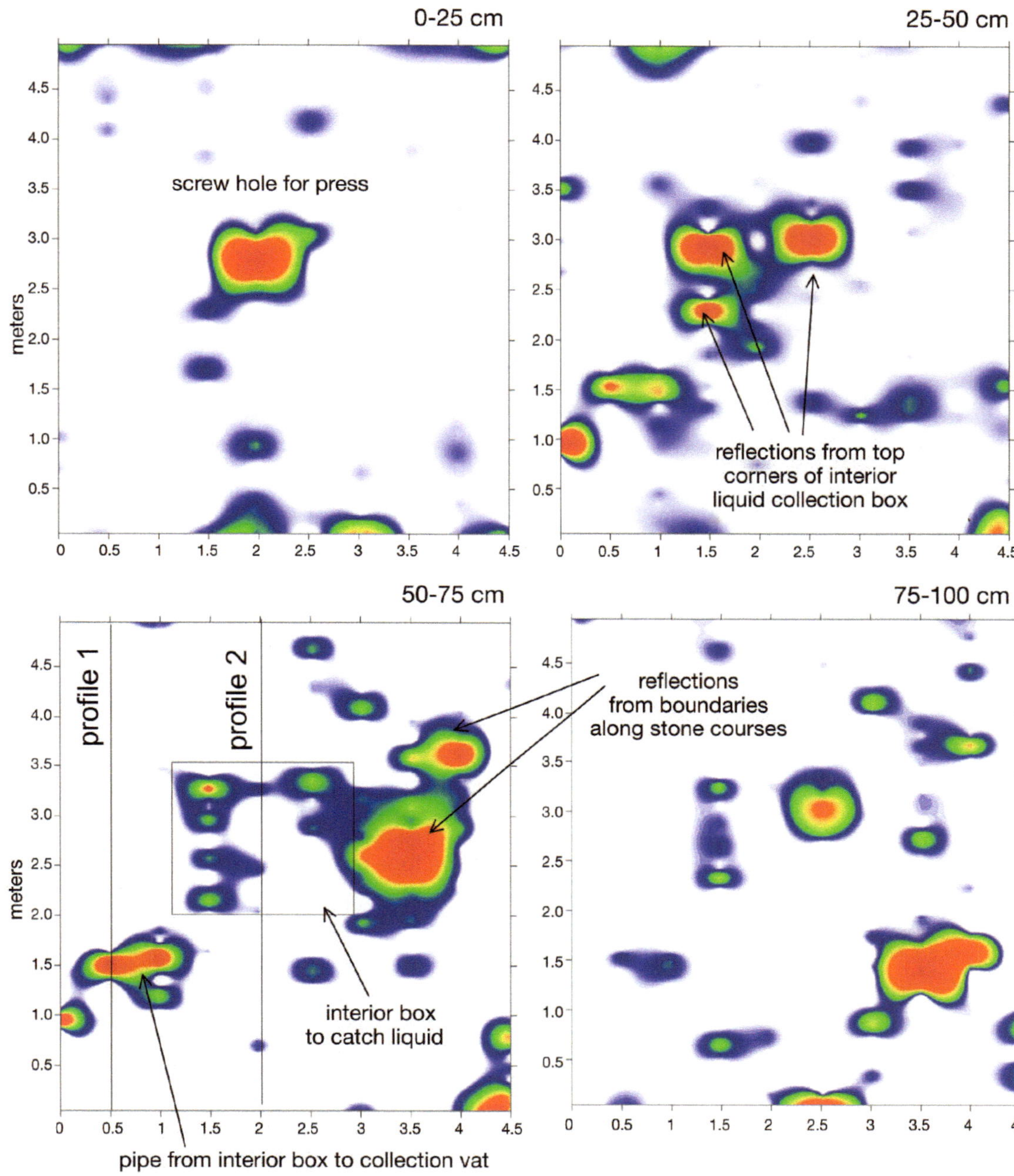

Figure 10-10: Amplitude maps of the collection box below a wine-press. Data collected at the Hippos-Sussita site, Israel, with 400 MHz antennas.

be mapped (Figure 10-11). These became very good models for void spaces of this sort in basalt flows.

The reflection profiles collected over a number of these lava tubes produced a wide variety of reflections, which were often difficult to interpret without taking into account all of the variables that produce reflections in voids, discussed above. The most common features, visible in all profiles, were reflections generated from the jagged edges of *pahoehoe* flows along the sides of the tubes (Figure 10-12). A number of complex reflections were recorded within the empty space of the tube, probably generated from subtle reflections recorded as energy that traveled to and from the surface antennas at an angle. Those reflections tend to be discontinuous and often lower in amplitude than those generated from the basalt bed boundaries more directly below the antennas. The ceiling of the tube is sometimes visible as a continuous reflection. Reflections from these tube features are often so complex that a reverse-polarity reflection generated from the void space (Figure 9-7) is not apparent.

Sometimes the ceiling of the lava tube does not produce a distinct reflection, perhaps because it is irregular, and energy reflected at the void-stone interface is diffracted and reflected away from the surface antennas and not recorded. Irregular ceilings of lava tubes are common, especially where there has been some collapse after the lava hardened and the tunnel was created (Figure 10-13). The key to locating the lava tubes in this type of complex reflection profile, where there are few truncated lava flow reflections along the edges of the void, is to search for the air waves created within.

Figure 10-11: Measuring the ceiling thickness and void space in a very large lava tube in Hawaii. This lava tube near Kona, Hawaii, contains rubble on the floor from ceiling collapse, which allowed us access to this long, sinuous tunnel.

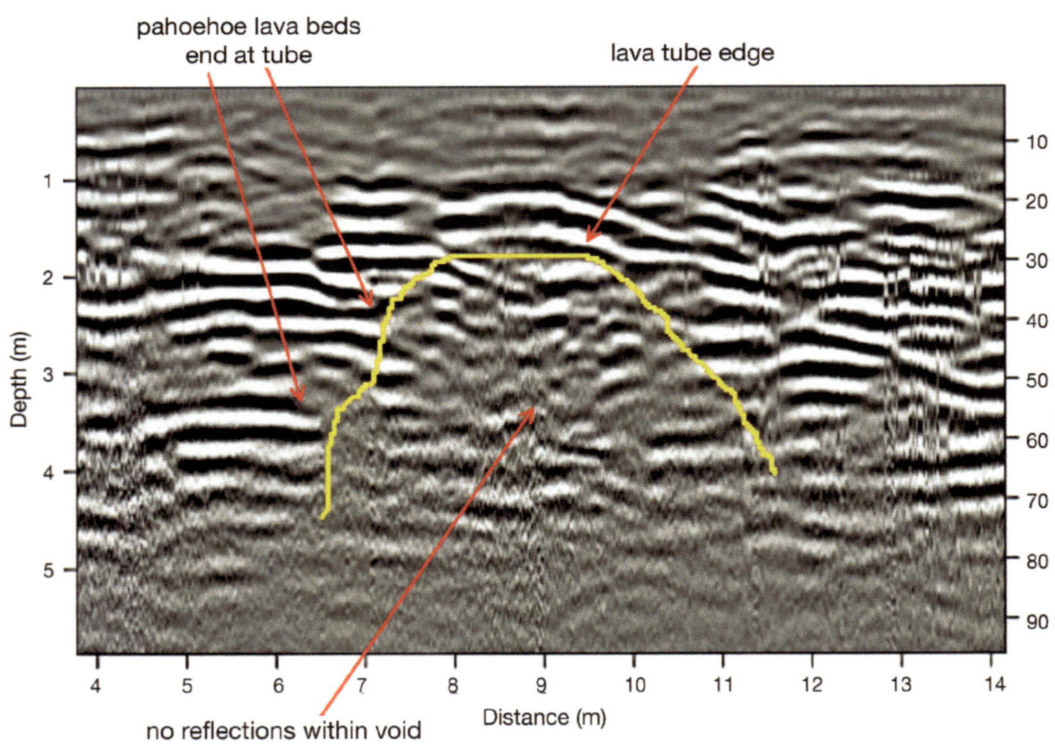

Figure 10-12: Reflection profile crossing a known lava tube. This tube, called O'oma Ahupua'a, is in the Kona area of Hawaii. There are no continuous reflections recorded within the void space, and the bounding basalt flows produce high-amplitude planar reflections. Data collected with 270 MHz antennas.

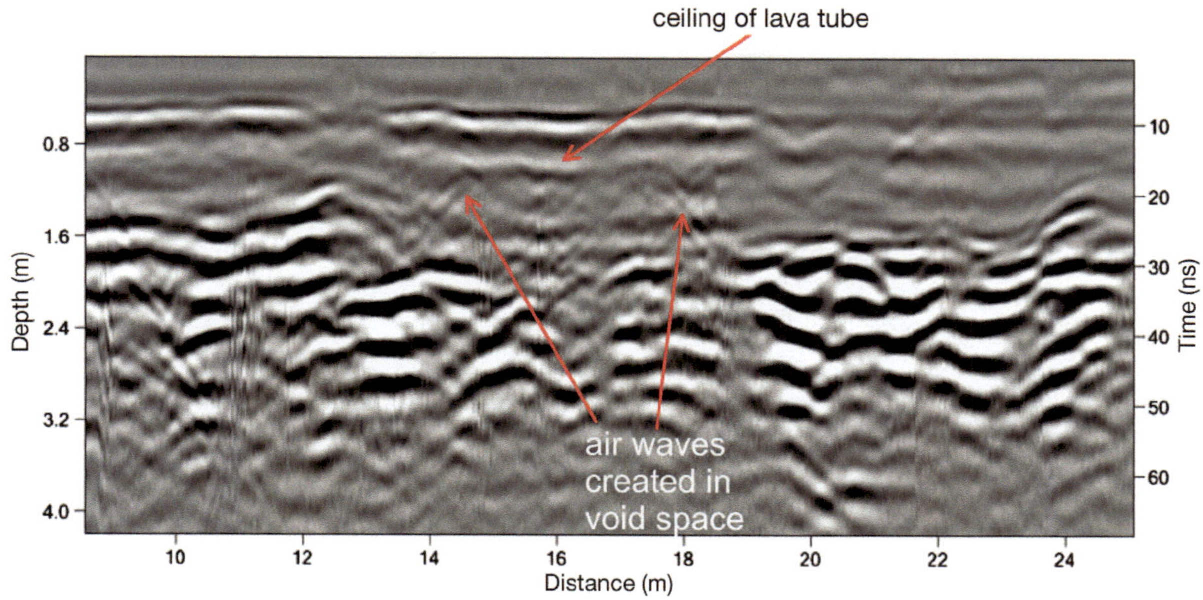

Figure 10-13: Reflection profile over a lava tube with air waves generated within the void space. This is the lava tube at the Pu'uwa'awa'a Ahupua'a site, Kiholo Bay, Kona area, Hawaii. Data collected with 270 MHz antennas.

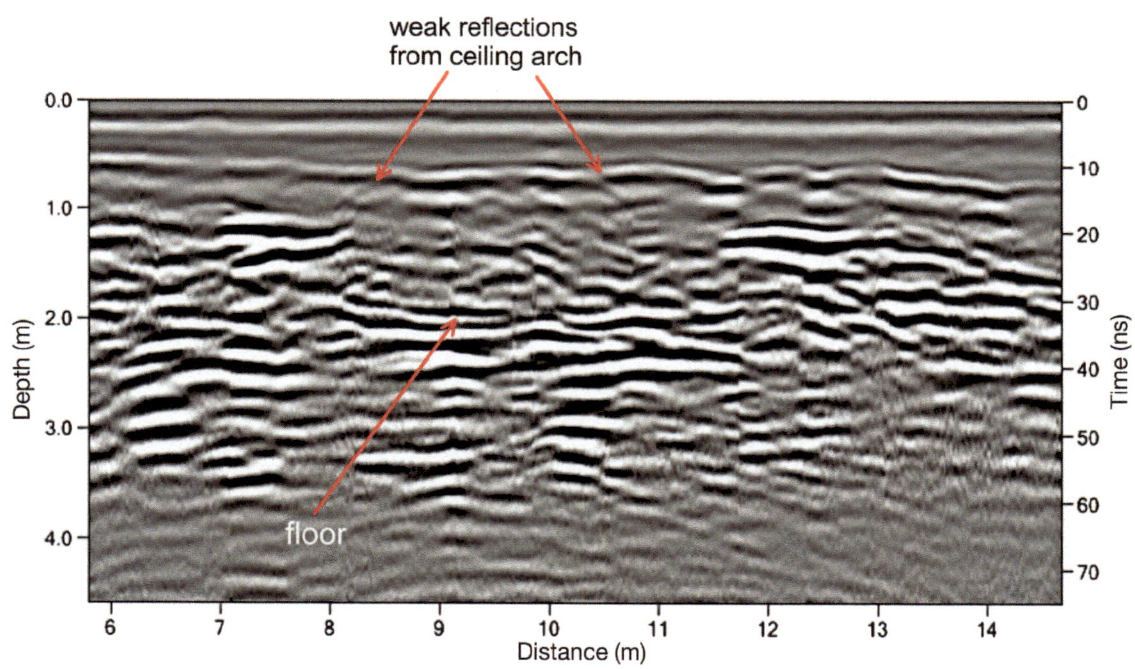

Figure 10-14: Reflection profile over a lava tube showing bed truncation. Data collected at the lava tube at the Pu'uwa'awa'a Ahupua'a site, Kiholo Bay, Kona area, Hawaii, with 270 MHz antennas.

Floor reflections are also visible in some profiles (Figure 10-14). The edge of the lava tube is indicated by basalt flows that are truncated on the edge of the void, while the ceiling produced only very weak reflections.

In an attempt to simplify the very busy reflection profiles common in basalt flows and lava tubes, reflection traces were stacked during collection, producing a more averaged profile (Figure 10-15). This was an effective way

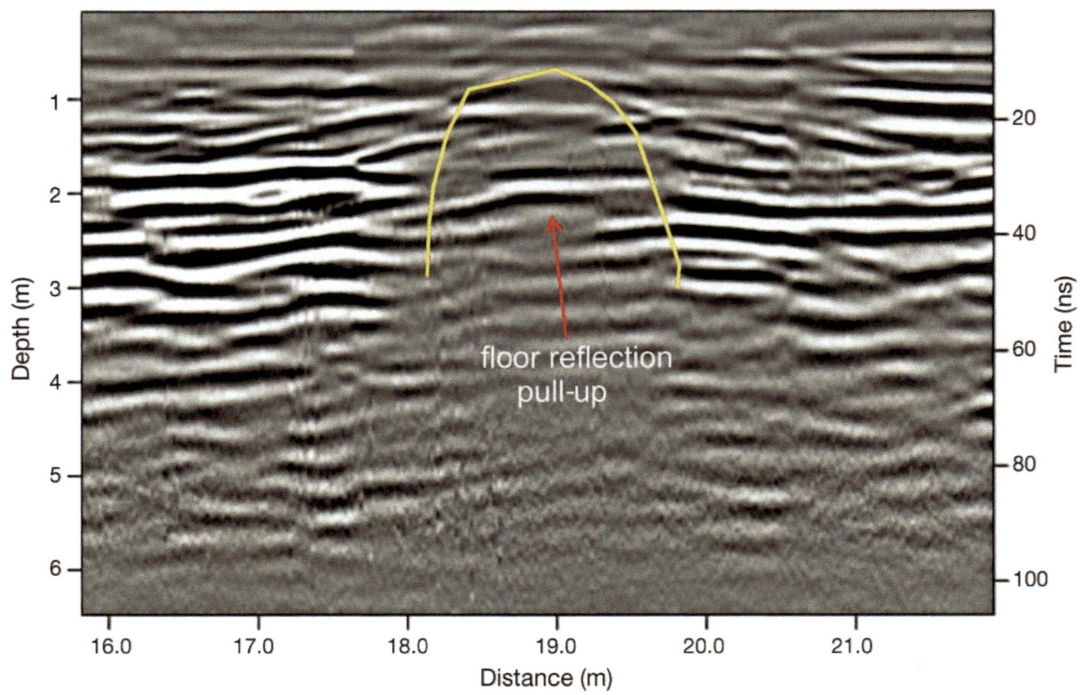

Figure 10-15: Processed reflection profile over a lava tube with a floor pull-up. This profile is from Pu'uwa'awa'a Ahupua'a at Kiholo Bay, Kona area, Hawaii, where traces were stacked with a four-trace average. The edges of the tube are visible by the truncated flow reflections, and the floor is pulled up under the void space. All the more subtle reflections, such as air waves, have been destroyed by this processing step. Data collected with 270 MHz antennas.

to show the edges of the basalt flows adjacent to the tube, but the resulting averaged reflections are still complex. In Figure 10-15, the floor reflection is clearly visible as a velocity pull-up, and the truncated flows are also apparent. By producing processed averaged trace profiles of this sort, all the air waves and other more subtle reflections that were generated by these complex features were destroyed. For this reason, I prefer to use unprocessed reflection profiles and then spend the time necessary for analyzing each, to locate the void spaces.

As much as I have tried to produce useful amplitude maps of lava tubes, the reflections are so complicated that they have proven to be almost useless. Air waves, out-of-plane reflections, velocity pull-ups of the floor reflections, and a variety of reflections generated from edges and ceiling irregularities produce only confusing amplitude maps when the profiles are sliced and reflections displayed in map form. The complexity of these features necessitates interpretation of these types of features using individual reflection profiles.

These examples of lava tubes are probably illustrative of the types of complexity that would occur in other types of caves. If the surrounding layers are horizontally stratified layers of sediments, I would expect the reflections to look much like these from lava tubes. If the surrounding rock were fractured or had other, more random discontinuities within it, the reflection profiles would take on a whole new type of complexity, which I have not addressed here.

11

Using GPR Interpretations to Understand People

The early practitioners of geophysical archaeology, who usually came not from archaeology or anthropology but were instead recruited from physics, geology, or other, very scientifically-based disciplines, were often not versed in the disciplines that were using their interpretations. The perceived usefulness of geophysics within the archaeology community was therefore often based on its ability to find interesting buried materials that could later be excavated using standard archaeological techniques. This historical use of geophysics, which emphasized discovery rather than interpretation and analysis, was probably occasioned by the limited experience of geophysicists in addressing the questions asked by the archaeological community. The use of GPR as purely an exploratory tool is still quite prevalent in the minds of many archaeologists, whose goal in using GPR is almost always to find buried cultural materials that they can excavate, often long after we GPR folks have gone home. Another use of GPR, by a minority of practitioners, is to use these powerful imaging tools to locate buried cultural materials so they can be avoided and preserved. Even fewer use GPR interpretations to generate a sampling strategy that relies on selective subsurface testing integrated with detailed GPR studies to test ideas about human behavior, social organization, or changes in cultures over time (Conyers 2010; Conyers and Leckebusch 2010; Kvamme 2003).

I have been making the case for a number of years that geophysical archaeologists who employ GPR should begin to use their interpretive techniques for more than mere prospecting, and incorporate them into research projects in a much more holistic way (Conyers 2010; Conyers and Leckebusch 2010). In my view, the integration of GPR images with standard excavation procedures is a good way to approach questions and test anthropological and historical hypotheses about the human past in ways not otherwise possible. This integration can be especially appropriate when site layout, feature size, and the orientation and other aspects of cultural remains are visible in GPR images and can then be related to human behaviors. Using GPR in this manner, the interpretations become a primary data source from which to study the human past, not merely an adjunct to standard excavation procedures. So far, my proselytizing on this issue has met with limited support by others who use GPR. I recently asked Christophe Benech, who specializes in magnetometry in the Middle East (Benech 2007), how many of his colleagues in Europe and Asia are using geophysics to study people and not merely to locate interesting buried features. He had a hard time coming up with more than two or three and expressed the same frustrations that I have. I have resigned myself to the fact that until more anthropologically trained geophysicists begin to understand the utility of GPR, the interpretive tools presented in this book will remain underutilized.

The key to using GPR as an interpretive tool to test hypotheses about people, history, social organization, and a wealth of other interesting subjects is to ask appropriate questions about the past first and then determine if there is a way that these can be tested geophysically. Just collecting data randomly and hoping to find something to dig up is a very "old school" technique, and one I am trying to see gradually phased out, at least in academic uses of GPR.

There will always be a need to explore for buried remains in the context of cultural resource management (rescue archaeology) during construction projects. That will remain a major reason for people to use GPR, and perhaps the main justification for many to invest in equipment and software. Good money can be made in such endeavors, which perhaps is justification enough for some practitioners. I also have concluded that there is nothing wrong with earning GPR profits in the context of finding, learning about, and also helping to preserve buried cultural resources. Even this type of project can produce data that can deepen our understanding of the past in some way, provided the right questions are asked during the GPR research, and an interpretation of the results is geared toward testing hypotheses related to those questions.

If buried cultural features contain spatial patterns within human settlements or cultural landscapes that can be related directly to human behaviors, the GPR method can have great utility. Such patterns might be house size, orientation, street layouts, construction and architectural techniques, and many other variables that can be imaged with GPR. The method's greatest asset is its ability to produce three-dimensional pictures of the ground with a resolution not possible using other, shallow geophysical methods. The results can then be directly related to excavation information and those data projected into a wider area in three dimensions. In this chapter, I present two such studies where GPR was used to test ideas about the past in ways not possible using any other method, whether geophysical or archaeological (perhaps barring hugely expensive and massive excavations that expose absolutely everything in the ground). One study involved using the size and function of circular kivas to test ideas related to cultural connections in the prehistoric American Southwest. The other relied on an analysis of street patterns and building layout in a complex three-dimensional grid in Jordan to interpret social organization and the behavior of early Nabataean people prior to the Hellenistic and Roman influence that transformed Petra into a monumental city.

TESTING HYPOTHESES ABOUT BEHAVIOR AND CULTURAL CONNECTIONS IN UTAH

In the American Southwest, cultural connections between widely spaced agricultural communities within the Chaco Canyon sphere of influence have been a topic of research for decades (Conyers 2010). I have touched on this subject in other chapters when discussing the great kiva (Figure 3-8) and the greathouse and its associated road (Figure 9-12) at the Bluff, Utah, site. At a site north of Bluff, called Comb Wash, an interesting set of buried features had long been considered evidence of important "outliers" connected culturally and perhaps economically with the central place at Chaco Canyon, New Mexico, far to the southeast (Conyers 2010; Conyers and Osburn 2006). These sites had been described as "great kivas," which are known elsewhere to have been ceremonial buildings where "Chaco-like" rituals took place. Connections between Comb Wash and the "Chaco center" were suggested by the presence of roads across large expanses of desert leading from southeastern Utah to Chaco, connecting other prehistoric villages and these great kivas. Other villages near Comb Wash contain the remains of Chaco-type greathouses and great kivas (Figure 3-8).

Great kivas are semisubterranean circular structures as much as 20 m in diameter, which appear to have been used for large communal ceremonies, presumably in a way that emulated similar important ceremonies at Chaco. Sites that contain great kivas have long been considered to have been under the influence of leaders and ideas from Chaco, and their presence suggests that people who lived near these outliers were within the "Chaco sphere." Long-distance trade is also known to have taken place throughout the Chaco sphere of influence, with people on its periphery and far to the south in present-day Mexico. Items that were obtained by trade from Mexico have been uncovered in the general vicinity of Comb Wash, lending support to these hypothesized connections of people who lived in southeast Utah.

At Comb Wash, which is many hundreds of kilometers away from the Chaco center, five large, circular depressions were described and mapped, with associated Chaco-style pottery found on the ground surface. The prevailing interpretation was that the Comb Wash people built great kivas in a very Chaco-like way in order to emulate important people and ideas related to this influential prehistoric "system" (the nature of which is still being debated). The five Comb Wash surface depressions were mapped by others (Conyers and Osburn 2006) and noted to be the greatest concentration of such structures outside of Chaco Canyon itself. This was taken as evidence that some kind of very important behavior had occurred at Comb Wash, unlike any documented elsewhere within the Chaco periphery. As none of these depressions had been excavated and studied in any great detail, we had no cause to doubt this interpretation. The reasons why they had never been excavated are (1) because they are located on U.S. government–managed lands and no permits were issued; and (2) because the nearby descendant communities had opposed excavations, owing to Native Americans' sensitivity about ceremonial buildings of this sort.

When one of my students proposed GPR research at Comb Wash, after permission had been granted because of the noninvasive nature of the proposed methods, we generated a number of working hypotheses related to a study of these five supposed great kivas. These multiple hypotheses revolved around stylistic and architectural differences within them and their placement and orientation on the landscape, which could potentially be determined with GPR. The idea was that there might have been different clans or other groups who for some reason chose to gather at Comb Wash periodically in "Chaco-like" ceremonies. One hope was that GPR could identify these hypothesized differences by looking at architectural variations in these five clustered great kivas. If those differences could be documented, it would help test a variety of anthropological ideas related to ethnicity, cultural connections, central places, and other ideas concerning the Chaco phenomenon and its power to persuade people located on the northwest peripheral edge of its influence.

Many hundreds of GPR profiles were collected in five grids over each "great kiva depression" using 400 MHz antennas (Figure 11-1). During collection, we were astounded to see that none of the kivas were "great" at all, as we could identify their walls instantaneously on the survey system monitor, and they looked to be very small. This was initially disturbing, as none of our working hypotheses appeared to be either supported or refuted by the data being collected. It is always difficult for the human brain to grasp the abrupt destruction of preconceived ideas. The data were processed into amplitude maps, and in three of the five grids processed, GPR images showed small, round kivas, as well as a number of other interesting buried features nearby, including room blocks, possible storage cisterns, and the floors of pit houses (Figure 11-2). None of the kivas was larger than about 5 m in diameter, and they were not "great" in any way. Two of the large depressions contained no buried architecture whatever and are likely the remains of water retention reservoirs from historic-era cattle-ranching operations.

Figure 11-1: Collecting GPR data in surface depressions at one of the Comb Wash depressions, southeastern Utah, using 400 MHz antennas.

The GPR maps at the three sites, which were found to contain buried prehistoric kivas, indicated they were of a size more consistent with "domestic" or "family" kivas built by people for long-term uses of various types, in relatively modest farming communities (Figure 11-2). We partially excavated one of them, which confirmed the GPR results, and it yielded pottery dates coincident with the period of Chaco influence. At Kiva 1, an earlier-used pit-house floor was found near the kiva, and a room block was located to the north, which was probably occupied at the same time as the kiva (Figure 11-2). The amplitude maps at Kiva 2 showed a similar buried kiva, but no associated room block or earlier habitation structures. Kiva 3 was somewhat more complex when viewed in the GPR amplitude maps, with many geological features visible. This kiva had been constructed by first digging into the sandstone bedrock before it was roofed over. The kiva center is distinct as an area of no reflections, owing to the homogeneous aeolian fill sediment. Surrounding bedrock layers produced complicated linear reflections generated from bedding planes.

These GPR results made it necessary to reformulate all our ideas on the prehistory of the Comb Wash area within the greater Chaco interaction sphere. Instead of this area being a regional ceremonial or other Chaco-inspired center of important activities, it appears to have been only peripherally influenced by Chaco. The Chaco-like ceramics demonstrate some influence from the regional center far to the southeast, and it is likely that some of the people who lived at Comb Wash might have traveled there sometime during their lifetimes. However,

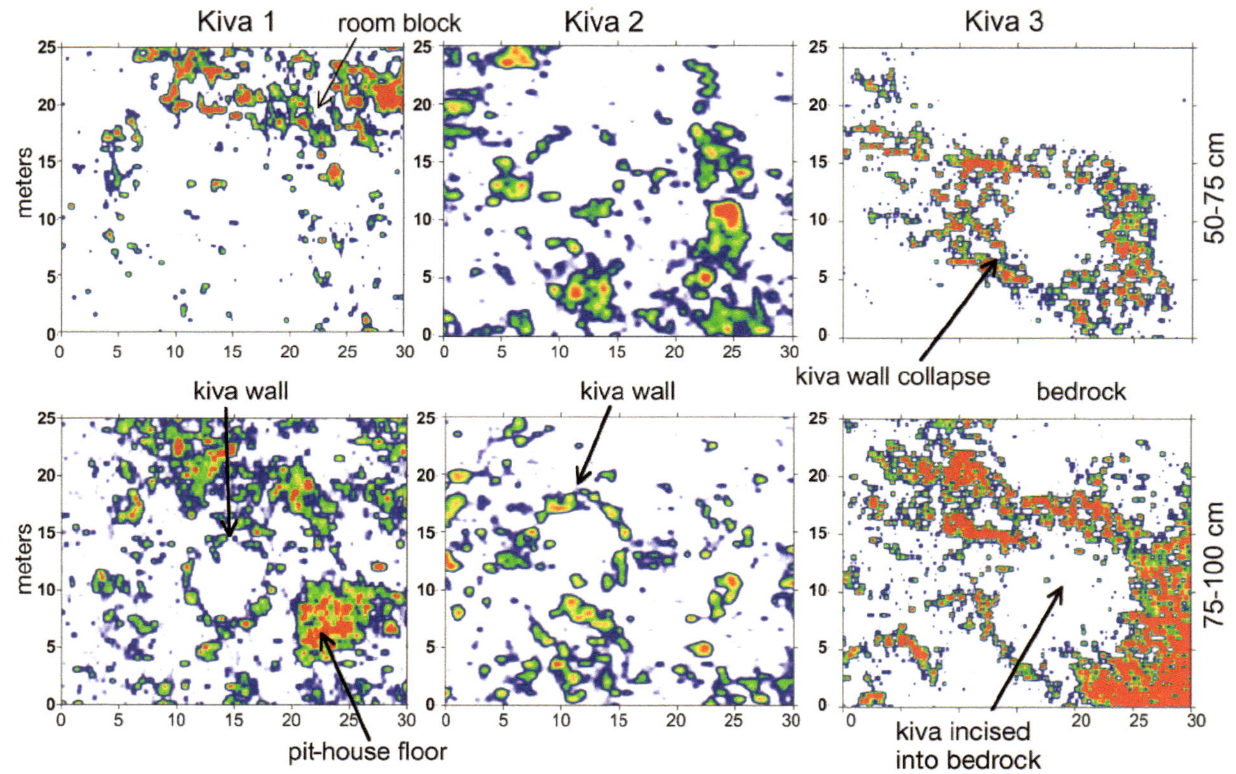

Figure 11-2: Amplitude maps of three kivas and other associated features. These images show three kivas at Comb Wash illustrating various "family size" kivas and other associated buried cultural and geological features. Data collected with 400 MHz antennas. *Courtesy of Tiffany Osburn.*

there was no evidence whatever from the GPR results that this area had been used as a large-scale ceremonial center or important gathering place of many different groups from some distance away. Instead, the interpretation indicates that prehistoric people at Comb Wash likely had lived in this area for generations, renovating existing pit dwellings and building new kivas, storage facilities, and associated above-ground room blocks while practicing everyday farming activities.

In this area, GPR was ideal for producing maps of the complex, three-dimensional aspects of sites that had architecture at different burial depths from different periods of habitation. One kiva was associated with bedrock and other stratigraphic layers, which would have made other geophysical methods impossible to use at that location. In these studies, the GPR method was excellent at testing hypotheses with a limited amount of invasive excavation. Two weeks of GPR collection and analysis, and a few more of excavations and artifact analysis, totally transformed and refuted many prevailing ideas about the past, to the dismay of people who had been working there for decades. This would have been impossible without using the GPR interpretations and, most importantly, would not have been accomplished without first posing ideas about the past that could be tested with GPR. As results were obtained that called for a reanalysis of preconceptions, new ideas were proposed and tested during the research as new data became available. This project shows the value of posing hypotheses in advance so that GPR data can be collected that have the ability to test ideas, as opposed to randomly collecting reflection profiles or grids of profiles in the hope of finding something in the ground.

Architectural Changes Related to Urbanization at Petra

At the site of Petra in Jordan, where we found the foundations of the Roman temple I discussed earlier (Figure 2-22), much is known about the standing architecture, and our GPR work was capable of mapping the monumental late Nabataean and Roman architecture near the surface. The GPR results were quite good at helping us understand aspects of the architecture built after the city became a wealthy trading center, which is what has drawn tourists and archaeologists to this site for more than a century. While that history of Petra is well known from decades of excavations, almost no research has focused on the earliest inhabitants of the valley, who were likely the ancestors of the people who made this city so important in antiquity. The archaeological remains of those simpler dwellings from the earliest inhabitants were either destroyed during later periods of urbanization or remain buried in the ground and unavailable for study. We therefore considered an analysis of these earliest inhabitants of Petra an appropriate task for GPR. The idea was that GPR images could be produced from the deepest layers, and the houses and earliest street patterns potentially studied. These are deeply buried below the shallower remains of what had always been the focus of almost all other archaeological research at Petra (Conyers et al. 2002; Conyers 2010).

While formal Roman annexation of Petra didn't occur until A.D. 106, monumental construction with a considerable Hellenistic influence began about a century earlier, perhaps as early as mid- to late first century B.C., by the Nabataeans, who had become wealthy by controlling land trade routes from Arabia to the Mediterranean Sea (Conyers 2010). There is subtle and potentially important evidence of the earliest Nabataean occupation of the valley from the second and early first centuries B.C. left by the people who originally lived in much more humble dwellings (Conyers 2010). These structures were likely razed and then covered over during a first-century urbanization campaign, which created the city layout that can be seen at or near the surface of Petra today. Some interesting questions had been posed but never adequately tested, relating to the social structure and community planning of those earliest Nabataean settlements; but because much of the archaeological evidence to test these ideas was either destroyed or buried, they remained nothing but interesting research questions.

One site at Petra conducive to the production of maps of the early Nabataean structures is called the Lower Market, just east of the Great Temple (Conyers et al. 2002; Bedal 2003). This open area had seen little archaeological work prior to excavations in the late 1990s and early 2000s (Conyers et al. 2002) and almost no deep testing of the earliest, deepest horizons, other than a tantalizing trench excavated into the area from the Colonnaded Street, just to the north (Figure 11-3). In that excavation trench, which abutted a large grid of GPR data, the possible foundations of early Nabataean dwellings had been exposed in the late 1960s and roughly dated by pottery. The stratigraphy in that trench appeared to show that an ancient living surface from earliest Nabataean times had been covered by a large amount of fill during the urban construction in the late first century B.C. That fill material was composed partially of architectural rubble derived from dwellings of an unknown age or function.

To test for the presence of preserved architecture below the present ground surface, one large grid of GPR reflection profiles was collected with 400 MHz antennas (Figure 11-4) (Conyers et al. 2002). Images of the shallowest buried architecture, within about 2 m of the surface, showed a number of temples, platforms, water lines, and possible water pools consistent with Bedal's (2003) interpretation of this area as a formal garden that included water features and temples.

The 400 MHz antennas were capable of transmitting radar energy to about 3–4 m in depth in the dry quartz sand ground (Figure 11-5), and the uppermost structures can be readily studied in amplitude maps and reflection profiles (Figures 2-17 through 2-22). A distinct high-amplitude planar reflection is also readily apparent in many of the profiles, dipping to the north toward what would have been the stream (wadi) in ancient times. This reflection was found to be at the same depth where evidence for an early pre-urbanization living surface had been documented in one small excavation. In all GPR reflection profiles, this dipping living surface was overlain by what was interpreted as architectural rubble, visible as jumbled point-source reflections (Figure 11-5). These reflection profiles indicate that this architectural debris had been used to fill and level a surface south of the wadi for the later monumental construction.

As GPR reflection data collected in a grid produce a three-dimensional block of information about the subsurface, it was determined that the earliest Nabataean features below the late first-century B.C. fill could also be mapped. It was hoped that the resulting GPR maps of the deepest layers would yield information on the first structures built in this area of Petra and, in doing so, help clarify aspects of social organization related to that built environment. The same reflection data set—which had been sliced horizontally and used to produce amplitude maps of the Roman and late Nabatean structures—was used to create another set of slices, a set that differed in the way the profiles were sliced. To obtain images of the deeper layers, each reflection profile within

Figure 11-3: Photo of the GPR grids at Petra, Jordan. The Great Temple is to the left, and the Colonnaded Street and the wadi are to the north. The small grid was collected to produce images of the temple, shown in Figure 1-1 and Figures 2-17 through 2-22.

the grid was interpreted to identify the sloping planar reflection, correlative to the buried living surface, which represents the ground surface along the southern flank of the valley prior to the first-century urbanization construction episode. The fill material above that surface produced an abundance of reflections, making manual interpretation too difficult, and therefore horizon amplitude slices had to be created.

All amplitudes of all GPR reflections in all profiles within the grid were digitized, gridded, and mapped in slices prepared parallel to the sloping ancient living surface (Figure 11-6). This task was necessary because horizontal slices would have cut across the buried sloping surface at many different angles, producing useless amplitude maps. Two slices were created, one directly on the living surface reflection, and another just above it (Figure 11-5).

Figure 11-4: Collecting GPR profiles in the GPR grid just east of the Great Temple at Petra, Jordan.

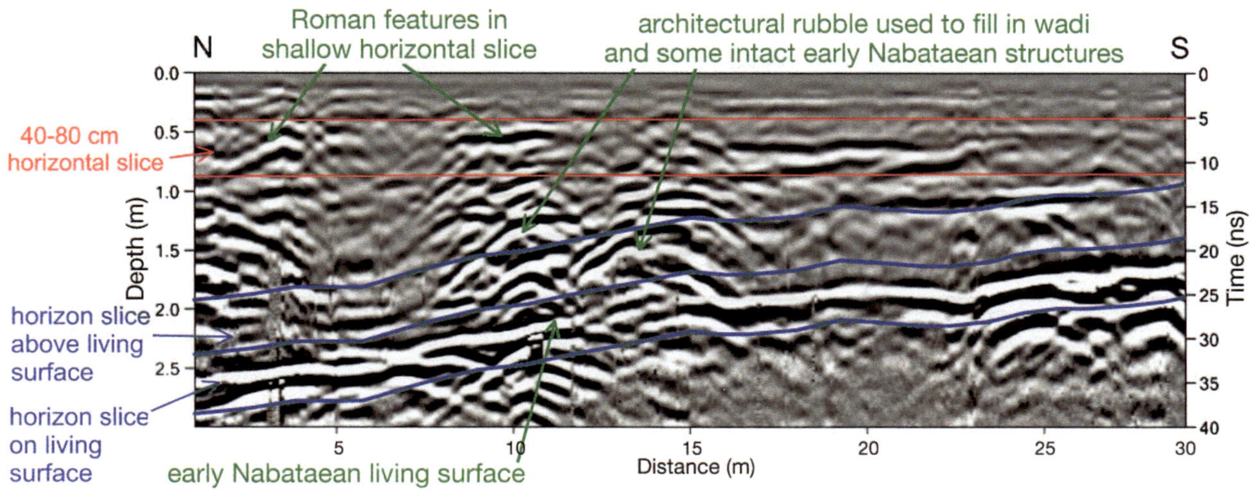

Figure 11-5: Reflection profile at Petra, showing the buried living surface. Rubble is visible on that surface and in the slices used to produce amplitude maps in Figure 11-6. Data collected with 400 MHz antennas. *Courtesy of Michael Grealy.*

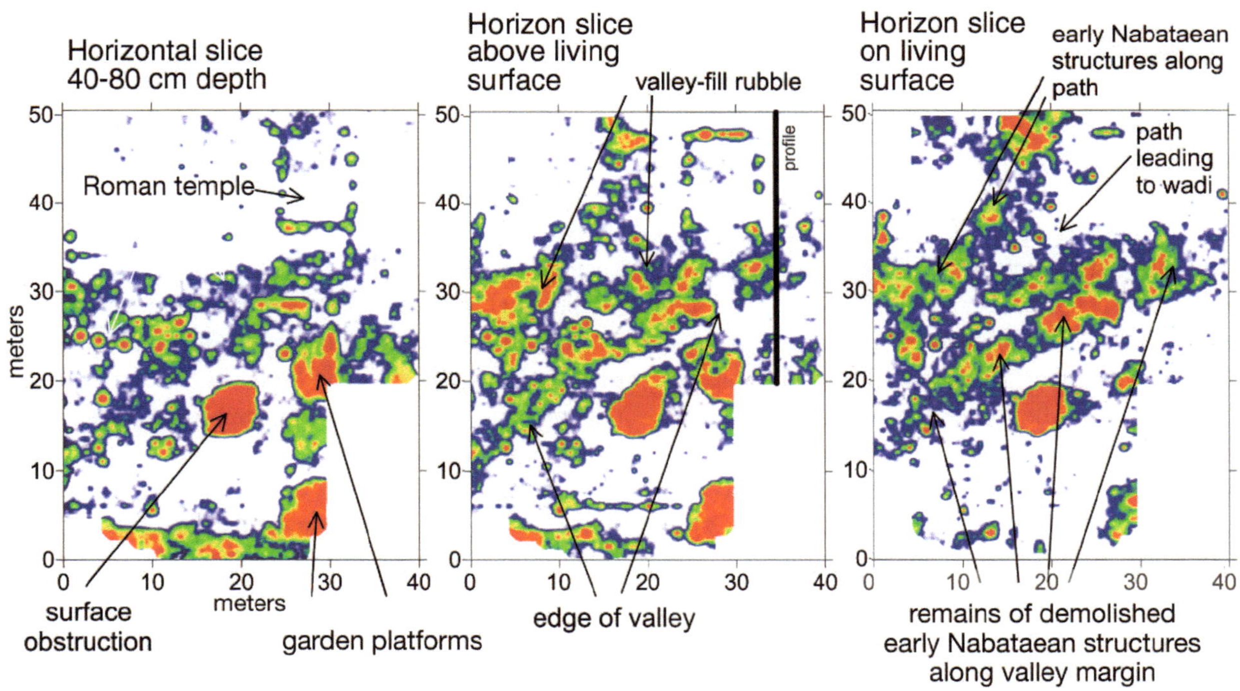

Figure 11-6: GPR amplitude slices from the Lower Market, Petra, Jordan. The left slice is from the 40–80 cm depth, which shows late Nabataean and Roman-period structures when the area was a formal garden. The middle slice shows rubble fill and the valley edge in a 25 cm thick slice produced from reflections generated directly above the horizon that was a living surface from earliest Nabataean times, occupied prior to the late first-century B.C. urbanization program. The right image is a 25 cm slice constructed directly on that earlier living surface, showing the foundations and remains of Nabataean structures built along the sides of the valley, with a pathway leading to the watercourse to the north. *Data courtesy of Michael Grealy.*

In the resulting amplitude maps, areas with no significant reflections are likely open spaces or pathways between buildings that had been constructed on that early Nabataean living surface. The high-amplitude areas are the remains of the buildings constructed along those pathways, which appear as high-amplitude rubble piles. They still retain some outlines of their original walls, even though they were largely destroyed during the filling episode (Figure 11-6). The pathways between these buildings led toward the wadi to the north.

The early-period pathways, visible as the linear zones of no reflection, led from the highlands in the south toward the water to the north (Figure 11-6). The positioning of the early Nabataean informal buildings on the margin of these pathways shows that they were built in various orientations and likely placed on whatever land was suitable for building at that time. This seemingly informal construction likely functioned as the part-time living quarters of people who were still involved in long-distance trade. The pathway to the water in the wadi was of prime importance for those inhabitants. Only later in the late first century B.C., when the Nabataeans had established control over trade routes from south Arabia to the eastern Mediterranean, did social differentiation and monumental construction take place. The remains of the later, more formal buildings (Figure 11-6) from that construction period appear in stark contrast to these earliest habitations. In the upper GPR slice, the structures within what became a formal garden were invariably oriented with the cardinal directions, consistent with Hellenistic-influenced building practices from later Nabataean and Roman times.

In this study, the architecture from two different periods within one small area of Petra demonstrates the very different types of city planning and social structuring practiced by the inhabitants of Petra over a period of three to four centuries. The GPR analysis of these two layers of the Lower Market indicate that the earliest Nabataean inhabitants built their structures along functional pathways leading to water, which is very much in keeping with their history as seminomadic people. In contrast, the upper GPR map shows that as the city grew and prospered beginning in the first century B.C., the built environment seemed to follow practices that suggest central planning and a much stricter and perhaps economically and politically differentiated social structure. The GPR images, when correlated to limited excavation information, produced images of these two buried landscapes and the structures built on them, in ways not possible with any other archaeological method except perhaps massive excavations.

CONCLUSIONS

Both these studies demonstrate how GPR methods in archaeological research have progressed to the point that they can be used for much more than exploration. While the resulting amplitude maps and reflection profiles must still be integrated with chronological and artifact information in order to be placed within a cultural history, they can often be used directly to test ideas about the past. Social change, technological innovations, and ecological adaptations can potentially be studied based on the orientation, layout, and change of buried features that are visible using GPR data. Ground-penetrating radar is a geophysical method especially effective in deeply buried or stratigraphically complex environments, because of the method's ability to discriminate important layers and map only certain levels or horizons. When GPR results are integrated with other data from standard archaeological methods, they fit nicely into the quantitative and scientific approach of studying many aspects of ancient life. When hypotheses about ancient people can be related directly to the extent, orientation, or geometry of buried architecture and associated anthropogenic features, geophysics can be used to directly support or refute hypotheses about the human past.

12

Interpretation in Collaborative Ventures

The Relationship between Geophysics and Archaeology

Almost all research work using GPR in archaeology is a collaborative effort, so I have given some consideration to the subject of how successes, failures, and interpretations are related to some interesting collaborations I have experienced. As geophysical methods have become a part of many archaeological projects, especially in the last few years, the integration of GPR interpretations with more standard archaeological information is now ever more important. In the early days of GPR, practitioners tended to work alone and just delivered the final products of geophysical analyses to others with whom they were working. To some archaeologists with little geophysical background, we were considered outsiders who brought "black boxes" to the field and collected strange-looking data. Our data collection procedures might have been interesting to watch for the first few minutes of fieldwork, but then quickly became boring as we moved back and forth across the ground, which is about as interesting as watching someone mow the lawn. To many of those collaborators, we were considered dispensers of esoteric knowledge that was mostly disseminated in incomprehensible images.

This less-than-flattering impression, recalled from my own past, is being transformed as GPR and other near-surface geophysical methods become more widely applied and as younger enthusiasts incorporate these methods into their research. In addition, with more informed interpretations and with images that look closer to what others expect objects and materials in the ground to look like, geophysical archaeologists are slowly becoming more mainstream. Of course, some in the more traditional archaeological community will continue to think of those of us who employ GPR as nerdy outsiders, as opposed to "real" archaeologists who dig things up. Excavation followed by analyzing artifacts that includes touching and looking at them remains the "real" archaeology for many. This intellectual divide must be bridged; and only by effectively interpreting GPR images and making those interpretations comprehensible and useful for others can we move out of the "black box" era. The best way to do this is to progress from the "finding anomalies" method of interpretation to the more complex interpretation methods discussed in this book.

I hesitated for some time to write this chapter, reluctant to sound unappreciative of my non-geophysical colleagues, many of whom tease me for being an operator of "gizmos" or think of me as someone who perhaps tells interesting stories based on incomprehensible images. I really do appreciate them very much, as they have given me and my students countless opportunities to work on interesting projects all over the world. And, for the most part, we remain friends and colleagues both. However, a message I just received from a collaborator settled my mind about writing this section of the book, and I began to compile a list of the many interesting but difficult projects where I have used GPR in collaboration with others.

The message that spurred me on was from one of my lovable collaborators regarding a project whose excavation results I wanted to add to my interpretations in one of these chapters. I had asked for details about what had been uncovered during excavations, which occurred many months after I completed my GPR work. My hope was that I could relate my GPR maps and profiles to the excavation data. For my part in the project, I thought I had satisfactorily explained, in both a long written report and in person, my interpretations and conclusions. However, my collaborator's answer to many of my specific questions about what features had been uncovered was that my "scans" had "read rocks." I am still puzzled how my detailed analyses of profiles and amplitude maps, which showed preserved architecture and a complex buried landscape, could have left her with the idea that I was "scanning" and only found "rocks." After a number of attempts to find out more about how the GPR information was used for planning the subsequent excavations, and how the two data sets could be compared and understood in context, it became apparent to me that my placement of reflection features in space and my interpretations about the origins of those reflections were mostly ignored. It appears that my numerous images and interpretations in many grids were used only in a general way for planning excavations, based perhaps on how enthusiastic I was in my report about what I could "see." Comments like "that was a hot spot" tend to back up this conclusion. My experience with other such collaborators suggests that my GPR interpretations were consulted only to get close to something of interest in the ground, and then the "real" archaeologists took over and dug wherever they thought was a good place. I am still in the process of trying to find out how the excavations in this particular project correlated with my GPR grids, and I worry that although all were surveyed into space, no one really paid much attention to the relationships between the GPR interpretations and what was actually uncovered in the ground. It is certain that no one on the excavation team was particularly knowledgeable about GPR, nor did they give much thought, if any, to what their discoveries in the ground might have looked like in the interpreted GPR maps and profiles. It appears that once the digging began, GPR results were the very least of their interests.

So, in my list of themes for this chapter regarding the interpretation of GPR with respect to collaborations, I'd put first and foremost the difficulty of effectively communicating GPR results to others. Also high on my list of frustrations is the reluctance of excavators, especially those with no GPR interpretive ability or inclination, to use GPR results in a constructive way. Surely my friends and colleagues who have invited me to become involved in their projects are not maliciously ignoring me or my GPR results. Perhaps my inability to correlate my results with excavation information is largely due to my becoming involved in other projects by the time excavations began, leaving me insufficiently connected with the project after my work was finished. Or it might have been that the GPR interpretations had served their purpose by the time the digging began, and so my usefulness to the project was at an end. How archaeologists perceive the usefulness of GPR long after the interpretive results are disseminated is a very good clue to whether collaborators really understood what was done. When someone uses such phrases as "when you are scanning the ground," I suspect they think that GPR is much like a metal detector that just "reads" something in the ground and perhaps makes a noise when antennas are over the top of something important. Perhaps they think of the amplitude maps as just colorful "readings" of those "beeps" when I am "scanning" the ground. I need to explore these types of perceptions by others more in the future.

A variation on this theme has come up when I have been asked onto a project to serve as a "visiting high-tech scientist from out of town." That is, occasionally my inclusion in a project has been requested because "high-tech" credibility was deemed necessary for funding or project legitimacy. What better way to do that than to bring in someone who uses "sophisticated" equipment to produce colorful and complex images of the ground? Such collaborations can be quite interesting, of course, and I am honored to be asked to take part in exciting and interesting projects. Some have even produced memorable and useful results. They can be difficult, though. One such project occurred early in my career, when David Hurst Thomas from the American Museum of Natural History in New York asked me to help out with research on a Spanish colonial church at the historic San Marcos Pueblo near Santa Fe, New Mexico. As we have continued to be friends and have worked together on other projects, I feel at liberty to mention his name here, as I assume he considers me more than a visiting nerd (or he would not have asked me to collaborate so often).

In this project in New Mexico, the GPR maps were very useful in finding the walls of a compound just south of the colonial church in what was hypothesized to have been the *convento*. Within those walls, the GPR amplitude maps showed a circular feature, which excited us both, and we immediately hypothesized that it might have been a "traditional kiva" built within a Catholic church's precincts (Figure 12-1). The presence of a circular structure of that sort, if confirmed to be indigenous architecture, would potentially tell us a great deal about "cultural assimilation and negotiations" between the native people who were still building traditional ceremonial kivas and the recently arrived Christian immigrants. David's excavations, based on my GPR interpretations, later determined that the circular feature inside the walls had been formed by adobe melt from the still-standing *convento* walls that filled up the corners of the enclosed rectangular area as their tops were slowly eroded. Then sometime in the 19th century, when this area was a cattle ranch, the local cowboys disposed of, and periodically burned, manure in this circular basin. When, for this book, I spent awhile looking at the profiles from which the amplitude maps were made, this usage and history of the circular feature became apparent (Figure 12-2). At the time, however, I had not yet discovered that adobe walls produced no reflections at all, and that the melted adobe adjacent to them was highly reflective. I was concentrating strictly on the circular basin so visible with GPR, which excavations later showed was filled with burned cow manure. I have never heard the end of my interpretations; and David facetiously tells me every time we meet how much he appreciated those interpretations during his weeks of digging through manure in a quest for interesting archaeological features! After reanalyzing the data and maps from San Marcos, it is now apparent that the circular amplitude features were caused by post-abandonment erosion, but I had never

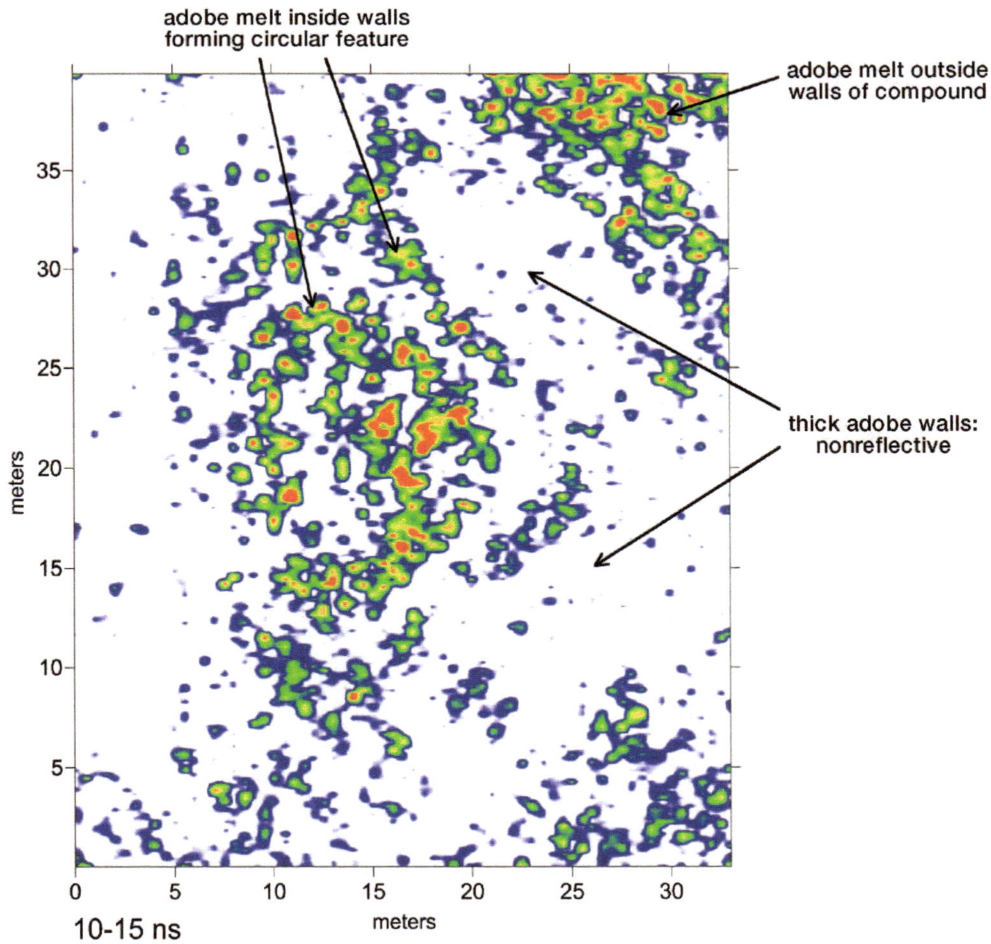

Figure 12-1: Amplitude map of the convento at San Marcos, New Mexico. This image illustrates the nonreflective adobe walls filled with adobe melt, which produced the somewhat circular feature within the walls. Data collected with 500 MHz antennas.

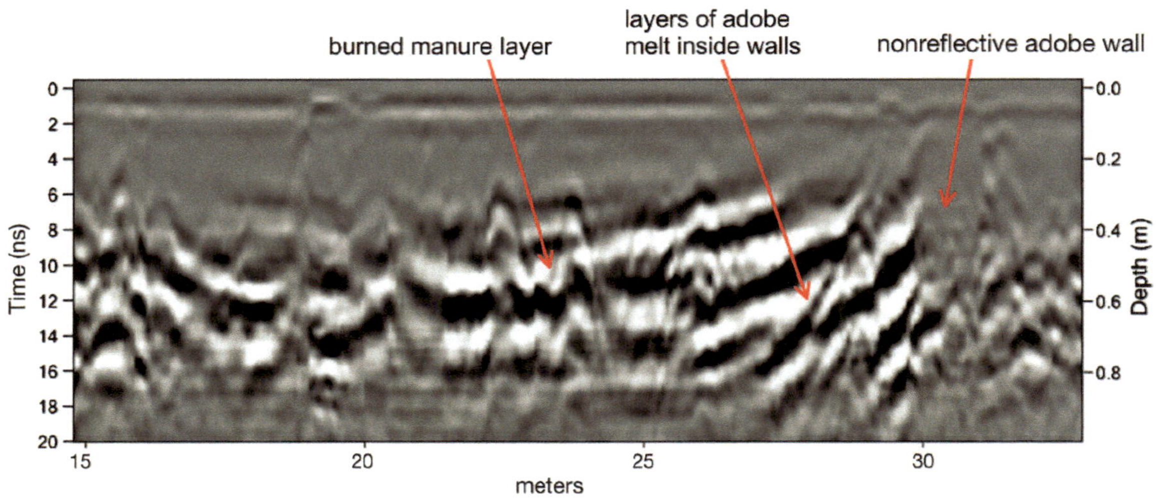

Figure 12-2: Reflection profile showing the adobe melt layers inside the nonreflective adobe walls. Data collected at San Marcos Pueblo, New Mexico, with 500 MHz antennas.

been confronted with this situation before, and I didn't include it in my hypotheses. I have worked with other collaborators who ended up using my GPR interpretations and found similarly uninteresting remains, but they have been much more reticent in their subsequent contacts with me.

A long-term collaboration, information sharing, and the ability on my part to integrate the excavation information into the GPR database were all very important in the convent project in New Mexico. Collecting GPR data, interpreting the resulting processed images, and then walking away from a project rarely lends itself to informative collaborations.

POOR COMMUNICATION ABOUT SITE CONDITIONS

A common refrain voiced by all my GPR friends is how poorly they were informed by collaborators about the conditions of a site prior to arriving there to collect data. All of us who routinely collect GPR data have learned to ask many probing questions of our collaborators about ground conditions and soil types prior to planning fieldwork. Nevertheless, their analyses are often sorely off the mark and devoid of the kind of site information we need to plan for collecting GPR data. We all laugh about sites described to us as "flat and open" that turned out to be more like "on the edge of a rocky hillside covered with impenetrable thorn bushes." I have conducted many surveys in a forest so thick with undergrowth that even small animals could not penetrate—landscapes initially described to me as "open" or "not too overgrown." Reflection profiles in areas such as this tend to start and stop at random locations in what can only be hoped is a rectangular grid, and the profiles tend to meander, with numerous antenna coupling changes throughout.

Once two of my students agreed to conduct a project in northern New Mexico where the ground was described as "open," only to find it covered with sagebrush (Figure 12-3). They attempted to get the proposed grid cleared of these bushes, but were told that this would constitute an "un-environmental degradation" of the landscape. So they attempted to delineate survey lines with string, supported on the top of the bushes, which they tried to follow with the antennas, very unsuccessfully. This was not only a very difficult survey to conduct, but the antenna tilts and coupling changes were so dramatic that the resulting reflection data were almost useless.

Figure 12-3: String used to delineate profile transects in brushy terrain. This survey, undertaken near Aztec Ruin, northern New Mexico, proved to be worthless. *Image courtesy of Jennie Sturm.*

Other challenging conditions, which cannot be guarded against in advance, are related to weather. In central Oregon, the week before a survey was to be conducted a big storm covered the project area in about 30 cm of snow (Figure 12-4). I had traveled a long distance to get to this site and had made commitments to provide results to my collaborators, so we had to proceed in this snow-covered ground anyway. Our interpretations were severely compromised because the somewhat compacted snow had partially melted around the small trees and bushes, creating many small air pockets just below the snow surface. Each of these void spaces produced a very high-amplitude air wave reflection, and the resulting reflection data were very difficult to interpret (Figure 12-5). It was also very difficult to collect data in the deep snow, and we neglected to bring skis or sleds to help us move around the grid. There are, however, very good reasons to conduct GPR in fresh snow, such as searching for avalanche victims (Modroo and Olhoeft 2004; Instanes et al. 2004).

Figure 12-4: Collecting 400 MHz reflection profiles in deep snow in central Oregon. The survey wheel slipped, and many air waves were recorded as energy moved within the snow layer and reflected off air pockets within the snow.

Other challenging field conditions were confronted elsewhere due to basic misunderstandings about what the capabilities of GPR are in different ground conditions. It was assumed by collaborators that I could collect GPR easily in rice paddies in Portugal, across boulder fields in the high Rocky Mountains, and in very small open areas within standing modern architecture. None of these projects worked out well. The rice paddies contained somewhat brackish water that attenuated energy almost immediately (Figure 12-6). Also, the local farmers would not let us wade across their fields (how did they expect us to collect the data, I wonder?). We tried to put the antennas in an inflatable canoe and pull them across the paddies, to no avail (Figure 12-7). In the boulder field, the transmitted energy coupled at variable depths below the surface and an abundance of high-amplitude point-source reflection hyperbolas were generated, but no coherent reflection surfaces that would indicate any structure of any sort (Figure 12-8).

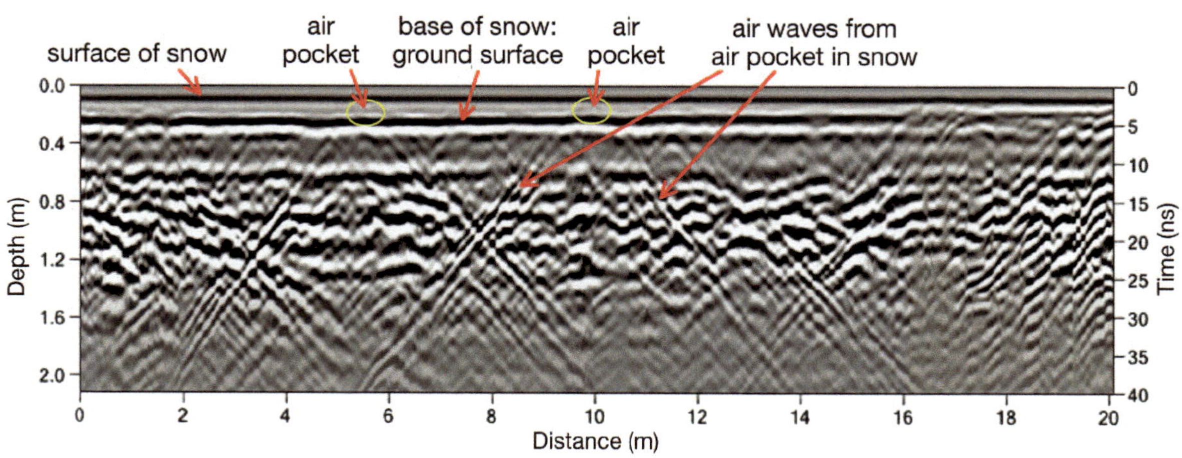

Figure 12-5: Reflection profile collected over snow, showing air waves. The air waves were generated from air pockets in the snow around small bushes and trees. Data collected with 400 MHz antennas near Klamath Falls, Oregon.

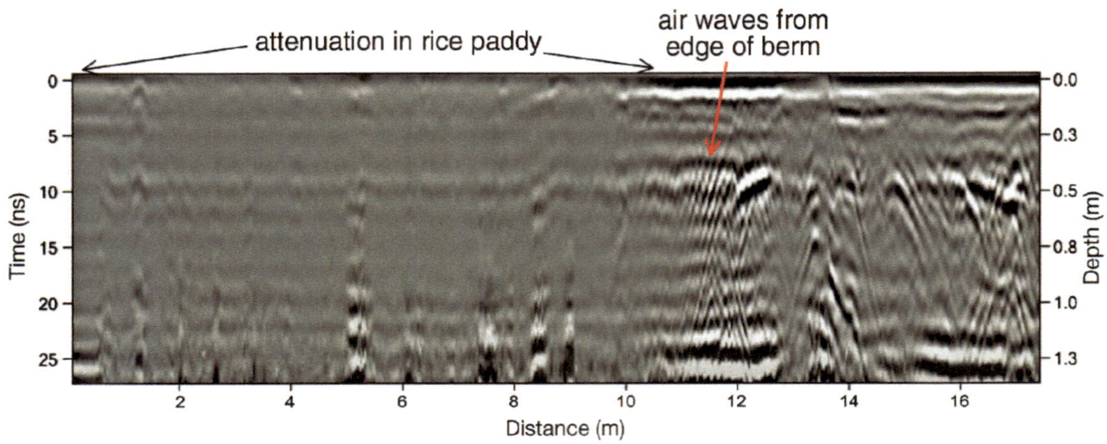

Figure 12-6: Reflection profile collected in an inflatable canoe across a flooded rice paddy in coastal Portugal. All energy was attenuated in the somewhat brackish water of the paddy, but good reflections were generated as soon as the antennas moved on shore. Data collected with 400 MHz antennas

Figure 12-7: Attempting to collect GPR profiles by placing antennas in an inflatable canoe. Profiles were then collected as the canoe was pulled back and forth across a flooded rice paddy, in coastal Portugal.

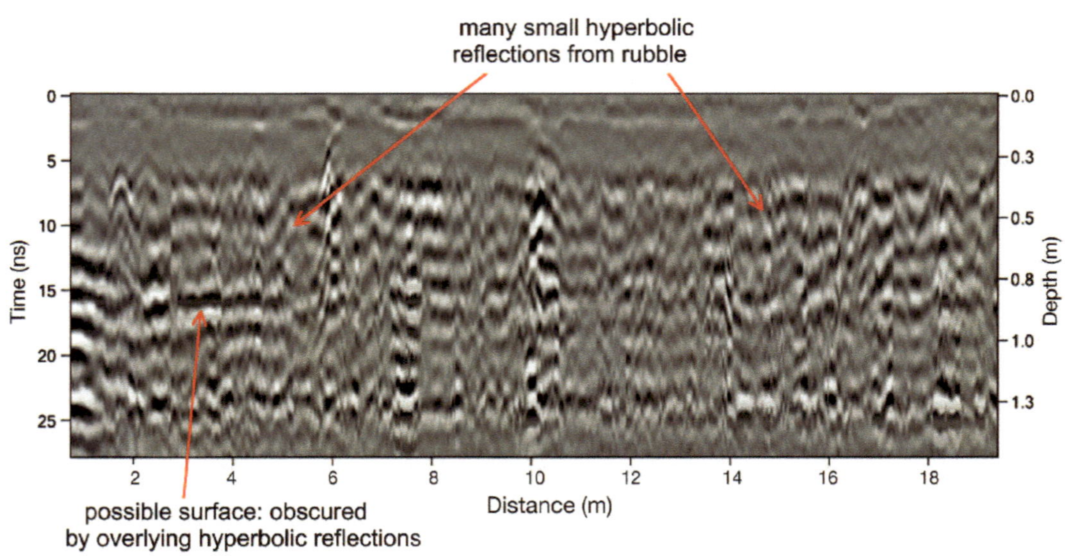

Figure 12-8: Reflection profile collected across a boulder field. The abundance of reflection hyperbolas and differential coupling made this data set less than useful. One possible surface below the boulders produced a discontinuous planar reflection, obscured by the abundance of reflections above it. Data collected using 400 MHz antennas near Black Hawk, Colorado.

A common lesson from all these situations is that poor communication between collaborators leads to a poor outcome. It must be taken as a given that non-geophysical collaborators cannot be expected to understand the difference between good and bad ground conditions for GPR, so it is our responsibility to become informed in advance. Some of the "ground condition shock" I have experienced when first arriving in the field might have been alleviated if I'd had, long before agreeing to conduct a survey, more detailed information about the ground, pictures of the site, maps, and satellite images of the Earth's surface, most of which are now easily available from the Internet. But even when I am more informed about site conditions than has been the case in some projects, I am hesitant to say no to some research that might be both fun and informative. It is always possible to modify one's collection and processing procedures and thus produce good results even in poor conditions.

UNREALISTIC EXPECTATIONS ABOUT THE PRECISION OF RESULTS AND THE SCALE OF SURVEY AREAS

Collecting many very small grids (under 2 × 3 m in size) tends to be a waste of time, as it is very difficult to produce images of features and their surroundings from a batch of small disconnected maps or profiles. I have found that in order to identify reflections and features produced from small grids, a larger grid that includes the surrounding materials is usually necessary for comparison purposes. Larger grids are almost always preferable to smaller ones. Nonetheless, a collaborator may insist "you must just scan this last small patch of open ground before you go." Perhaps they think that by moving the antennas over the surface of an isolated 2 × 1 m area, GPR can "detect" something of interest. I have never found any small grid of this sort to be even marginally useful unless it can be somehow incorporated with many larger adjoining grids.

Working with engineers is something I have learned, the hard way, to avoid. They tend to ask different questions than archaeologists and are much more comfortable measuring aspects of their world in millimeters rather than in the larger, often less precise units we usually deal with. Often the best that archaeological GPR can do with interpretations is to calculate both depth and location within a grid to a few centimeters or tens of centimeters. As much as I try to make engineers understand some of the inherent errors with GPR, they always insist on very precise interpretations. If their results are just a little bit different from an interpretation, they tend to become quite upset. In contrast to engineers' expectations of millimeter accuracy, archaeologists are often happy if a GPR interpretation can get close, and tend to be much more forgiving of the inherent errors in the GPR technique. I continue to work with engineers, however, as they pay well and often require only the "immediate results," as opposed to the long, detailed written reports expected by archaeologists. Being able to put a mark on the ground where something is located (and then being paid handsomely for it) can sometimes feel more rewarding than having to write up lengthy reports that are ignored anyway. However, if there is potential liability involved in how the interpretations are used by engineers, the risk can far outweigh the financial benefits. This happened to me once when I gave an inexact location for a high-tension cable within a concrete floor slab on the ninth floor of a high-rise building. I learned later that cables of that sort hold the infrastructure of the entire building together! My interpretation was about 2 cm off, and when the construction workers used my GPR maps as a guide, they cut the cable, necessitating a total rebuilding of one floor of the building. I have never been asked back to work with the company that hired me, and that is fine with me.

Similar imprecision or inexact interpretive results have occurred when conducting GPR in cemeteries, which can also lead to uncomfortable results. In one project, we were asked to find areas within a very crowded cemetery so that the mortuary could sell new plots and bury additional "customers." We delineated what appeared to be an "open area" with no burials, which was then sold for interment. When the gravediggers later turned up human bones in those "open" areas, it was not a good advertisement for the GPR method. There were no casket remains in those areas and therefore no void spaces or other physical materials that could be detected in the GPR reflection profiles. Those "open" areas we saw were probably used to bury indigents many decades ago with a minimum of expense, and if there had been coffins, they decomposed rapidly along with the bodies in them. No records of those burials had been kept. Over time in many wet or acidic ground conditions, all materials decompose swiftly, and in the space of a decade or so there is no physical or chemical difference remaining to differentially retain water and produce a radar wave reflection. As shown in Chapter 8, if the ground does not contain natural layering that would be visible with GPR if truncated by grave digging, the burials will remain undetectable with GPR. For this reason, it is always good to add a disclaimer in all reports conducted in cemeteries about the possibility of not being able to locate some remains due to conditions of this sort.

In my early years of GPR, I felt fortunate to have located a buried pit-house village in southern Arizona, which would have taken weeks and a great deal of money to find and excavate using traditional methods. I was asked to collaborate with a prominent consulting archaeologist on this project as my first test of the commercial viability of GPR mapping. My interpretation concluded that in a 30 × 30 m grid there were 12 prehistoric house floors below alluvial sand and gravel. My collaborators then excavated the complete GPR grid with heavy equipment and exposed all of the floor features, uncovering 13 houses. I had missed one of them in my interpretation, whose floor was composed of the same material as the geological matrix of the site and therefore did not produce a high-amplitude reflection (Conyers and Wallace 2004) (Figure 12-9). Having just missed it in my interpretation then, I subsequently reanalyzed the data from this project, only to find that its floor still did not produce a visible reflection no matter what types of processing I applied to the same data set. When I asked the archaeol-

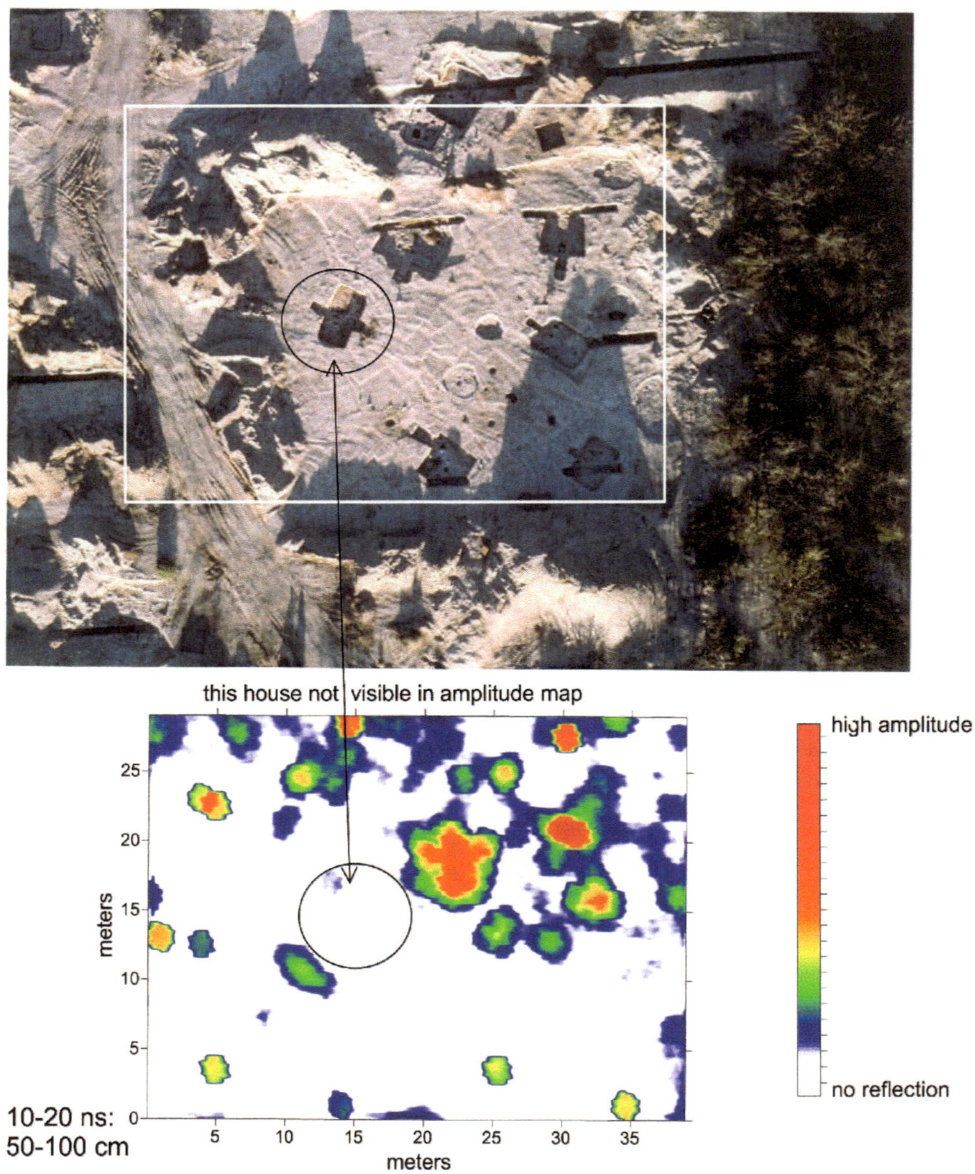

Figure 12-9: Comparison of an amplitude map of many Hohokam pit-house floors and an aerial photo of the grid after it was exposed. Data collected in southern Arizona using 500 MHz antennas.

ogists involved in this project if they were ever going to use GPR again for other projects of this sort, the answer was "your results were not good enough for us." For them, a success rate of 12 out of 13 was not competitive with their method of exposing everything to the human eye by completely stripping the area with heavy equipment. I can appreciate, if not agree with, this type of conclusion, especially when consulting profit is taken into account. The project leader must have felt that it was better for his firm to keep many employees on the payroll and charge out the costs of massive excavations in order to find 100% of the buried cultural remains than to pay one outside consultant and identify only 92%. The only reasonable response I have for this kind of thinking (which is mostly about money) involves the nondestructive nature of GPR mapping as compared with the total excavation and destruction that goes along with those large-scale ground-stripping projects. Of course, there are many other good arguments to be made for GPR, including finding the buried village and preserving it completely, or mapping it with GPR and then testing only those features that can add new knowledge within a research plan, and preserving the others.

Unrealistic expectations by people who have never used GPR before are often amusing, but troubling as well. I was asked once to go to Guatemala to map Mayan sites in a huge sugarcane field. Soon after I arrived, it became apparent that my collaborator thought I could somehow collect GPR data over many tens of square kilometers in about a week or so. When I gently told him that in the allotted time I would perhaps be able to map much less than 1% of his study area, he was crestfallen. I will never forget the look in his eyes when I was transformed from a potential magician, who would tell him everything he wanted to know about what was under the ground, to someone much less informative. Neither of us has ever gotten over that misunderstanding based on very different expectations of what is possible with GPR.

I have tried to think of ways that expectations by all prospective collaborators can be adjusted and modified prior to starting a project. Once again, everything comes down to communication. Today, when many forms of worldwide messaging can be instantaneous, there is no excuse for not discussing all aspects of a project in advance. A detailed list of "deliverable products from interpretation" and an accurate estimate of how much area can be covered per day should be included in some type of "memorandum of understanding."

SKEPTICS, TRUE BELIEVERS, AND THE MASS MEDIA

Worse than having skeptical collaborators is working with "true believers" of the GPR method, people who consider you to be some Superman with X-ray vision, with an ability to see absolutely everything hidden in the ground. The true believers are never those with any experience in GPR interpretation, or they would have been dissuaded from that idea long ago. There is nothing worse than the disappointment of people who hold strong opinions of this sort when things do not work out as anticipated. Once I was asked to map a buried stagecoach road that was known to be only about 10 cm below the ground surface near Palm Springs, California. Try as we might, we could not produce images of that road. I could not see any reflections from it in reflection profiles or in the amplitude maps. Later, it was apparently quite visible to the archaeologists who excavated it, and our negative results disappointed a number of collaborators who were initially great GPR supporters. I have since given this data set to others to process and interpret, in the hope that I missed something in my interpretation. They have also failed to produce images of the road. I can only conclude that the roadbed was not different enough from the ground matrix to be visible, or was located within the near-field of the 400 MHz antennas and therefore more difficult to detect (or both). These collaborators quickly changed from true believers to skeptics as a result of this failure.

I have been involved in a few television programs that tempted me to try to become a TV star using GPR. Unfortunately, my lack of on-screen charisma and a tendency to say the wrong things on camera have mostly made such collaborations less than satisfactory. I should have realized that TV shows these days must deliver short, entertaining scenes that can hold viewers' attention lest they quickly change the channel. Lengthy talks on the wonders of GPR do not seem to meet these criteria. In the process of recording these shows, I have been asked if I could please add more colorful images and sounds to my data collection procedures, to make the presentation more compelling. I have never been able to satisfactorily accommodate these wishes.

One TV program I became involved in included screaming teenage girls, who were coaxed into spending the night sleeping in a tent over some unmarked graves I had found with GPR earlier that day. I should have suspected what was going to happen, as the show was scheduled for broadcast on Halloween night. I foolishly believed that the program was going to focus on the science of GPR and how it can be used to make images of features in the ground. Instead, everyone involved in the program knew all along that I would be used to find the graves, give legitimacy to the so-called plot, and then allow the "action" to take place after I went home. Some of my students saw this program and thought it was really great. I thought it was horrible. Fortunately, my friends and colleagues tend not to watch programs such as this, and I have never heard any negative comments or chuckles from them about how I was exploited in this way. I do not recommend this type of collaboration for anyone wishing to give legitimacy to real science.

In another program involving "real" archaeologists who do research in Indiana Jones-like ways, I was coaxed into declaring on camera that some of my female students are just as "ditzy" as the actresses who played the lovesick students in one episode of those pseudo-archaeology movies. I don't remember saying these words and, of course, I don't think this about my female students, but I have been chastised often for those comments. Neither of these television collaborations has enhanced my scientific career or the relationships with my female students.

UNUSUAL AND DOWNRIGHT CRAZY REQUESTS FOR GPR

When I was in Hawaii in 2005 working with the U.S. military at their human remains identification laboratory, Sam Connell and I were asked to help out one morning at the Iolani Palace in Honolulu during King Kamehameha Day celebrations. The hope was that we could identify the location of a time capsule placed in the palace walls in 1872 by King Kamehameha V during its construction. This capsule was reported to contain a copy of the constitution of the Kingdom of Hawaii (before annexation by the United States) and numerous photos and mementos from that time. It was known to have been put in the northeast corner of the building, which was the standard location for time capsules placed by those in Freemasonry lodges, of which the king was a member. We used 900 MHz antennas to collect reflection profiles within the walls, holding the antennas outside the building on the wall's exterior faces. I was unsure what the velocity of the building material was and therefore didn't know if the high-amplitude reflections I was seeing were generated along stone faces within the wall or perhaps from the inside surface of the wall. To test whether radar energy was penetrating through the stone into the void space inside the building, we performed a test where the antennas were kept stationary on the outside wall and pointed horizontally into a room in the palace that was cleared of bookshelves. We turned on the system to collect 20 traces per second and kept it stationary (Figure 12-10). A very large Hawaiian woman in a flowery dress, called a *mumu*, was enlisted to move the back of her body along the interior wall directly over the antennas. We could see her through the open window, and she notified us with a sharp yell when she was positioned directly in front of the transmitting antennas, with the wall in between. A distortion of the high-amplitude reflection from the interface between the interior, the wall, and the woman's body was visible as she moved in front of the antennas (Figure 12-10). Velocity was then readily determined, and we were able to calibrate the reflection profiles to depth and adjust the time-window to maximize resolution within the walls. The capsule was then quickly located in one area of the wall near where it was thought to have been put (Figure 12-11).

This time capsule project turned out to have some interesting political repercussions. To our surprise, our discovery—having been announced on all the Honolulu evening news programs and in articles in the local newspapers—set off a controversy over Hawaiian sovereignty, and was seized on by groups actively proposing Hawaii's secession from the United States. Apparently, the inclusion of the pre-annexation Hawaiian constitution in the time capsule, and Sam Connell's position with the U.S. military, raised questions about whether we were part of a conspiracy to destroy evidence that might somehow hasten the reestablishment of the Hawaiian monarchy. Fortunately, threats made against us by these militant groups were not known to me until much later, after I had left Hawaii for the mainland; and luckily my name was misspelled in all the newspaper articles.

Complex ground conditions need not always be a deterrent to good data, as we discovered at Ashkelon, Israel, where we were tasked with determining if a very steep slope outside the ancient city walls might contain the Islamic-age or much older Philistine ramparts. Elsewhere along the well-preserved city walls, these flat, paved surfaces had been excavated and were visible for study. Our study area was a very steep slope at one area of the site, where we positioned 17 profiles in areas between trees to test this idea (Figure 12-12). Many of the reflection profiles (Figure 12-13) showed evidence of these rampart surfaces under the ground, which were confirmed with later excavations.

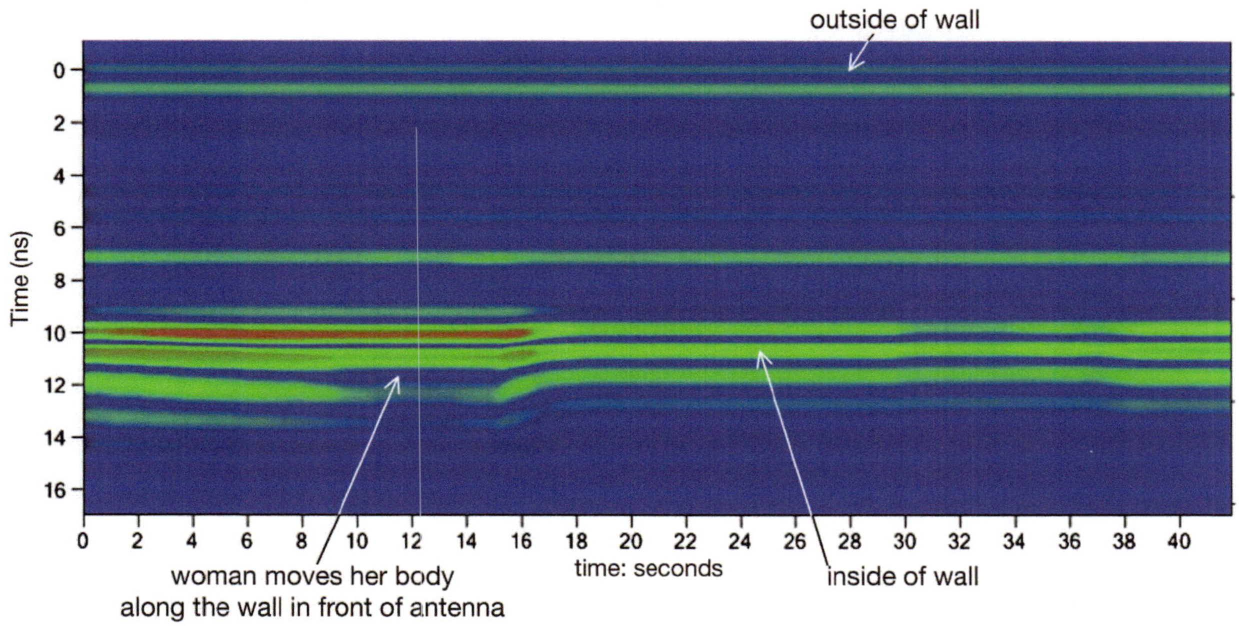

Figure 12-10: Reflection profile generated from reflections collected in time from a stationary antenna held to project energy horizontally through the wall of the Iolani Palace, Honolulu, Hawaii. The body of a volunteer wearing a flowered mumu can be seen moving in front of the antennas, which was measured in time and used to calibrate distance. Data collected with 900 MHz antennas.

During my forays into collecting GPR in lakes, I had always been fortunate to have access to motorized boats or fiberglass canoes. However, when attempting to collect reflection profiles in the Dominican Republic, no boats of this sort were available, and we were relegated to using the absolutely worst boat imaginable for this type of endeavor. It had been constructed from mismatched pieces of wood and leaked constantly, with the tops of sideboards only a few inches above the water surface (Figure 12-14). The only propulsion device was a long pole pushed into the lake bottom by a slightly inebriated fisherman standing in the back of the boat (Figure 12-15). It was a hair-raising experience, and to say I was apprehensive about putting many thousands of dollars of electronics in a boat of this sort is an understatement. The wood, however, turned out to be an excellent surface for the transmission of radar energy into the water, and we produced an excellent set of reflection profiles (Figure 4-24). Later on during that project, we found someone in a nearby town who owned a fiberglass boat with a motor, which proved to be much better for this project.

My success with GPR in lakes was a topic of discussion one evening in Ethiopia, where I had been using GPR for more standard mapping of buried remains in cave floors and fields. We thought it would be interesting to try this method in Lake Chamo and set out the next morning to collect some profiles in that very large lake

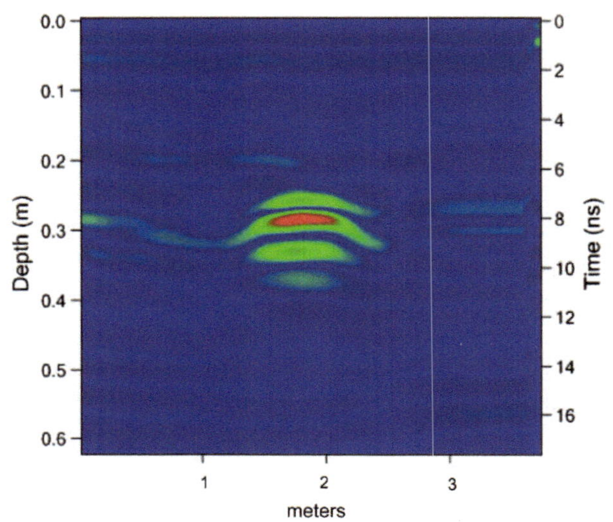

Figure 12-11: Reflection profile showing King Kamehameha V's time capsule, Iolani Palace, Honolulu, Hawaii. Data collected with 900 MHz antennas.

Figure 12-12: Location of the very steep slope where GPR reflection profiles were collected along the rampart of Ashkelon, Israel.

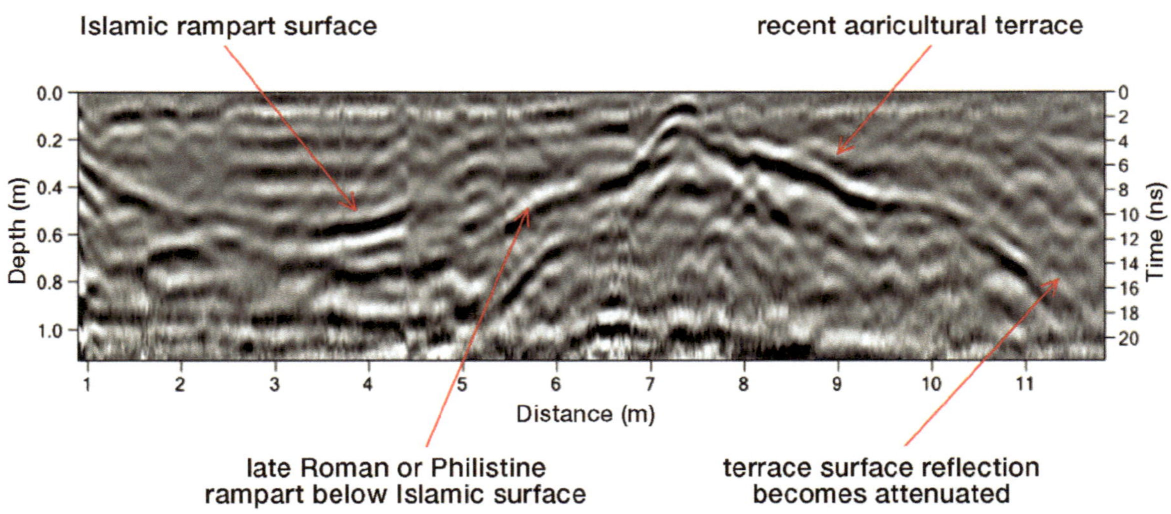

Figure 12-13: Reflection profile, uncorrected for topography, showing two rampart surfaces along the eastern edge of the city walls of Ashkelon, Israel.

within the African Rift Valley. There were no nonmetallic boats to be found, and we had to put the antennas in a plastic bag, holding them against the side of the steel boat in order to get energy into the water (Figure 12-16). Dodging hippos and Nile crocodiles, we collected many kilometers of totally worthless reflection profiles. The very weak reflections obtained from the floor of the lake and reflections from any sediment layers deposited

Figure 12-14: GPR equipment placec in a very leaky homemade boat. This very scary project was in Lake Saladilla, Dominican Republic.

Figure 12-15: Collecting reflection profiles on Lake Saladilla, Dominican Republic, from a very leaky and poorly constructed boat. *Image courtesy of Sally Horn.*

there were completely overwhelmed by the huge reflection, and multiple reflections from the side of the metal boat. It was a very fun day otherwise, where we became intimate with the hippos, crocodiles, and tsetse flies that abounded on this lake.

CONCLUSIONS

Perhaps it has been my good fortune to have been subjected to less "GPR dogma" than others during my years of doing GPR. As a result, I have had fewer reasons to say no to possible collaborators and their requests, mostly due to my ignorance. Even today, when I think I know what I am doing (always a fatal type of conceit), I become involved in many more projects than I should. When people ask me to collaborate in their research, or to help out with interesting problems, I am usually flattered to have been asked, and if it is not too dangerous or time-consuming, I usually oblige. I regret very much not taking my GPR system with me on some early projects I collaborated on in Costa Rica during the rainy season, as I was still under the impression that wet ground

Figure 12-16: Attempting to collect reflection data holding 400 MHz antennas wrapped in plastic, along the side of a metal boat in Lake Chamo, Ethiopia. No reflections were recorded except for those from the metal boat side.

was bad for GPR. Those never-collected data sets would no doubt have been spectacular, as I now know that many tropical soils are excellent for GPR. Even the wet manure project in California (Figure 6-9), which was grim and one the worst days in the field of all time, allowed me to learn something about ground conditions and radar attenuation. So I am not suggesting that GPR users avoid projects just because the conditions are known to be difficult. They might just work out, and if so, we might all learn something.

I do not, however, suggest repeating my attempt at collecting data from a leaky boat, as just one false move would have destroyed tens of thousands of dollars of equipment. Innovating in the field is also great fun; the use of the body bulk of the woman in the flowery *mumu* turned out to be an extremely useful way to calculate velocity (although a metal plate could have been used to produce even better results).

Communication with collaborators before, during, and especially after projects is extremely important, although few people seem to be very good at it. I hear story after story about a "disconnect" between perceptions and reality with regard to what GPR can really accomplish. Perhaps many people are just not very good at listening, or we GPR folks are not very good at communicating the realities of GPR. Maybe we just don't ask the correct questions, or when we do, we fail to persevere until we get the answers we need before going into the field. Descriptions of soil types, ground conditions, vegetation type, and density are useful only to the extent that the person doing the describing (or interpreting the descriptions) has the necessary background to provide (or understand) a certain level of specificity. To say that the ground material at a site is "real sandy" could mean anything from "it is all sand" to "it has some sand in it," "someone told me there was some sand nearby," or "there is sand at the surface, but it is all clay below about 10 cm." Conversations with prospective collaborators therefore need to be lengthy and detailed and should be followed up with an exchange of pictures and further descriptions. Sending a list of soil types that are "good" or a published paper related to GPR will do no good whatsoever, as it is unlikely that the recipient will ever read it or pay much attention to it.

The time to overcome communication problems is before you arrive in the field, especially if someone else is paying your way there. Once you get there, it is usually too late. Great attention should be paid in advance to survey condition details, and an understanding agreed upon about what can actually be accomplished in the time allotted. Of course, even when you manage to do this, you may still be asked, "As long as you're here, how about surveying every other farm within 10 miles?"—after you have completed the agreed-upon work. This problem of being asked for more can be overcome more easily if the GPR project is being done for commercial rather than for academic purposes. The way to deal with collaborators requesting additional "as long as you're here" work in commercial projects is to tell them you will add a huge sum to your final invoice. If they actually agree on that sum, then the additional work can become a "win-win" situation. It is best to get those agreements in writing.

Communicating with collaborators after a project has been completed is the most difficult problem of all, as we are all busy, and it takes a real effort to follow up on excavation results that only become available months after the GPR reports have been completed. As soon as we have gone home and left collaborators alone with our reports, those non-GPR friends of ours may tend to change (or misconstrue) the conclusions to fit their own needs or ignore them completely. Your presence in the field with GPR equipment will soon seem little more than an interesting interlude into "nerd-dom" unless you make yourself overtly available and proactive by following up with results after you leave the field. It is extremely difficult to get results from collaborators after you are no longer a physical presence, and it is naive to expect them to keep you informed—unless you have somehow been cursed with a collaborator who constantly asks you to reprocess and reinterpret your data time and time again until you "get it right" or at least "meet their expectations." Those very needy collaborators who are never satisfied and consider you their "tech-slave" should be avoided at all costs.

A Checklist of How to Approach and Be Successful in Collaborative Ventures

It is helpful to keep a checklist of a few basic items to be considered and perhaps checked off by all concerned when planning GPR work and collaboration with others. Whenever I have neglected one or more of these, the project suffered and was more likely to produce less than optimal results:

- Communicate before, during, and after conducting GPR work with all collaborators. Try to educate others about what is possible and what your results will consist of when you are finished. Explain to collaborators who have no geophysical experience what to expect during collection and when reviewing the results obtained.

- Find out as much as possible about study sites in advance with respect to topography, soil, and sediment types and the depth of features to be resolved. This can be done by studying photos, online satellite images, topographic maps, and published soil surveys. Vague comments that the ground surface is "clear" or that the soil is "sandy" are not enough.

- Study in advance what others have published and reported about the study area, especially the composition of buried features, their depth, orientation, and complexity. Before you begin collection, use this information to produce models of what GPR profiles might look like when visualizing these features so that you are prepared for what you may encounter during collection.

- Try to agree on the size of areas to be surveyed, and make sure all participants understand how much data can be collected and processed each day. It is especially important to communicate how long it will take to process, interpret, and then write up findings on each grid, depending on their size and complexity. I usually say I can collect 50 × 50 m in a day. This is usually a low estimate, and with good helpers and uncluttered ground, much more area than this can be surveyed. The time to process and interpret each grid is often the limiting factor, and I usually stress that there is a day of interpretation necessary for each day in the field at a minimum.

- If possible, bring a computer to the field and be ready to generate useful maps and profiles every evening from that day's collection. Then present those preliminary interpretations to all members of the team to stimulate a discussion about what can be done the next day. Sometimes it might be beneficial to repeat collection in one area if the setup parameters were not optimal, or you may want to change the transect orientation, or make a more detailed analysis of a smaller area within a large grid that was just collected. Often, if a large amount of ground needs to be surveyed, I will collect profiles at 50 cm or even 1 meter line spacing and process those data to produce low-resolution images first. Then I can interpret what is visible in those images and return the next day to collect smaller grids with 20 or 25 cm line spacing at perhaps a different grid orientation to confirm and enhance interpretations made the day before within the larger, less well-defined areas.

- When beginning collection at a new site, do not be lured into collecting huge amounts of data soon after arrival in the field. Spend a few hours calibrating equipment, determining which antenna is best for the depth of penetration and resolution necessary. Then try to find a trench or outcrop where the materials are exposed, and collect profiles where the visible units can be directly compared with GPR reflection profiles. Do velocity tests along these exposures, if possible, by placing a metal bar into the face of the exposure and doing direct time-depth conversions. Use those velocities to help determine the time-window to use in collecting data and determine what units are producing reflections.

- Move the antennas around a grid before collecting profiles, and look at the collected reflection traces on the computer screen. Look for depth of attenuation, and changes in stratigraphy and other ground conditions that might affect final calibrations.

- Make sure that all the GPR grids are tied in space to many different, immovable objects, and use GPS to place them within a universal surveying grid. The GPS placements might have errors, your flags marking grid corners can be lost in a matter of hours after leaving the field, datums may be moved or destroyed; therefore, it is warranted to have as many different placement measurements as possible. There is nothing worse than generating a set of interpretations that are "floating in space."

- Don't hesitate to try new collection procedures in unusual areas or under conditions that are difficult to manage with standard procedures. As long as it is not dangerous or might result in a loss or damage of equipment, be adventurous in data collection, as you just might be successful and learn a good deal. Be adaptable.

- Before leaving the field, make sure that all collaborators have maps, profiles, and at least a cursory report of results that they can use after you have gone. Do not presume they will actually read those reports or understand the images, so spend some time explaining your interpretations in non-geophysical terms to the non-geophysical members of the team. Paper copies are best if there is a color printer available. Politely and subtly threaten all concerned about failing to keep you informed of what takes place during excavations after you have left. In today's e-mail world, there is no excuse for not keeping in touch daily.

- Ask to have digital images of excavations, profiles, photos, and interpretations by field members e-mailed to you daily after you have returned from the field. A final report on the GPR results will be enhanced if you can receive immediate results that can be tied directly to your GPR data. Use that information for your final report.

- Back up all data and images in many different places as soon as possible. I do not sleep well after returning from the field unless that day's data are in at least three different storage locations.

- Never be too sure of yourself with interpretations that come from a limited amount of data analysis and processing. Be conservative in making pronouncements about successes or failures. You will probably be wrong if you are too hasty. Spend time with your data. The software needed to try different processing methods and image creation steps is now commonly available. Once profiles and amplitude maps are created, return to the basic raw data over and over again to determine what in your data is producing the images you are producing. Never be happy with just creating "anomalies" in your images. Anomalies are being produced by something in the ground (unless they are background noise or other disturbances), and you must determine what it is about the ground that is producing the features visible in images. The best way to do this is to keep looking at the reflection profiles in two dimensions, adjusting scales, zooming in on some features, processing them in different ways, and comparing those with the amplitude maps at different depths.

- Finally, when you produce new and different images that are found to be both meaningful and useful, let all the rest of us know about them. Publish your results, put them up on your personal website, or present your results publicly at conferences so that everyone can learn from new and different methods.

13 Conclusion

During the writing of this book, I sent draft copies of some chapters to a few friends and colleagues who use GPR frequently in their research. The comments I received back are revealing about where the subject of geophysics, and GPR in particular, stands today within some disciplines. One colleague who offered feedback on this book uses GPR almost solely in a for-profit enterprise, and his company does very well locating many types of buried materials for engineering purposes. He dabbles in archaeology, but his primary focus is finding and mapping buried pipes, determining the thickness of roadbeds, and understanding how fractures might have weakened retaining walls and other structures of that sort. His message to me on the subjects presented here was, "What do you think you are doing giving away all the GPR secrets?" This comment shouldn't have surprised me, as I would expect someone who does GPR on a commercial basis to look on the ability to interpret collected data accurately as "esoteric" knowledge—akin to the knowledge retained within medieval craft guilds. Those groups had a very selective membership and passed down "the tricks of the trade" only to those who were part of the group. Skills were taught to a select group of apprentices, journeymen, and masters of the craft, and knowledge was guarded from anyone outside the guild. Membership in guilds was very controlled, and the trade secrets were passed down from generation to generation among that select group, protecting the benefits they garnered from the products generated. While not overtly exclusive in this same way, GPR knowledge is passed down in a similar fashion, mostly through universities, but also by a few practitioners who teach classes for the general public and by instructors hired by the GPR system manufacturers and software developers. To some in the GPR profession, I might therefore be considered a person who is attempting to "break" the guild's power to control profitable and important information.

In the opposite vein, I had a few comments that were versions of "it is about time someone wrote something on GPR interpretation." Those comments were edifying and seemed to support my contention that interpretation of GPR—as the next step after basic collection and processing skills have been learned—needs to be enhanced. There are only a few ways to make that happen. One is to open up the "guild" to more apprentices. In my personal situation, I try to do this with my students, but it is very difficult to broaden my group of GPR researchers much wider, as I teach mostly my own graduate students and a few undergraduates. For a time, I taught a three-day class on GPR once a year to anyone interested, and hundreds of people attended. But in only three days, newcomers to the subject can learn just enough to be dangerous. Those classes can barely touch the surface of how to interpret results.

One of the motivations for writing this book came out of an afternoon workshop I was asked to lead at the University of California, Berkeley, in 2011. The workshop organizers wanted me to lecture on the usual topics of GPR methods and then to show some results. I was somewhat tired of giving that kind of talk, so when I found out that the workshop was to be restricted to just a few people who had used GPR before, I quickly changed the proposed agenda. Instead, I had the participants load the GPR processing software programs onto their laptop computers and then told them I wanted each person to interpret the data sets I presented to them. The workshop was broken up into "themes," much as I have done in this book. I had profiles collected on beaches, in dry ground, wet areas, on lakes, and many others. On the screen, I showed them pictures of someone collecting the data at each location so they could get a feel for ground conditions, and also identified where the data had been collected and for what purpose. Everyone was given a few minutes to produce their own images of the data, and we jointly analyzed what we were seeing on our computers and then generated interpretations. It turned out to be a great learning experience for all of us, and in the process of thinking through the differences in data types and processing procedures, I learned a good deal. That small workshop showed me that the next obvious step in learning about GPR—once the basic collection and processing skills have been acquired—comes from thinking about GPR images with respect to ground conditions and the nature of the geological and cultural reflections collected, and then having to make interpretations based on knowledge of what produces the data set. When the workshop was over, all of us were very enthusiastic about this interactive experience of focusing strictly on GPR interpretation. (Only the prospect of cold beer at a nearby tavern motivated us to stop.) This book is a direct offshoot of that experience, and I have expanded the themes much beyond those initial topics.

Next on my "to-do" list is to produce an online GPR interpretation website where participants can use their own software and process their own data, and then make their images and interpretations available for a large community of GPR scientists to see and comment on, perhaps giving us all instructions on the different ways these can be done. The site could perhaps start with some of the data presented here, but expand to include anyone's data collected in many interesting areas of the world. This type of project is my next planned endeavor, which would hopefully make GPR more widely used and interpreted in more informed ways that have real meaning in different projects. In such a process, perhaps a wide community of GPR users can effectively "break the guild" of "esoteric knowledge" about GPR.

I keep coming back to my mantra that revolves around the use of GPR to test ideas regarding people, history, and cultures instead of being used solely as a way to prospect for buried remains. The themes and examples discussed here are only a small sampling of what might be possible when very precise three-dimensional images of the ground are interpreted in ways that can produce interesting insights into human behavior. Once a researcher has learned how to operate GPR systems and collect good data, and has access to software for processing and producing images of the ground, there are no limits to what can be asked and potentially answered about complex buried sites and ancient landscapes. All one needs is a basic proficiency in the fundamentals of GPR, and knowledge of how radar reflections are generated and processed, and any number of interesting and very useful projects can be initiated. There is no reason to end a GPR project as soon as interesting buried features are discovered and then let the excavators have all the fun in interpreting the remains with regard to ideas regarding the past. If GPR results can be used to drive excavations, and then the excavation data can be integrated into the aerially extensive data set produced with GPR, research topics can be broadened in ways not even realized by most researchers.

The hardware and software used to collect GPR data sets will continue to generate more advanced, larger, more precise, and expansive information about the ground. But it is still the responsibility of informed, knowledgeable, and experienced GPR interpreters to determine how those data are used and how informed, and informative, the interpretations of these impressive three-dimensional packages of information will be.

References

Baker, Gregory S., and Harry M. Jol (editors)
2007 *Stratigraphic Analysis Using GPR*. The Geological Society of America Special Paper 432. Boulder, Colorado.

Bedal, Leigh-Ann
2003 *The Petra Pool-Complex: A Hellenistic Paradeisos in the Nabataean Capital (Results from the Petra "Lower Market" Survey and Excavation, 1998)*. Gorgias Dissertations: Near Eastern Studies 4. Gorgias Press, Piscataway, New Jersey.

Benech, Christophe
2007 New approach to the study of city planning and domestic dwellings in the ancient Near East. *Archaeological Prospection* 14: 87–103.

Bristow, C. S., and H. M. Jol (editors)
2003 *Ground Penetrating Radar in Sediments*. The Geological Society Special Publication 211. London.

Cameron, Catherine M., and Phil Geib
2007 Earthen architecture at a Chacoan great house. *Journal of Field Archaeology* 32: 1–14.

Conyers, Lawrence B.
1995 The use of ground-penetrating radar to map the buried structures and landscape of the Ceren Site, El Salvador *Geoarchaeology* 10: 275–299.
2004a *Ground-Penetrating Radar for Archaeology*. AltaMira Press, Walnut Creek, California.
2004b Moisture and soil differences as related to the spatial accuracy of amplitude maps at two archaeological tests sites. *Proceedings of the Tenth International Conference on Ground-Penetrating Radar, Delft, The Netherlands*, June 21–24, 2004, pp. 435–438, edited by E. Slob, A. Yaravoy, and J. Rhebergen. Delft University of Technology.
2006a Ground-penetrating radar. In *Remote Sensing in Archaeology: A North American Perspective*, edited by Jay K. Johnson, pp. 131–160. The University of Alabama Press, Tuscaloosa.
2006b Ground-penetrating radar tests in Hippos. In *Hippos-Sussita*, edited by Arthur Segal, pp. 99–107. Zinman Institute of Archaeology, University of Haifa Press, Haifa, Israel.
2006c Ground-penetrating radar for archaeological mapping. In *Remote Sensing for Archaeology*, edited by James Wisemen, pp. 329–344. Kluwer Academic/Plenum Publishers, New York.
2006d Ground-penetrating radar techniques to discover and map historic graves. *Historical Archaeology* 40: 64–73.
2009 Ground-penetrating radar for landscape archaeology. In *Seeing the Unseen—Geophysics and Landscape Archaeology*, edited by Stefano Campana and Salvatore Piro, pp. 245–256. CRC Press/Balkema, Taylor and Francis Group, London.
2010 Ground-penetrating radar for anthropological research. *Antiquity* 84: 175–184.
2011a Ground-penetrating radar mapping of non-reflective archaeological features. In *Proceedings of the Ninth International Conference on Archaeological Prospection, Sept. 19–24, Izmir, Turkey*, edited by Mahmut Drahor and Meric Berge, pp. 177–179. Archaeology and Art Publications, Istanbul.
2011b Discovery, mapping and interpretation of buried cultural resources non-invasively with ground-penetrating radar. *Journal of Geophysics and Engineering* 8: 13–22.

2012 Advances in ground-penetrating radar exploration in southern Arizona. *Journal of Arizona Archaeology* 3: 80–91.

Conyers, Lawrence B., and Catherine Cameron
1998 Finding buried archaeological features in the American Southwest: New ground-penetrating radar techniques and three-dimensional computer mapping. *Journal of Field Archaeology* 25: 417–430.

Conyers, Lawrence B., and Samuel Connell
2007 The applicability of using ground-penetrating radar to discover and map buried archaeological sites in Hawaii. *Hawaiian Archaeology* 11: 62–77.

Conyers, Lawrence B., and Dean Goodman
1997 *Ground-Penetrating Radar: An Introduction for Archaeologists*. Altamira Press, Walnut Creek, California.

Conyers, Lawrence B., and Juerg Leckebusch
2010 Geophysical archaeology research agendas for the future: Some ground-penetrating radar examples. *Archaeological Prospection* 17: 117–123.

Conyers, Lawrence B., and Jeffrey Lucius
1996 Velocity analysis in archaeological ground-penetrating radar studies. *Archaeological Prospection* 3: 312–333.

Conyers, Lawrence B., and Tiffany Osburn
2006 GPR mapping to test anthropological hypotheses: A study from Comb Wash , Utah, American Southwest. In *Proceedings of the 11th International Conference on Ground-Penetrating Radar, Columbus, Ohio, June 19–21, 2006*, edited by Jeffrey J. Daniel, pp. 1–8. Ohio State University, Columbus, Ohio.

Conyers, Lawrence B., and Hartmut Spetzler
2002 Geophysical exploration at Ceren. In *Before the Volcano Erupted*, edited by Payson D. Sheets, pp. 24–32. University of Texas Press, Austin.

Conyers, Lawrence B., and Henry D. Wallace
2004 Ground-penetrating radar techniques and three-dimensional computer mapping at the Valencia Site. In *Archaeological Excavations at Valencia Vieja*, pp. 135–143. Technical Report No. 2001–11. Desert Archaeology, Tucson, Arizona.

Conyers, Lawrence B., Eileen G. Ernenwein, and Leigh-Ann Bedal
2002 Ground-penetrating radar discovery at Petra, Jordan. *Antiquity* 76: 339–340.

Doolittle, J. A., F. E. Minzenmayer, S. E. Waltman, E. C. Benham, J. W. Tuttle, and S. D. Peaslee
2007 Ground-penetrating radar soil suitability map of the conterminous United States, *Geoderma* 141: 416–421.

Ernenwein, Eileen G.
2006 Imaging in the ground-penetrating radar near-field zone: A case study from New Mexico. *Archaeological Prospection* 13: 155–158

Gaffney, Chris, Vince Gaffney, Wolfgang Neubauer, Emonn Baldwin, Henry Chapman, Paul Garwood, Helen Moulden, Tom Sparrow, Richard Bates, Klaus Löcker, Alois Hinterleitner, Immo Trinks, Erich Nau, Thomas Zitz, Sebastian Floery, Geert Verhoeven, and Michael Doneus
2012 The Stonehenge Hidden Landscapes Project. *Archaeological Prospection* 19 (2): 147–155.

Goodman, Dean
1994 Ground-penetrating radar simulation in engineering and archaeology. *Geophysics* 59: 224–232.

Goodman, D., and Y. Nishimura
1993 A ground-radar view of Japanese burial mounds. *Antiquity* 67: 349–354.

Goodman, D., Y. Nishimura, and J. D. Rogers
1995 GPR time-slices in archaeological prospection. *Archaeological Prospection* 2: 85–89.

Grealy, Michael
2006 Resolution of ground-penetrating radar reflections at differing frequencies. *Archaeological Prospection* 13: 141–145.

Instanes, Arne, Ida Lønne, and Knut Sandaker
2004 Location of avalanche victims with ground-penetrating radar. *Cold Regions Science and Technology* 38: 55–61.

Kvamme, Kenneth L.
2003 Geophysical surveys as landscape archaeology. *American Antiquity* 68: 435–457.

Modroo, Justin J., and Gary R. Olhoeft
2004 Avalanche rescue using ground-penetrating radar. In *Proceedings of the Tenth International Conference on Ground-Penetrating Radar, Delft, The Netherlands, June 21–24, 2004*, edited by E. Slob, A. Yaravoy, and J. Rhebergen. Delft University of Technology.

Neubauer, W., A. Eden-Hinterleitner, S. Seren, and P. Melichar
2002 Georadar in the Roman civil town Carnutum, Austria: An approach for archaeological interpretation of GPR data. *Archaeological Prospection* 9: 135–156.

Olhoeft, G. R.
1981 Electrical properties of rocks. In *Physical Properties of Rocks and Minerals*, edited by Y. S. Touloukian, W. R. Judd, and R. F. Roy, pp. 257–330. McGraw-Hill, New York.

Sheets, P. D.
1992 *The Ceren Site: A Prehistoric Village Buried by Volcanic Ash in Central America.* Harcourt Brace Jovanovich, Fort Worth, Texas.

Thompson, Victor D., Philip J. Arnold III, Thomas J. Pluckhahn, and Amber M. Vanderwarker
2011 Situating remote sensing in anthropological archaeology. *Archaeological Prospection* 18: 195–213.

Viberg, Andreas, Trinks, Immo, and Kerstin Lidén
2011 A review of the use of geophysical archaeological prospection in Sweden, *Archaeological Prospection* 18: 43–56.

Index

NOTE: Page numbers in *italics* indicate pages with term referenced in illustration or caption.

About the Author

LAWRENCE B. CONYERS is a professor of anthropology at the University of Denver, Colorado. He received B.S. and M.S. degrees in geology and geophysics from Oregon State and Arizona State Universities respectively. His Ph.D. degree is in anthropology from the University of Colorado, Boulder. Before working with ground-penetrating radar in archaeological applications he spent seventeen years in petroleum exploration and development using seismic techniques.